*Landman's
Handbook on*
*BASIC LAND
MANAGEMENT*

i

Landman's Handbook on BASIC LAND MANAGEMENT

Edited By

Lewis G. Mosburg, Jr.

IED Publishing House, Inc.
Oklahoma City, Oklahoma

First Printing: November, 1978

Library of Congress Catalog Card
Number: 78-61347

ISBN Number: 0-89419-027-X

Printed in the United States of America.

Editor's Dedication

To Dwain Schmidt....

.... without whom the books never would have been written, or the institutes instituted.

About the Author

Lewis G. Mosburg, Jr. is senior partner in the Oklahoma City law firm of Mosburg & Day.

Born on March 2, 1933, in Tulsa, Oklahoma, Mosburg is the son of a former assistant chief geologist of Stanolind Oil and Gas Company (now AMOCO Production Company).

Attending the University of Oklahoma, Mosburg received his B.A. degree (with distinction) in 1954 and his LL.B degree (with distinction) in 1956, being elected to membership in Phi Beta Kappa and Order of the Coif. Upon graduation, he was employed in the legal department of Stanolind Oil and Gas Company until entering the private practice of law in Oklahoma City in 1958.

A frequent lecturer and writer in the fields of oil and gas, real estate, and tax sheltered investments, Mosburg teaches annual institutes on Contracts Used in Oil and Gas Operations, Oil and Gas Leases, Petroleum Land Titles, and Land Support Personnel, as well as serving as a special lecturer in law at the University of Oklahoma Law Center and moderating and serving as a faculty member for numerous other seminars. He also serves as special counsel (and former general counsel) of the Oil Investment Institute, and is a member of the California Oil and Gas and Real Estate Advisory Committees and the Committee on Direct Participation Programs (formerly the Committee on Tax Sheltered Programs) of the National Association of Securities Dealers, Inc.

In addition to cassette series and articles in the areas of energy, real estate, and taxation, Mosburg is the author or editor of numerous books, including the *Landman's Handbook on Petroleum Land Titles* (now in its third printing); *The Tax Shelter Desk Book* (The Institute for Business Planning, 1978); *The Tax Shelter Coloring Book* (The Institute for Energy Development, 1977); *Real Estate Syndicate Offerings Handbook* (Property Press, 3rd ed., 1976); *Financing Petroleum Exploration: Private Placements Under Rule 146* (Energy Concepts International, 1976); and *Real Estate Syndicate Offerings: Law & Practice* (RESD, Inc., 1974).

ILLUSTRATIONS

TABLE OF CONTENTS

Chapter Two
PETROLEUM LAND TITLES

Chapter Three
THE OIL AND GAS LEASE

Chapter Four
CONTRACTS USED IN
OIL AND GAS OPERATIONS

Chapter Five
CANADIAN PETROLEUM AND NATURAL GAS LAND OPERATIONS

Chapter Six
BASIC TAX FOR THE LANDMAN

Chapter Seven
EXPLORATION TECHNIQUES

Chapter Eight
ELECTRICAL WELL LOG

INTRODUCTION

ON BEING A LANDMAN

The Landman is a critical member of the exploration team. While it is up to the geologist to "originate" the prospects to be drilled, from that point the Landman must secure the leases necessary to permit the company to undertake its exploration effort and must insure that the lessor's — and the company's — title is sufficiently "clean" to justify commencing a well. The Landman must negotiate the contracts on which the company will rely in conducting its operations, and be able to explain the lease the company wishes to take to a potential lessor, as well as interpreting the company's rights and obligations under the leases he has been successful in securing. And he is required to have an understanding of the Federal income tax consequences of these transactions.

The Landman is often the only representative of the company whom the lessor will meet. Thus, the Landman must be able to explain to lessors, and prospective lessors, how the oil and gas search is conducted.

The *Landman's Handbook on Basic Land Management* will introduce you to your role as a Landman in each of these areas, and show you how to do your job. It is by no means exhaustive, and other volumes in the *Landman's Handbook* series will go into far greater detail on how to interpret a lease, check a title, or understand how your company actually finds and produces oil and gas. This book is intended to get you off to a good

1

start as a Landman, and to serve as a handy single volume reference for the more experienced Landman.

It is difficult to understand the oil and gas industry, or your role in it, without knowing something about its history and the evolution of our system of land ownership. In Chapter One, "Evolution of Land Ownerships", you will see how European and colonial land grants and concepts of land ownership have influenced modern oil and gas title law; how a system of "public domain", offshore, and state land ownership is complemented by the private ownership of the majority of U.S. land; how the United States varies from most of the world in its concepts concerning ownership of minerals; and how today's oil and gas lease evolved.

Having gained this historical perspective, Chapter Two, "Petroleum Land Titles", will consider your role as a title "examiner" in those numerous instances where you, rather than an attorney, are called upon to pass upon the status of your lessor's title, and what to look for in reviewing title sufficiency. After reviewing the basic concepts of land ownership and what to look for in checking a title, this chapter will show you how to determine the validity of deeds and other "conveyances" of land; the other means by which title may pass from an old to a new owner; how to draft and interpret mineral conveyances and reservations; and how to cure defective titles.

Chapter Three, "Oil and Gas Leases", will overview how the various clauses of a lease fit together into an interlocking pattern, and will then review, clause by clause, your company's rights and obligations under the lease.

Chapter Four, "Contracts Used in Oil and Gas Operations", deals with the various contracts with which a Landman must be familiar. After considering the basic principles which apply to all contracts (as well as what is a contract), the chapter will look at support letters, farmout agreements, joint operating agreements (with particular emphasis on the "Model Form Operating Agreement" used in most areas of the United States), and division orders and gas sales contracts.

Canada continues to increase in importance to oil and gas operators. American companies are becoming more involved in the oil and gas search in Canada, both through direct efforts and operations conducted with Canadian "partners". Even though you may be working in Texas or the Rocky Mountains today, tomorrow may find you transferred to Canada, reviewing the operations of a Canadian joint venture or advising on whether or not a Canadian operation should be considered. Thus, Chapter

Five, "Canadian Petroleum and Natural Gas Land Operations", discusses land ownership in Canada, including Canadian survey systems; development of the Canadian oil and gas industry; the various types of petroleum and natural gas leases and other agreements granted in Canada; how to acquire necessary surface rights; how oil and gas operations are affected by other areas of governmental regulation in Canada; and the Canadian tax system. Also covered is the Canadian system of government.

Chapter Six, "Basic Tax for the Landman", discusses the various federal income tax laws that affect oil and gas operations. Among topics explored are the depletion allowance, including the effects of the Tax Reduction Act of 1975; the tax consequences of farmout agreements, including the impact of 1977's catastrophic Revenue Ruling 77-176; the deductibility of intangible drilling and development costs, including prepayments; and methods of avoiding unnecessary classification as an "association taxable as a corporation".

Chapter Seven, "Exploration Techniques", considers the origin of hydrocarbons; how oil and gas is "trapped" in reservoir rocks; surface and subsurface maps; the types and use of well "logs"; the use of geophysical exploration techniques; and how to convert this information into a decision to drill.

The subject of well logs is so critical to an understanding of petroleum operations that our final chapter, Chapter Eight, is devoted entirely to "Electrical Well Logs". The many uses of electrical logs and the many variations of logging techniques which are available are explored in this chapter.

Throughout the book illustrations, charts and sample documents are provided to make understanding easier. In addition, each chapter is introduced by "Editor's Comments", to tie in the material that follows with your Landman's role.

— *Lewis G. Mosburg, Jr.*
Oklahoma City, Oklahoma — June, 1978

3

Chapter One

EDITOR'S COMMENTS

"Know the past to understand the present and predict the future!"

There is no way for you, as a Landman, to understand our system of land ownership, or to know what clauses can and cannot safely be deleted from the Oil and Gas Lease, without understanding how these aspects of petroleum land operations evolved.

In this chapter, "Evolution of Land Ownerships", Jeff Womack traces the evolution of our U.S. system of land and mineral ownership and of the Oil and Gas Lease. Womack also gives useful tips (§ **1.11**) concerning the conduct of the Landman's "record check" (also see Chapter Two, "Petroleum Land Titles").

You should find the following points discussed by Womack of particular interest:

* The division of land ownership between Federal, state and private individuals, including the ownership of the Continental Shelf (§ **1.04-1.07**).

* The meaning of the "law of capture", which, in the absence of statute, governs a landowner's right to explore for oil and gas, and the problems created by this rule (§ **1.09**).

* The various types of interests that can be created in oil and gas (§ **1.09**).

5

- The development of the modern day Oil and Gas Lease (§ **1.10**).

— L.G.M.

Chapter One

EVOLUTION OF LAND OWNERSHIPS

By J.F. Womack *

§ 1.01 Introduction

Scientifically, it has been estimated that the earth is about four billion years old. The scientific members of the Exploration Team have accumulated a backlog of information about the earth. They have learned that Antarctica was once a tropical climate. The North and South Poles have shifted at times during the geologic past. The new Continental Drift theory expounds the idea that there was once one large land mass that divided into continents known today and that they are constantly shifting. They have discovered that the geologic rocks found in Florida are similar to the rocks found on the bulge of North Africa; that new land is being formed in the Atlantic and Pacific Oceans; and that land along the Pacific Ocean and along the Mediterranean is disappearing back into the mantel. This has opened up new areas of interest for new prospects for oil, gas and other minerals.

* *Editor's note:* Womack is manager of Landowner Relations for the Exploration Department of Exxon Company, U.S.A. Houston based, Womack was the 1978 president of the American Association of Petroleum Landmen.

In the geologic time scale, the age of man and in turn the ownership of land and the evolution of land law has occurred in very recent time. If we used a 24-hour day as the time scale for the four billion years of existence of the earth, the age of man would be the last second of the 24-hour period. The discovery of oil would be a part of the last one-quarter of a second.

In spite of this contrast, it has taken a millenium to develop the laws that we have today. Just like the shifting of continents, these laws are constantly changing and developing. It has been a change of evolution rather than revolution.

I think it is important for a Landman to know something of the history of the land since land is the basis of all wealth. It is certainly the basis for a Landman's wealth or at least his livelihood.

§ 1.02 European Law and Its Influence

With the evolution of land ownership and land law, it is not necessary to go back to the tribal and nomadic type of land usage. Since English law or the common law became the predominant force here in the United States, we need to begin with the Norman Conquest of 1066. This conquest gave to England a new dynasty, a new ruling class and a new system of landholding.

Grants were given as rewards for those associated with the conquest and consisted in part of land for use and occupancy and in part of rights and privileges for services to be rendered to the lord, baron, or landholder under the sovereign. Some were free except for allegiance to the sovereign and some unfree because they were always owned and administered by the lord, baron or other holder of the grant or use.

The catalog of estates evolved by the common law were as follows: the fee simple, fee tail, life estate, estate for years, periodic estate, estate at will and estate at sufferance. Some of these you will recognize in our present land laws.[1] So it is true that the imprint of the past is discernible in the present.

---

[1] *Editor's Note:* See **§2.02(b)**, "Transfer of Title — In General".

One way to transfer property in those days was by "livery of seisin". Feoffment was the name of the ceremony of transfer. In this ceremony the buyer and seller walked upon the land with witnesses. The seller would take a handful of dirt and pour it into the hand of the buyer and then walk off the land leaving the buyer in possession of the land.

Through these hundreds of years these estates grew in usage, changed, were legislated away, or the customs changed where they were no longer applicable and usable. The Statute of Uses in 1536 introduced new methods of creating and transferring estates, made possible creation of new types of legal future interests, and gave rise to our modern trusts. The Rule in Shelley's Case was another landmark and affected interests of remaindermen. Through these old laws came the joint tenancies and tenancies in common whereby undivided interests were created, became transferrable, and subject to the heirship and laws of descent and distribution. These laws are the basis of interests being researched, leased, and administered every day by Landmen. Much of this English common law prevails in most of the states of the United States today.

In addition to the common law, we have the influence of the Spanish Civil Law. During the 16th century when Spain held domain over most of the western world, Cortez and Coronado and other explorers brought back gold, silver and other precious gems. Charles III of Spain issued a Royal Mineral Ordinance in 1783 which stated in part that "all mines, gems and minerals and *earth juices* belong to the Crown." When Mexico and Texas were republics they adopted constitutions which honored the laws of Spain on grants of property.

The ownership by France is reflected in the laws of the State of Louisiana under the Code of Justinian or better known as the Napoleonic Code. You will find that Louisiana laws differ considerably from those in the other states.

So we do have all of these different histories incorporated into our present laws even though they have changed. The influence of other European nations is reflected in local customs and laws where the settlers from those nations made their homes.

§ 1.03 The Colonial Period

During this country's colonial period, we had Spain holding lands in the west, southeast and southwest. France owned the strip generally along the Mississippi north and south including the Ohio Valley. The British Crown granted to various groups and individuals and later to the colonies themselves land grants extending from the Atlantic Coast to the Pacific Ocean. No one at the time knew just how far it was to the Pacific Ocean. Nonetheless, the grants were given under varying terms and some extended to the Pacific Ocean. The states of Massachusetts, Connecticut, Virginia, North and South Carolina and Georgia held such charters, but they did not mean much with the French holding the Ohio Valley, Spain claiming west of the Mississippi, and the Indians trying to hold on to their tribal lands.

The early land titles during this time of history were somewhat confused. In general, the philosophers granted the Indians the right of possession based on occupancy, but the discoverer nation of England, France or Spain had an option, so-called, vested in the king to purchase these titles through solemn treaty. They tried to bestow upon the Indian tribes their idea of real property like they had known in Europe. In any event some lands were purchased through solemn treaty and sometimes direct from the Indians without a formal treaty, and the titles were then passed on to individuals by such means as to encourage the spread of settlement. There were many speculators in land at that time, and this continued for many, many years. Try to place yourself in the position of the early settlers who conceived that land was a status dream. They had not been allowed to own land in England and Europe, and here they could afford to own large quantities and gain the status never realized in the old country.

During this period England, France and Spain were having problems in Europe and land was traded during this imperial struggle. Then in 1763 the Peace of Paris ousted the French from the Ohio Valley and Indians were defeated in many small battles. With these events the colonies reasserted their old claims and it opened the door to expansion to the Mississippi. There was a proclamation issued in the same year that expansion was not to extend beyond the Alleghenies until such time as the Indians agreed to an adjustment of the border. But the movement west could not be stopped even with some tremendous victories by the Indians during the Revolutionary War. During all of this period there were struggles between France and Spain and France and England, between Indians and settlers, between the colonies with overlapping claims to the land, and between speculators and new settlers.

Georgia was the only state who refused to relinquish her western lands to the United States when she ratified the Constitution. She sold over 25,000,000 acres for $207,580 to three groups of land companies. That would have been a pretty nice lease block to acquire at less than one cent per acre, wouldn't it? Of course it was not too good a purchase at the time near the Mississippi with Spain claiming part of it and the Indians still in the neighborhood. Another good buy at the time was the land near Nashborough, now Nashville, which sold in 100-acre lots for five cents per acre. The land promoters had all kind of schemes such as giving 500 acres to the first woman to settle and 500 acres for the first child born in the area. It kind of sounds like present day promoters who give you a set of dishes or some other gift if you will come to look at the property.

The Scioto group had obtained its option on five million acres in the Ohio Valley. To raise money for meeting their payments, the Scioto speculators sent a man to Paris to sell stock. He fell into the hands of an English schemer ironically named William Playfair. Together they formed the Compagnie de Scioto and flooded France with glowing tales: the Ohio country was "the most salubrious, the most advantageous, the most fertile" ever seen. Five hundred American "cultivators", they said untruthfully, were already erecting houses and planting this "garden of the universe" in preparation for the coming of settlers fortunate enough to secure acreage.

The Bastille had fallen; mob rule threatened France's established order. People besieged the Compagnie offices with their money. During February and March, 1790, five ships carrying more than 500 hopeful men, women and children plowed across the Atlantic toward what the passengers visualized as a Utopian heaven of gleaming white houses, to be named Gallipolis.

Playfair absconded. When the French landed at Alexandria, Virginia, they learned that the Scioto Company had not met its payments and therefore owned no land. The originator of the Scioto scheme made amends of a sort. He helped secure lands for the bewildered newcomers opposite the mouth of the Great Kanawha, inside the Ohio Company tract. He arranged for John Burnham to recruit 50 men at 26 cents a day to clear a few acres and build barracks — several rows of drab log buildings divided into 80 rooms. There most of the shockingly unprepared immigrants settled down to rebuild their lives.

As was mentioned, after ratifying the Constitution, the states gave up western lands to the U.S. government. One of Thomas Jefferson's land committees, searching for an orderly method of putting the public domain into private hands, faced two main problems: the price and the measurement. The minimum price was set at $1 per acre and the minimum acreage at 640 acres or a square mile. This proved too costly to most settlers who found that they could not clear that much forest for cultivation. They ultimately forced the issue so they could buy 320 acres and pay for it in four equal installments.

There were two systems in existence at the time in the bounds of a piece of land. The Virginia system prevailed in the south where a man had simply to step off whatever he wanted, regardless of shape, guiding himself by natural landmarks — hollow oak trees, creek beds, and the like. Afterward he had the plot surveyed and registered it in the land office, but disputes arose over overlapping boundaries and generally it was a chaotic system. If you have had any experience in Kentucky and West Virginia, you will find these types of descriptions today which are very cumbersome in determining the boundaries between land and ownerships.

The Congressional committee decided to end the chaos by surveying the western lands in a precise pattern before permitting occupancy so the land system in vogue in New England was adopted. The basic unit of measurement was a square, six miles to a side, called a township. Each township was to be divided into 36 pieces, called sections, each a square mile. The north-south boundary lines of the townships were called "range" lines and the east-west boundaries, "township lines". Southerners were slow to agree to this, but finally this pattern prevailed and became the official surveying method of nearly every civilized nation in the world.

All of these laws of course did not become law without a lot of discord, but ultimately the famed Ordinance of 1787 was passed so that at last there was workable machinery by which the western territories could develop as their populations grew toward full and independent statehood. During this period of history no one had heard of an oil and gas lease, and during this period whale oil was used.

There have been many explanations by experts on the various aspects of the energy crisis, but it is interesting to know that this is not the first oil crisis in the world. There was one about 100 years ago. Before oil was discovered in Pennsylvania, sperm oil or whale oil used for lamps soared from 23 cents a gallon to $2.55 per gallon in 1866. After oil was discovered, the price dropped to 40 cents per gallon and finally out of the market entirely. Given a free market system, this will happen again as the world discovers new sources of energy to meet the demands.

From all of the confusing years of the colonial period, it is obvious that private ownership of land was encouraged to more adequately develop this new vast area of wilderness. The trend toward private ownership is becoming a thing of the past as shown by the tremendous public domain still owned by the United States.

§ 1.04 The Public Domain

After the Revolutionary War, the states ceded their claims west of their traditional boundaries to the national government. The successive purchases and annexations by which the nation expanded to the Pacific and its present limits should be a familiar story.

In 1803, the Federal government acquired in excess of 500 million acres of land through the Louisiana Purchase. The following story is a bit of an oldie and you may have heard it before. It appeared in the *Houston Chronicle* and was submitted by Newell Choate of Hitchcock, a land title expert.

"The Post Office Department at Washington, D.C., searching the titles to post office sites in Louisiana, was dissatisfied with one because it went back no further than 1803. To the department's request for earlier information, the attorney for the owners replied as follows:

"Please be advised that the government of the United States acquired the title of Louisiana, including the tract to which your inquiry applies, by purchase from the government of France, in the year 1803. The government of France acquired title by conquest from the government of Spain.

"The government of Spain acquired title by discovery of Christopher Columbus, explorer, a resident of Genoa, Italy, who, by agreement concerning the acquisition of title to any land he discovered, traveled under the sponsorship and patronage of her majesty, the queen of Spain.

"The queen of Spain had received sanction of her title by consent of the Pope, a resident of Rome, Italy, and ex-officio representative and vice-regent of Jesus Christ. Jesus Christ was the son and heir-apparent of God. God made Louisiana.

"I trust this reply complies with your request."

In 1819, the purchase of the Floridas added nearly 43 million acres more. In excess of 334 million acres were acquired from Mexico under the Treaty of Guadalupe Hidalgo in 1848. Nearly 79 million acres were purchased from Texas.

The Oregon Treaty in 1846 brought an additional 181 million acres, and the Gadsden Purchase in 1854 added nearly 19 million acres more. And finally in 1867, we purchased Alaska and acquired 365 million acres. (See Figure 1 for a pictorial review of these acquisitions.)

The annexation of Texas in 1845 did not add to the Federal lands since Texas, being an independent nation, was able to negotiate the terms of its admission to the union and retain its public lands.

There are 2.2 billion acres onshore in the United States, and in spite of the system of private ownership, the Federal government still owns 755.3 million acres or about one-third of the total. Several years ago, the Public

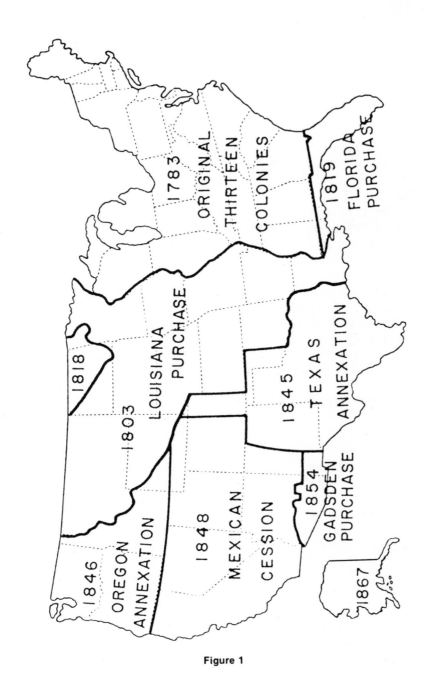

Figure 1

15

Land Law Review Commission was established, and a very lengthy and comprehensive report was made on the future possibilities for use of this land. This is available from the Government Printing Office in Washington, D.C.

The treatment of the public domain by Congress, to whom control is given by the Constitution, has been divided into three major phases: (1) a period of sale of lands for revenue purposes, (2) a period of disposal of lands for internal improvements, and as a means of promoting settlement and mineral developments, and (3) a period of retention of public lands for the stated purpose of conservation, involving the withdrawal or reservation of large areas from disposition.

By 1950 through sales, homesteads, grants to railroads, grants to states for reclamation purposes, and for support of education, etc., over one billion acres had been disposed of, but there still remains about one-third of the total under Federal ownership and control.

§ 1.05 The Continental Shelf

In addition to the 755.3 million acres onshore, the United States owns most of the Continental Shelf, which, bordering the Atlantic Coast contains 127,000 square miles; the Pacific Coast, 18,500 square miles; and the Gulf Coast, 144,000 square miles. I do not have the number of square miles off Alaska, but nearly all of the Bering Sea is Continental Shelf.

The ownership of this Continental Shelf is interesting. As far back as 1667 the English common law had considered that the marginal sea and the arms thereof belonged to the Crown. After the worldwide explorations started by Columbus, nations claimed whole oceans and seas, but these claims could not be enforced. Then a new doctrine appeared, first started in 1702, which in effect stated that the *effectiveness* of occupation and control of the seas was the basis of ownership. This proposal was that the sea might be appropriated to the distance that a cannon shot could reach. Thereafter an Italian jurist, Galiani, proposed a fixed range for cannon shot of three miles. This is the basis of the three-mile limit.

Just how far a nation may now claim is not a settled matter in international law. Our Federal government has claimed to the edge of the Continental Shelves, which have a maximum width off the Atlantic Coast of 100 miles, off the Pacific of 25 miles, and off the Gulf of Mexico of 140 miles.

You may be familiar with the Tidelands Controversy. Prior to about 1930 it had been generally considered, even by the U.S. Supreme Court, that the several states owned their marginal seas, at least to their traditional boundaries or historical limits, usually three miles. Texas, however, whose boundary had been set at three leagues or 10½ miles from its coasts, was an exception.

There was much agitation over these marginal seas which culminated in the decision in the case of *The United States vs. California* on June 23, 1947. The Supreme Court held that California is not the owner of the three-mile marginal belt along its coast and that the federal government rather than the state has paramount rights in and power over that belt, an incident to which is full dominion over the resources of the soil under the water area, including oil. In subsequent decisions the court held that neither Louisiana nor Texas stood on different footings from that on which California stood. Finally the matter was settled by Congress for the most part ceding to the several states the parts of the Continental Shelf within their traditional boundaries by the Submerged Lands Act. Generally the ownership has been determined.

§ 1.06 State Ownership

It is important to recognize that the various states are proprietors and charged with the administration of about 100 million acres. As mentioned, the original 13 colonies retained land not previously granted to private owners within their boundaries when the United States was formed. Most of this land eventually found its way into private ownership.

More than 225 million acres have been from time to time granted to the states by the Federal government for various purposes: to promote public education, to promote internal improvements and, in some cases, to help build railroads. During the settlement period the states, including Texas, pursued policies designed to place the ownership of the land in private hands to a large extent, but later the policy, like that of the national government, tended to swing to that of retention.

As in common law where the sovereign owned bays, inlets and other arms of the sea affected by the tide, the states with few exceptions have usually retained ownership of lands beneath waters affected by the tides, and usually also the beds of navigable streams. In Texas there is a total of about 1¾ million acres of lakes, bays, rivers and submerged lands owned by the state.

§ 1.07 Private Ownership of Land

Unlike the situation in most countries of the world, the private ownership of a parcel of land by grant or patent from the sovereign in this country ordinarily carried with it ownership of the fee; that is, everything pertaining to the land, from the center of the earth to the zenith of the sky. This was in keeping with the common law tradition in the original colonies and territories. As to lands acquired from France, Spain, Mexico and Texas, the minerals were by legislation vested in the private owners of the surface, even though not originally granted under the laws of the previous sovereign.

So of all the land onshore, about two-thirds is owned by individuals, and this could be a significant reason for the fact that most of the oil found to date has been on this privately-owned land. It has been more available for exploring and leasing under the free enterprise system. Governments, as a rule, have not been greatly interested nor particularly successful in exploration, and in the case of the United States this idea was coupled with the broad doctrine that the government held public lands only in the interest of the people, and that its people were entitled to secure these lands for private ownership with the least possible restriction.

That is a pretty brief history of the last 1,000 years, but what we here are really interested in is, of course, mineral law.

§ 1.08 Ownership of Minerals

Who owns the oil and gas which is found under the surface of the land?

Who has the right to explore for and produce these hydrocarbons? This varies throughout the world. In old England, under the old regime in Russia, in the United States and in a few other parts of the world, mineral titles remained with the owner of the land. The government did not exercise the right of eminent domain, which right gave the government the right to take the minerals away. But under the leadership of France, most of the countries of western Europe and in other parts of the world have appropriated to their governments the undiscovered mineral resources, particularly those beneath the surface. Some countries like Canada allowed private owners to retain the rights where previously they had been specifically conveyed before certain dates.

During this period in our history, there were really two basic concepts of conflicting considerations:

(1) There was the assumption that mineral resources, which are wasting assets, accumulated through long geologic periods, are peculiarly public property — not to be allowed to go to private ownership, but to be treated as a heritage for the people as a whole and to be transferred in the best possible condition. Some of the early minerals to be developed were either for money or war purposes, leading to the acceptance of the idea that these belonged to the government or to the sovereign.

(2) The other assumption was that the discovery and development of mineral resources require a free field for individual initiative, and that the fewest possible obstacles are to be put in the way of private ownership. Therefore, in framing laws of ownership, concessions have been made to encourage private initiative in exploration and development.

Worldwide there has been emphasis on these two underlying considerations. In the United States, at one extreme, the laws have usually been such as to give the maximum possible freedom to private initiative and to allow easy acquirement of mineral resources from the government. At the other extreme in South Africa, Australia and South America, it is impossible for the individual to secure title in fee simple from the government; he must develop the mineral resources on what amounts to a lease or rental basis, the ownership remaining in the government.

Most of the nations are tending to this latter emphasis. Even in the United States under the Mineral Leasing Act, the emphasis has changed for the remaining public domain for coal, petroleum, gas and a number of other minerals.

In spite of this trend, the majority of the minerals in the United States are privately owned and, of course, this is of interest to Landmen.

Large use of mineral resources is of comparatively recent date. Some of the mineral industries are not more than a decade or two old and the greater number of them are scarcely a century old. In the United States the mineral industry dates mainly from the gold rush to California in 1849. The formulation of laws relating to the ownership of minerals has on the whole *followed* rather than *preceded* the development of the mineral industries.

§ 1.09 Property Rights in the Minerals[2]

After discovery of oil at Titusville, Pennsylvania in 1859, rapid strides have been made in understanding the occurrence, nature, and behavior of oil in the ground since it became commercially important. The laws relating to the property rights in oil were developed at a time when the lawmakers, the litigants and the courts had, at best, only a vague and imperfect understanding of these matters. Some courts drew an analogy that oil and gas were similar to rivers of water underground. Other courts compared oil and gas to percolating waters. Probably the largest group of courts took the view that oil and gas are like wild animals. Legal principles and precedents relating to wild animals were therefore applied to oil and gas, frequently with unfortunate results.

The migratory nature of oil and the application of the wild animal principle led to the so-called law of capture. Under this rule if one landowner drilled an oil well on his land, even though it drained oil from under his

---

[2] *Editor's note:* For a further discussion, see § **2.03**, "Ownership and Transfer of Title to Minerals".

neighbor's land, the oil belonged to him. The neighbor could protect himself only by drilling his own well and capturing his own oil. This widely accepted principle has been justly and severely criticized because of inequities created and the wasteful drilling and overproduction it causes. Proration in most states has supplied a partial answer to the problems created, and unitization, both of drilling units and of entire fields, is receiving more and more attention as an even more effective solution of the problem.

The courts in the United States are divided on the theory of the landowner's interest in the oil and gas underlying his land. Some courts consider that he owns these minerals in place. Texas, West Virginia and Mississippi courts take this view. Other courts take the more widespread view that the landowner does not own his oil until he reduces it to possession. They do ascribe to him the sole right to explore for and produce minerals on his land, so that while differently arrived at, under normal conditions both approaches can lead to reasonably satisfactory results. Under either theory in most oil states the ownership is qualified to the extent that the police power may be exercised in the public interest to prevent waste. The owner of the fee simple title has numerous separate rights in his minerals. With these he may deal separately in an infinite number of ways.

It may be helpful to distinguish a few of the interests that are bantered around in the industry with which the landowner can deal.

Fee Simple Ownership. Usually this is simply called the fee. If you own all of the land, the surface and the minerals, you own the fee simple title just like under the English common law. When you have this kind of ownership, you have the right to the surface; you own the minerals; you have the leasing rights and are entitled to the lease bonus payments; and you own the royalty.

Surface Owner. A surface owner owns only the surface. Sometimes it is referred to as the "fee" in the surface which stretches the exact definition of fee, but he is not entitled to lease the land for oil and gas because he does not own the mineral rights. However, he is affected because the surface is used for access roads and well sites.

Mineral Owner. The mineral owner owns the minerals separate from the surface. Again, usage has developed this term into the mineral fee. He has the right to remove the minerals from the premises, and he has a superior right over the surface owner since the mineral estate is the dominant estate. The owner of the minerals is the only one who has a right to lease the minerals for oil and gas.

Royalty Owner. A royalty owner owns a portion of the oil or gas that is produced from the land. The royalty owner does not have the right to lease the land for oil and gas nor is he entitled to any of the bonus or consideration money. He is entitled to his royalty only after production of the oil and gas. He owns this royalty even if the lease comes to an end. It is sometimes called a "perpetual" royalty.

Overriding Royalty Interest. The overriding royalty is like a royalty in that it is an interest in oil and gas after it is produced from the land. It differs from the royalty interest in that it comes out of the working interest or that part of the production that goes to the person who has acquired the lease rather than to the landowner. This interest terminates when the Oil and Gas Lease terminates and it is not a perpetual royalty.

Payments out of Production. A production payment is also carved out of the production attributable to the working interest or lessee interest under a lease, but differs from the overriding royalty and royalty interests in that it terminates when a specified amount of money has been received out of production.

§ 1.10 The Lease and The Producers 88

You could almost trace the history of the law of oil and gas through the various changes that have occurred in the lease forms. The very early leases were a one paragraph letter type of a contract (see Figure 2). The original lease on the Drake lease at Titusville, Pennsylvania, provided for a term of 15 years with the right of renewal. Rental, which was really a royalty of one-eighth, specified that the one-eighth could be purchased at 45 cents per gallon. Incidentally, that comes to over $18.00 per barrel which is over 100 years ago. Possibly that will make you a little happier about today's

Dated December 30, 1857
Deed Book P, p. 357
$1 in hand.

Pennsylvania Rock Oil Company
               to
E. B. Bowditch and E. L. Drake

'Demise and let' all the lands owned or held
under lease by said company in the County of
Vanango, State of Pennsylvania, 'To bore, dig,
mine, search for and obtain oil, salt water,
coal and all materials existing in and upon
said lands, and take, remove and sell such,
etc., for their own exclusive use and benefit,
for the term of 15 years, with the privilege
of renewal for same term. Rental, one-eighth
of all oil as collected from the springs in
barrels furnished or paid for by lessees.
Lessees may elect to purchase said one-eighth
at 45 cents per gallon, but such election,
when made, shall remain fixed. On all other
minerals, 10 per cent of the net profits.
Lessees agree to prosecute operations as early
in the spring of 1858 as the season will
permit, and if they fail to work the property
for an unreasonable length of time, or fail to
pay rent for more than 60 days, the lease to
be null and void.'

By agreement dated February 12, 1858, and recorded in Deed Book P, p. 441, the above lease
is amended so that 12 cents for every gallon
of oil shall be in full of all other rental;
also giving lessees the privilege of renewal
for 25 years.

Figure 2

prices. This lease also called for drilling in the early spring as the season would permit, the lease being dated in December. All of the early leases provided for immediate or early drilling because it was thought at that time that the oil was fugitive like a river and if you did not drill in a hurry, the oil would move to adjoining land. The idea was to capture it and reduce it to possession in a hurry. So all of the early laws with this fugitive nature in mind created some of the laws still on the books today.

The early leases also followed agricultural leases and the salt leases prevalent at the time. The early settler who did not have access to the sea found that salt water lay beneath the surface, and they needed this salt for their cattle and their own consumption. In fact, some feel that the first discovery of oil was after boring for salt water. These salt leases provided for the owner to be entitled to a share of whatever was produced so the oil leases naturally followed this pattern.

Salt has some history in Texas. During the Civil War salt became in short supply, and one large source was located in Hidalgo County. Disputes arose over the ownership of the salt. After a thorough research of the records, it was determined that salt was a mineral and was therefore reserved to the sovereign by that Ordinance by Charles III of Spain. Therefore, title passed to Texas and Texas had title to the salt. As a result of this controversy, the new state constitution, adopted in 1866 after the war, contained a provision which quitclaimed title to all minerals to the owners of land in Texas.

There were generally two periods of development of the lease. One was the immediate need to drill to capture the oil or to terminate the lease. Later, the second period came into being when more was known about the nature of oil and gas and that is the period when land was held for future development with delay rental provisions and continuance of the lease while it was producing. There were many intervening leases that occurred as the second period of development came into current use. Court decisions at the time were based on two ideas, the fugitive nature of oil and gas and the immediate development necessary for the owner to get his consideration for signing the lease. There was no cash bonus, no delay rental clause, and no thereafter clause to keep the lease in force at the end of the primary term, but these decisions became painful for the operators. The modern lease form resulted after a decision in an Oklahoma case styled *Brown v. Wilson* in which the lease was attacked by the landowners suing to cancel the lease. Although the decision was reversed two years later, attorneys representing some of the major oil companies at the time got together and drafted a lease which for the most part prevails today in the basic and standard provisions. They took the lease form to Burkhart Printing Company in Tulsa, Oklahoma for printing, and the printer put "Producers 88" in the upper left hand corner. This title is a subject of some debate. Some think the "Producers" refers to the Producers Oil Company and others think it was just the printer's idea since it applied to producers generally. Some think the "88" resulted from the 88th form printed from an index ledger. Others think it was just an arbitrary number or it may have been just to distinguish it from a Producers Oil Company Form 8 which had been printed and which resembled the new form. In any event, the name "Producers 88" has become a standard term in the industry and you still find that terminology probably followed by the word "Revised". It has been revised many times and is still undergoing changes through court decisions and new governmental regulations.

It is somewhat amazing that basically the form adopted has prevailed. All of the leases will have four basic provisions: (1) The lease provides for a consideration or cash bonus, (2) it has a fixed term, (3) it has a royalty, and (4) it has a delay rental provision allowing for drilling or paying to delay drilling within the primary term. Of course, there are exceptions with paid up leases or leases for a term of one year, but generally the leases in existence today have three basic provisions adopted many years ago.

§ 1.11 "Take Off" of the Records

Getting down to a specific geological prospect for oil and gas, the Landman works with a land map covering the area of interest. From this map an ownership "take-off" is made to see how the property is owned and if it is available for leasing.

On private ownership lands, he will check the records in the county courthouse to see who owns the minerals under the land and whether there is presently a lease on record covering such minerals. Many times, a Landman may use the records of an abstractor or commercial lease ownership maps, or perhaps an on-the-ground investigation in lieu of county records in an effort to determine whether a tract is leased. Most major companies, and many smaller companies and individuals, subscribe to a commercial lease ownership map service such as Tobin, Pomco and the like. A reference is made on the map to the landowner's name at the bottom of a particular subdivision of the map. If such tract is leased, the name of the company or individual who acquired the lease will appear in the middle or upper portion of the tract, together with the date on which the lease expires. This type of map is quite valuable to a Landman since it saves him considerable time in checking county records to see if a particular tract is leased. These maps are updated monthly, and generally are very reliable.

If it is necessary to check the records in the courthouse, there are available grantor and grantee indices. If you have either the grantor or the grantee name available, these records can be searched. They are arranged by years, and it can be a time-consuming job tracing a title down to date.

On public lands, including state and Federal, the same type of information is available from commercial lease map services as for private lands. Information may otherwise be obtained on Federal lands from the appropriate office of the Bureau of Land Management. On state lands, the State Land Office is generally the source of such information.

Chapter Two

EDITOR'S COMMENTS

An understanding of the evolution of land ownership, and the development of the Oil and Gas Lease, is important for an understanding of your role as a Landman. However, where "Petroleum Land Titles" are concerned, you are expected to be able to actually review title records to determine from whom the lease should be taken and what must be done to render title acceptable for the company's present purposes. This requires that you understand:

• What you are looking for when you "examine" a title (§ **2.01**).

• The difference between "marketable" and "business risk" title (§ **2.01**).

• How to get title out of the Federal or a state government, or a restricted Indian (§ **2.02(a)**).

• The necessary elements for transfer of title by conveyance, judicial action, adverse possession, and other means of title transfer (§ **2.02** et seq.).

• When adverse possession *will not* cure a title (§ **2.02(e)**).

• How properly to draw a mineral conveyance or reservation (§ **2.03(b)**).

• What to look for in examining a mineral conveyance (§ 2.03(c)).

• What type of curative steps to take in curing various title defects, and pitfalls to avoid as you attempt to cure a title (§ 2.04).

— L.G.M.

Chapter Two

PETROLEUM LAND TITLES

By Lewis G. Mosburg, Jr.

§ 2.01 Basic Concepts

No operator can drill a well without first having secured a valid oil and gas lease which gives the operator the legal right to extract the oil and gas. No matter how thoroughly the geologist or the geophysicist has done his job in locating a promising prospect, if leases cannot be secured on the prospect area the work has been in vain. Similarly, if the leases are taken from the wrong persons, or prove not to cover the anticipated interest in the minerals, the operator will find that he has taken all the risk of loss in drilling the well, but is now entitled to only a portion, or none, of the fruits of its efforts if the well is successful.

It falls on you, as a Landman, to make numerous decisions involving petroleum land title law in determining from whom to take these leases. Often the legal department can be consulted where sticky title questions arise. However, in the normal course of business, time and other practical considerations necessitate that you serve as a "mini-lawyer" in preliminarily resolving many title questions.

It is not the purpose of this chapter to make you a legal expert, or to suggest that you will not have to call on your company's legal counsel in answering difficult legal questions concerning the petroleum land titles. Also, the laws of each state vary as to the details of its law concerning land titles. However, this chapter does seek to present basic guidelines, applicable in most areas of the United States, to serve as ground rules concerning petroleum land titles, and to help you both recognize the common legal issues affecting petroleum land title law, as well as aiding you in reaching decisions in your day-to-day leasetaking.

The Anglo-American legal system permits private "ownership" of land, i.e., permits an individual to acquire the right to use and enjoy property to the exclusion of the world at large, and to transfer this "ownership", or right of use, to others. The process of determining who is the "owner" of a given tract of land is referred to as "examination of title".

Title examination may involve the examination by an attorney of law of an abstract of title and the preparation of a formal "title opinion". However, unless the drilling of a well is imminent, or a large bonus is being paid for the lease, it is often you as a Landman who will serve as the "examiner", studying either a certificate of title or the title records themselves to reach your conclusions as to the ownership of the land, and from whom you should secure a lease. This act of "title examination" calls for you to determine whether or not title has been properly acquired from the government (the "sovereign"); to study the various subsequent transactions affecting the title to determine their validity and legal effect; and to ascertain whether or not the land is burdened by "encumbrances", such as outstanding mortgages, easements, unreleased Oil and Gas Leases, or restrictions concerning the uses to which the owner may put the property. The act of title examination also requires that you be alert for "danger signals" which would impose a duty to inquire outside the record concerning the existence of other possible claims, or which serve as a warning that certain non-record claims may exist which would prevail over the rights of one purchasing from the apparent "record owner".

In the course of your examination, you must have a standard to use in determining whether or not your lessor's title appears to be acceptable for the company's purposes. While the attorney in preparing a formal title opinion will normally apply the standard of "marketability" — usually interpreted as a "perfect title of record" — you, as a Landman, will normally

30

seek to determine whether or not the title appears a "reasonable business risk".

"Business risk" is not synonymous with "safe" and "good" title: this is asking the impossible. Certain titles may prove defective even though they are "perfect" of record, and neither an attorney nor a Landman can guarantee that the title will be absolutely safe from attack. Similarly, what is a reasonable business risk will be affected by the purpose of the examination and the philosophy of the company. Obviously, a risk that is "acceptable" when you are "checkerboarding" leases for a dollar an acre may not be acceptable when a decision is reached to drill a multi-million dollar test well. Also, many production purchasers insist on a marketable title reflecting "perfect title of record" for division order purposes. However, when leases are being acquired based upon certificates of title or record check, or where decisions are being reached as to whether or not to waive requirements raised on a "marketability" basis in an attorney's title opinion, a more practical approach is taken, utilizing the "reasonable business risk" standard.

§ 2.02 Ownership and Transfer of Title to Land

§ 2.02(a) Inception of Private Title. As a result of various treaties, annexations and wars during the 18th and 19th centuries, the Federal government of the United States became an owner of all the land now comprising the "lower" 48 states and the state of Alaska. The pattern of these acquisitions is outlined in Chapter One.

As discussed in Chapter One, as a result of various grants our country has three major classes of owners: the Federal government, the states, and finally, private individuals who own over two-thirds of this country's land.

In most parts of the world, private ownership of land does not include title to the minerals. However, in the United States, private ownership can include mineral ownership. And it is on this privately-owned land that the substantial part of the oil and gas exploration of the United States has been conducted.

Private title must commence with a grant from an appropriate "sovereign". In the United States, the original source of title will usually be a grant from the United States of America. (An exception is except as to those lands lying with the original 13 states or the state of Texas.)

The necessity of a Federal grant applies not only to the lands granted directly by the Federal government, but to most state and Indian lands as well.

Congressional grants could be made directly to various private owners; they usually were made in that way where the grantees were public bodies or railroads. However, most Federal grants to individuals were made pursuant to general acts, such as the Federal Homestead Act.

Under the Homestead Act and similar statutes, the "entryman" earned his right to a "patent" from the Federal government by physically entering on the land, and occupying, cultivating and improving it in the manner called for by the statute. Once his right to a conveyance from the Federal government has been "perfected" by these acts, the entryman was entitled to his patent.

Prior to perfecting his right to a patent, the entryman's rights were not subject to sale by him, and could not be inherited or devised by will. However, once he had perfected his right to a patent, he was issued a final receipt or certificate by the registrar or receiver of the appropriate United States Land Office. This receipt or certificate vested an equitable title in the entryman; and he could now sell the land and grant mortgages or leases. The land also now formed a part of his estate, and thus could be inherited or devised. However, for *marketable* title, it was still necessary for the entryman to secure a patent from the Federal Land Office.

The patent, once issued, was not subject to collateral attack by the Federal government; thus you, as a title examiner, do not have to look behind the patent to determine the propriety of its issuance so long as it is regular on its face.

Unlike most conveyances, a patent could be issued in favor of an entryman who was deceased at the time the patent was issued. Similarly, if the "chain of title" for the land in question was dependent on conveyances executed pursuant to an original conveyance from an entryman who was depending upon his final receipt, and had never secured a patent, the new patent would still be issued in the name of the original entryman, and not the present owner.

What if the conveyances in the chain of title predate the patent? If the conveyance is by *warranty deed*, and a patent or final receipt has subsequently been issued, the title is sufficient subject to the securing of a patent. However, if the conveyance did not contain covenants of warranty, the examiner must determine if the final receipt or certificate issued to the entryman *predated* the conveyance. So long as a final receipt or certificate had been issued prior to the conveyance, the conveyance would serve to transfer the entryman's equitable title to the grantee. However, if the receipt or certificate was issued *after* the date of the conveyance, and the conveyance lacked covenants of warranty, the grantee acquired nothing by the conveyance.

A particular problem arises where the patent from the United States is dated subsequent to July 17, 1914, the date on which provision was first made for a discretionary reservation of minerals by the United States upon the issuance of a patent. Subsequent grantees are bound by the *actual* contents of the patent as reflected in the Federal General Land Office. Since reservations may have been missed when the patent was entered into the county records, as a careful examiner you should request a verbatim copy of the original patent from the General Land Office to insure that the patentee acquired title to both the surface and minerals from the United States.

Where the lands border a river, the patent from the United States to such "riparian" lands covers the land underlying the river bed to the center of the stream if the river is not navigable. However, where navigable waters are involved, title to the river bed passed to the state upon statehood, so long as the land had not been previously patented. (As to the effects of subsequent changes in the location of the river, see § **2.02(c)(5)**, *infra*.)

Often land was not granted directly to private individuals, but was granted by the United States to a state. The state may still own such land, or

33

may have subsequently conveyed title to a private owner.

The 13 original states and the state of Texas entered the Union as owners of most of the lands within their boundaries. The remaining states were granted millions of acres of public lands by the United States upon statehood. Additional lands may have been acquired by a state by purchase, mortgage foreclosure, or condemnation.

Where a lease is being acquired from a state, or a chain of title depends upon the state grant, you must first determine the validity of the state's title. Where title has been acquired by the state through private deed, mortgage foreclosure or condemnation, the same rules govern as would apply to a similar acquisition by a private individual. However, unless you are dealing with the lands of the 14 states that entered the Union as independent sovereigns, the state's title will depend upon grants from the United States, often made as a part of the declaration of statehood.

Frequently, the Congressional act itself serves as a conveyance. If the land under examination was specifically described in such a statute, no further conveyance from the United States is required. However, the majority of lands covered by such Congressional granting acts merely provided for the *selection* of such lands; and where the state claims title to a particular tract as a part of such "selection lands", you must secure a copy of the selection list filed with the General Land Office covering such land, together with the Federal government's certificate or endorsement of approval.

What as to the ownership of lands underlying the coastal waters of the coastal states? By legislation, it has been settled that the states' ownership extends three miles from its "coastline" into the Atlantic and Pacific Oceans, and nine miles into the Gulf of Mexico. The balance of the coastal waters — the Outer Continental Shelf — is owned by the Federal government. However, still unresolved is the exact location of the "coastline" — the low water line of the ocean — for each of the various coastal states.

After determining the validity of the state's title, you as an examiner must determine whether the prior or proposed conveyance from the state was, or will be, regular. This requires a determination as to which state agency or subdivision has been charged with the duty of administering such lands for the state, and whether or not this agency has the power to dispose of the land. The applicable state statute must thus be studied to ascertain:

(1) Was *ownership* of this land or interest authorized by the governmental agency?

(2) Was *disposition* of the land authorized?

(3) Were the proper statutory *procedures* followed, or will they be followed, in *selling* the land, and in *executing the conveyance?*

In many instances, the state may have been authorized, or required, to reserve some interest in the minerals upon sale. The statute and the sale proceedings must be studied to see if either provided for such a reservation; either might control even though the certificate of purchase, or possibly even the state patent, contained no reservation. Also, in some states, you as an examiner will be required to go behind the certificate of purchase, or even the patent, to determine that the sale from the state was properly conducted.

Special problems exist where the lands under examination involve lands presently or formerly owned by an Indian tribe or a restricted Indian. Title to the lands of the Indian tribes was held by the United States as a guardian for the tribe. However, it was the frequent practice of the United States to "allot" such land to the individual tribal members.

When examining title to lands so allotted, the examiner must first document the allottee's title by determining whether a proper Allotment or Trust Patent has been issued by the United States.[1] The title acquired by the allottee would then be alienable, except to the degree alienation was restricted by Congress. However, alienation normally *was* restricted, to a greater or lesser extent, by some Congressional act.

No general, uniform restrictions apply to the alienability of allotted

---

[1] The type of patent would depend upon the Congressional statutes applicable to the particular tribe.

lands. Instead, restrictions varied depending upon the tribe involved, the age and degree of Indian blood of the allottee, the source of the owner's title, and certain other factors, such as the designation of the land as "homestead" or "surplus". The applicable restrictions also were dependent upon the Federal statutes in effect at the time of the purported conveyance, which were changed with frightening frequency.

Restrictions, where applicable, could amount to an absolute prohibition against alienation, or a provision that there could be no alienation without the approval of the Secretary of the Interior or some designated Federal or state court. Any conveyance in controvention of these restrictions, or without the necessary approvals, was absolutely void, and, in the absence of a Federal statute, could not be corrected by state curative statutes or adverse possession pursuant to state statutes of limitation.

If an allottee died owning an allotted tract, the succession to his allotted land was also subject to Federal regulation, both as to the applicable laws of inheritance and the right to devise by will. The deceased's estate, as to his allotted lands, must also be administered, and any will probated, by the tribunal designated by Congress.[2]

In addition to the regulation of allotment, alienation and inheritance of Indian lands, Congress also regulated court proceedings involving Indian lands, and was a necessary party to any judicial actions affecting them; neither the United States nor the affected Indian would be bound by a court decree in an action where the United States had not been properly joined as a party. Such joinder required compliance with specific statutes enacted by the Congress to permit the United States to be joined as a party to actions involving Indian lands and required compliance with a specified, and detailed, statutory procedure.

In addition to allotted lands, each Indian tribe also possesses tribal lands which have not been allotted to individual tribal members. The sale or

---

[2] The limitations did not apply to non-allotted lands acquired by the Indian in his individual right.

lease of such lands is subject to applicable Federal statutes and regulations.

§ 2.02(b) Transfer of Title — In General. Title to land can be transferred by conveyance, by judicial action, or by various involuntary acts of title transfer such as adverse possession, death, after acquired title, or inurement. The rules governing the validity and effect of such transfers varies with the type of "title transaction" by which title is transferred.

Ownership rights in real property may be divided in various ways among various owners. Ownership may exist in *divided ownership* of adjacent tracts of land; or in *undivided ownership*, either as co-owners (tenants in common), joint tenants with the right of survivorship, or tenancy by the entireties (an unseverable joint tenancy between husband and wife). Ownership may also be divided between ownership of the surface and the minerals; and mineral ownership may be divided as to the substances covered thereby, as to depths and as to the duration of the ownership, as well as the size and legal nature of the various rights in the minerals.

The division of ownership as to the duration of the interests — division of ownership along time lines — applies to all forms of ownership of real estate. The person entitled to the immediate use and enjoyment of the property are said to own the *"present interest"*. Persons entitled to the immediate use of the property at some time in the future are owners of *"future interests"*.

Just as title to land cannot be abandoned, there can be no gap in the perpetual ownership of the land. Thus, whenever an interest in land ownership *may not* be capable of perpetual duration, such present interest must be subject to ascertainable ownership of the future interest.

If the owner of the land owns an interest which has no limit on its duration, he is said to own the *fee simple absolute*. If the owner owns a present interest which *may* endure forever, but is capable of being terminated by some event which may or may not occur, such as the cessation of use of the land for a specified purpose, he is said to own a *defeasible fee*, which will be followed by a *reversion*. Termination may be automatic *("so long as...")* or subject to a condition giving the future interest holders a right to claim the land ("but if... then grantors shall have the right to re-enter such land."). If the duration of the present interest is measured by some person's life, the present interest is said to be a *life estate*, which will be followed by a *remainder*. If the present interest is limited to a specified period of time, it is a *tenancy for years*, followed by a *reversion*.

§ 2.02(c) Transfer of Title by Conveyance.
§ 2.02(c)(1) In General. The most common method by which private title is transferred is by the execution of a presently-operative instrument intended by the grantor voluntarily to pass ownership of some interest in land to the transferee. Such an instrument is referred to as a "*conveyance*".

The key that distinguishes the conveyance from other attempts voluntarily to pass title are the words "presently operative"; a conveyance must be intended to take effect immediately to pass *some* title to the grantee. The grantee's interest may be limited to a future interest in the property; in other words, his right to the *use and enjoyment of the property* may be intended to be deferred until some time in the future. However, unless the interest is being transferred to him immediately and irrevocably, whether as a right presently to occupy the property or to occupy it in the future, the document will fail as a conveyance.

The rules concerning passage of title to land by conveyance are based upon two basic premises. First, the transfer should be by a *written instrument* executed in accordance with certain *formalities*. Second, to protect subsequent good faith purchasers, the conveyance should be recorded as a part of the public records. However, as between the parties, the conveyance will normally be binding if otherwise valid even though it has not been recorded, or even acknowledged.

The formalities required for a valid conveyance include that it be by a *written instrument*; that there be appropriate *parties grantor and grantee*; that the document contain *words of grant*; that it properly *describe* both the land and the interest being conveyed; and that it be properly *executed*, by being signed by the grantor and properly delivered.

§ 2.02(c)(2) Necessity that the Conveyance be in Writing. A "conveyance", by definition, is a written instrument. Statutes patterned after the English "Statute of Frauds" have uniformly been passed in all states and require that voluntary transfers of title be in writing. The adoption of recording acts in each of the states also creates a practical requirement for written conveyances. Similarly, where title to real estate is held in trust, local statutes normally require that the trust be created by an instrument.

§ 2.02(c)(3) Parties to the Conveyance. A conveyance must be executed by a party with *capacity to convey*, in favor of a party with *capacity to hold title* to real estate. If not, it is defective, and may even be totally void, even as to bona fide purchasers relying upon the apparent regularity of a recorded deed which does not reveal the incapacity of a grantor or a grantee.

Each party who is to be bound by the conveyance must be named in it as a party grantor, either by specific designation or by a general reference to "we" or "the undersigned". Minority or lack of mental capacity may also void a conveyance as to a particular grantor, or render the deed voidable.[3]

Just as the grantor must be designated in the deed, so must the grantee. Errors in properly naming the grantee will not void the conveyance, so long as he or she can be identified. (Such errors may affect the marketability of the title.) However, the grantee must be an existing legal entity such as an individual, a corporation or partnership, or a trust.

If the designated grantee is not in existence as a proper legal entity at the time of the conveyance, the deed is void; thus, conveyances to unborn or deceased persons, dissolved corporations, unincorporated associations, or the estate of a deceased person, create defects in the title.

§ 2.02(c)(4) Words of Grant. Every conveyance must contain *operative words* showing an intent on the part of the grantor presently to transfer

---

[3] What constitutes incapacity to convey, and the effects of such incompetency, is a matter of local law.

some present or future interest in the property to the grantee, or else the conveyance is void. Typical language is the classic "grant, bargain, sell and convey". However, any language showing the requisite intent presently to transfer title, such as "set over and assign" or "transfer", is sufficient.

§ **2.02(c)(5) Description.** A conveyance, to be valid, must contain a legally adequate description of both the land affected and the interest intended to be conveyed.

In the majority of states, the point of departure for a land description is the *rectangular survey system* established by the Congress in the 1780's by passage of the National Land Act. Under this survey system, six-mile square "townships" were established, each township contained 36 640-acre "sections", each one mile square. Boundary lines for these townships were lines running north and south ("meridians"), crossed by lines running at right angles to the north-south lines ("parallels").

Due to the curvature of the earth and human error, not all 36 sections within a township will contain exactly 640 acres. Accordingly, provision was made for irregular "lots" along the north and west of each township.

For examples of the operation of the rectangular survey system, see Figures No. 3 and 4.

Not all lands within the United States were surveyed under the rectangular survey system. The 13 original states, plus certain other eastern states, were surveyed under the Colonial Surveys prior to the enactment of the National Land Act. Many of the southeastern states were surveyed by "metes and bounds", and many other portions of the United States were subject to foreign grants and the survey systems on which these grants were based.

One of the most important exceptions to the rectangular survey system is the State of Texas. In Texas, grants were made and surveys conducted, by four different sovereigns — the Spanish government, the Mexican government, the Republic of Texas and the State of Texas — each using the Spanish "vara" as the unit of linear measurement. Most of these surveys, which are usually identified by the name of the surveyor, are noted for their substantial inaccuracies.

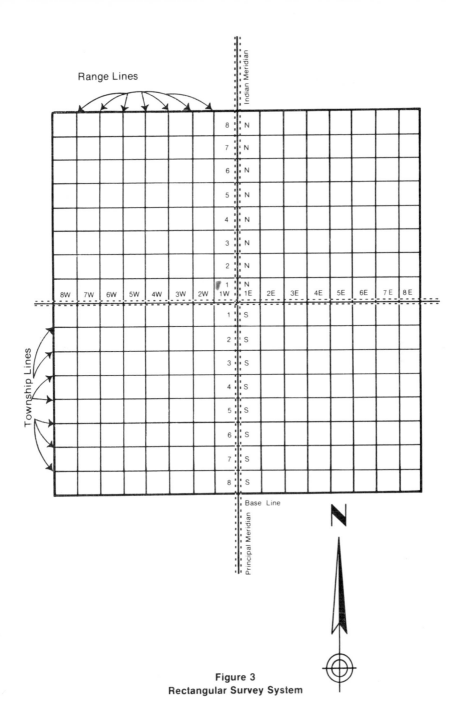

Figure 3
Rectangular Survey System

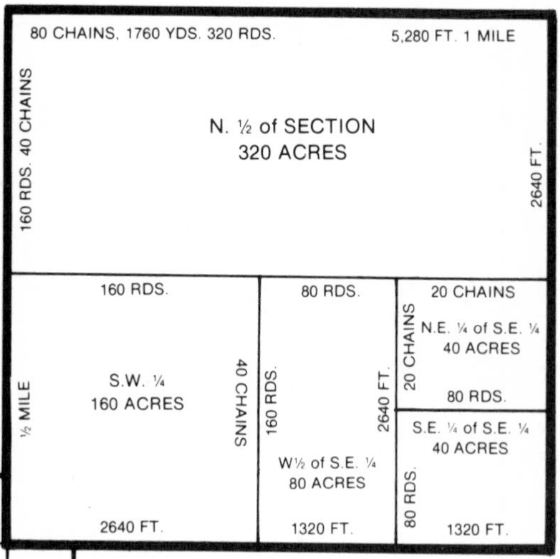

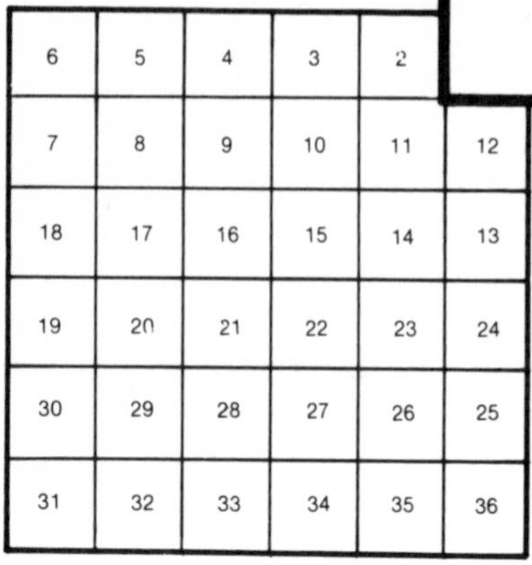

Diagram showing a Congressional Township of 36 Sections and plan of numbering same.

Section 1 is enlarged to show a regular Section divided.

LAND MEASURE

12 inches	make 1 foot
3 feet	make 1 yard
7 92/100 inches	make 1 link
25 links, 5½ yds, 16½ ft.	make 1 rod or pole
100 links, 66 ft., or 4 rods	make 1 chain
80 ch., 320 rds., 1,760 yds., 5,280 ft., 8,000 lks.	make 1 mile
144 sq. inches	make 1 sq. ft.
9 sq. feet	make 1 sq. yd.
30¼ sq. yds.	make 1 sq. rd.
40 sq. rds.	make 1 rood
4 roods, 160 sq. rds. or 43,560 sq. ft.	make 1 acre
640 acres make 1 sq. mile or 1 regular Sec.	

60 sec.	= 1 min.
60 min.	= 1 degree
30 degrees	= 1 sign
90 degrees	= 1 right angle or ¼ of a circle
360 degrees	= 1 circle

Figure 4
Rectangular Survey System

42

Once surveyed, how is an individual tract of land described for conveyancing purposes? The easiest and most accurate method of description is to describe the land by reference to its location under the rectangular (or other) survey system, either by conveying a specified section or a fractional part of a section, usually by reference to quarter section, quarter quarter section or half section.

If a tract of land to be acquired or leased cannot be described by rectangular reference, another description method must be used. Accordingly, the "metes and bounds" of a tract are frequently used as a description method: mathematically determined lines, with the "course" described by a reference to its angle from an established line, and its "distance" by its length or by reference to its points of beginning and end.

The metes and bounds description must have a fixed and ascertainable starting point which can be located on the ground, either in the form of a natural or artificial "monument" (which needs to be reasonably permanent), or, preferably, a point established by a prior survey.

Where the starting point or an intermediate point is a monument, the line will run to the center of the monument, unless it is extensive in nature — a river, mountain, etc. — where the line will run to the nearest side.

Boundaries of a metes and bounds description should "close" by returning to the starting point.

Where land is to be extensively subdivided, a "plat" or map of the subdivision is often prepared, dividing the tract into "lots" and "blocks", usually based on metes and bounds measurement. The lots can then be conveyed by reference to the lot and block as established on the plat rather than by the substantially more lengthy metes and bounds description. However, as in the case of a metes and bounds description, the plat must contain an ascertainable starting point, plus internal accuracy of measurement.

The methods of description just discussed are not exclusive. In addition, land can be described by reference to the lands or monuments adjoining the land; by a commonly accepted name such as "the Brown Farm"; by the size or acreage content of the tract; or by a "blanket" conveyance of "all land owned by me" in a specified county or survey. However, such description methods often create title irregularities.

Unless the description permits location of the land solely by reference to facts established by the public records, without uncertainty or reference to extrinsic facts, title is not marketable. However a description, though less than perfect, is not so fatally defective as to render the conveyance void if it can be shown, even by oral testimony, that there is only one tract of land to which the description could reasonably relate. However, if the description could refer to several tracts, the conveyance is void.

A conveyance of a riparian tract will cover all "accretions" to that tract — lands added by slow, imperceptible changes in the river's location. Similarly, changes in the river's location will change the boundaries of the adjoining tracts by "accretion" of that portion of land added by such slow and imperceptible change or similar reduction or "reliction", leaving the river as a dividing land between the tracts. However, where the change is sudden and perceptible — "avulsion" — such changes will not cause a change in tract boundaries.

Where land is conveyed subject to a right-of-way or other easement, or by reference to an adjacent road, easement or stream, land underlying this adjoining monument owned by the grantor normally passes to the grantee (or to the center of the monument if the grantor retains the adjoining tract). Similarly, exception of such an easement from the description normally excludes the easement only, and not any interest in the underlying land or minerals. However, interpretation of such conveyances is quite complicated; and it is wise to consult an attorney to resolve the exact extent of the land conveyed.

Problems concerning the description of the *nature and size* of the interest conveyed will be discussed later in this chapter.

§ 2.02(c)(6) Execution. A final requirement for a conveyance is that it be "executed" — *signed* and *delivered* by the grantor.

The requirement that a deed be signed normally requires personal subscription of the instrument by the grantor. However, under certain circumstances, a deed may be executed by an attorney-in-fact or by mark, so long as such execution strictly complies with established statutory procedures of the state in which the land is located. In the case of a conveyance executed by an attorney-in-fact, this requires a study of the power of attorney, to insure that execution of such a conveyance was within the authorized powers, and verification that the agency has not been revoked by the death or incompetency of the principal. (Death or incompetency might revoke the power of attorney even where there was nothing of record to alert you.)

"Delivery" refers not merely to a manual transfer of the signed instrument, but to some physical manifestation of the grantor's intent that the deed be immediately operative to transfer a present title to the grantee. Thus, handing over a signed instrument for inspection, without intending that it take effect, is not "delivery".

A key problem in delivery is the so-called "dresser drawer deed", where the owner, in an attempt to avoid probate, prepares deeds to his family but retains them in his possession to be discovered and recorded after his death. Such instruments are void as conveyances since there has been no delivery — they were revocable at any time up to the owner's death.

§ **2.02(c)(7) Other Conveyancing Technicalities.** *Acknowledgment and recording* are critical if constructive notice of the conveyance is to be given to subsequent purchasers and to cut off the rights of prior purchasers whose conveyances have not yet been placed of record. Acknowledgment and recording are likewise a prerequisite to the operation of most curative statutes.

Acknowledgment and recording are not usually required for the conveyance to be valid as between the parties.

The form of acknowledgment normally will be prescribed by state statute. The notary should be a disinterested party, and should not be an officer or shareholder of either the grantor or the grantee. However, the lack of "disinterested party" status will not invalidate the conveyance or its recording so long as the defect is not apparent from the face of the instrument.

Many states have enacted "local compliance" statutes, which provide that an acknowledgment taken in the form prescribed by the statutes of the state in which the conveyance is executed will be satisfactory even though not in compliance with the form established by the state in which the land is located and in which the deed must be recorded.

Many states have special requirements for conveyances of land in which either spouse may claim a *homestead* right, is community property, or which is subject to a spouse's dower or courtesy rights. A frequent requirement is that such an instrument be signed by both husband and wife (often, both are required to sign the same instrument), or that both be joined in any judicial action affecting the land.

Recitals that the grantor is single, or that the land is not homestead, will offer no protection if the recitals are not true. However, you will usually rely on recitations as to marital status (though not on recitals of non-homestead status).

§ **2.02(d) Transfer of Title by Judicial Action.** Often, judicial proceedings have the effect of transferring title to land. Thus, one part of a chain of title may be a mortgage foreclosure, a deed of trust, a probate proceedings, or an action to quiet title.

For a judicial proceeding validly to transfer title, the proceeding must comply with applicable statutory procedures, as well as constitutional requirements concerning "due process of law". In addition, the court must have jurisdiction over the land, the parties, and the type of action. Otherwise, the proceeding is void.

A court only has jurisdiction of land within the same state — a Texas court cannot probate a deceased Texas resident's estate as to his Oklahoma land. And to have jurisdiction as to the type of action, state statutes must authorize the court to hear this kind of case.

More complications arise in testing jurisdiction over the parties.

The primary problems in jurisdiction over the parties are whether or not the proper parties have been joined, proper service secured on the parties, proper pleadings filed and proper notices given.

A judicial proceeding will not serve satisfactorily to transfer title unless all possible claimants, including subsequent grantees or successors to the original owner, have been joined as parties. If possible homestead claims are involved, the owner's spouse must be joined. If the prior owner is or may be deceased, special statutes providing for service on the deceased's unknown successors must be complied with.

Even if all necessary parties have properly been designated as defendants to the proceeding, they will not be bound unless proper service has been secured upon them. Where the defendant resides within the state and can be located through a reasonable inquiry, "personal service" — delivery of a summons to the defendant by a court officer — is normally required. If the defendant is a non-resident or cannot be located, "service by publication" — publishing the summons in a newspaper — may be authorized; however, the procedure set forth in the statutes governing such service must be followed and a copy of the notice by publication mailed to all defendants whose addresses can be discovered by reasonable inquiry in order for due process of law to be satisfied.

Certain proceedings are looked at under the law as involving a determination of the rights in specified property as against the entire world ("in rem" proceedings), rather than a determination of the relative rights of named adverse parties ("in personam" proceedings). In such in rem proceedings, jurisdiction over the parties is gained by giving "notice" rather than by personal or other "service". However, such notice must follow the specified statutory procedure; and, as in the case of service by publication, a copy of the notice must be mailed or personally served upon all possible claimants whose identities and locations can reasonably be discovered.

Relief cannot be granted against a party unless such relief has been sought against that specified party. Likewise, the effect of a judgment will only be to bind those parties specifically covered by the final order and only as to the relief granted against them by the order. In addition, particularly in in rem proceedings, additional notices may be required at various stages of the proceeding for the final order to be binding.

Judgments by a court which lacks jurisdiction are always subject to vacation. In addition, default judgments and judgments against a minor may be subject to vacation for a specified period, although subsequent purchasers relying on the apparent regularity of the judgment in many instances take free of such claims.

§ 2.02(e) Transfer of Title by Adverse Possession and Other Involuntary Means.

Acquisition of title by *adverse possession* is the primary means of transferring title without the owner's consent and without judicial action. And even a bona fide purchaser relying on an apparent "record" title will not be protected if his grantor's title has in fact been divested by adverse possession.

Acquisition of title by adverse possession as prescribed by statute requires occupancy of the land in the same manner that an owner would make use of such land for the period prescribed by statute. The acquiescence of the true owner does not convert the possession into a permissive one.

Adverse possession is not a cure-all to all ancient title defects, since many interests cannot be barred by adverse possession. These unbarrable interests include rights of the government (including the United States, the states, and their various subdivisions and agencies) and public utilities; Indian lands in the absence of a Federal waiver; rights of servicemen; rights of persons under legal disability, such as minority or mental incompetency; and future interests, severed mineral interests and rights of co-tenants, so long as these rights were created prior to the time the adverse possession commenced. Non-possessory interests, such as liens and easements, are likewise not barred by adverse possession.

The "unbarrable" nature of rights of co-tenants and holders of severed mineral interests are particularly serious where adverse possession is being relied on to cure a defective title. No amount of surface "possession" can divest the rights of a prior severed mineral owner whose interest may have been improperly joined in a mortgage foreclosure. Similarly, a quit claim deed from one of two owners whose rights may (or may not) have been barred by a tax sale may result in a "co-tenancy", and stop adverse possession from running.

"After acquired title", also referred to as *"estoppel by deed"*, is another means of involuntary transfer of title. If a grantor has purported to convey a tract of land or a specified interest in the land by warranty deed, and did not in fact own the interest in the land which he purported to convey, any subsequent acquisition by the grantor will inure to the benefit of the grantee by operation of law to the extent necessary to make up the deficiency. Estoppel by deed will likewise apply in some states where title to the interest subsequently fails, such as upon the foreclosure of a mortgage which was not excepted from a deed's covenants of warranty, if the grantor subsequently reacquires title to the land.

§ 2.03 Ownership and Transfer of Title to Minerals

§ 2.03(a) Nature of Mineral Ownership. In most states other than Texas, "mineral ownership" actually constitutes a right to explore for and produce the oil and gas rather than ownership of the minerals in the ground (see Chapter One). (In Texas, the landowner is considered to own the minerals "in place".) However, whether the state follows the "in place" (absolute ownership) or "right to explore" (qualified ownership) theory, this "ownership" can be granted or reserved apart from a transfer of the surface ownership. The interest so created is referred to as a "separated" or "severed" mineral interest.

Often, more than one person will own an interest in the minerals. Where various persons own undivided interests in the minerals, no one of these co-owners has the exclusive right to their use. While under the majority view, any co-tenant may extract the minerals without the consent of the others, and the extracting co-tenant must account to the remaining co-owners for their proportionate share of production, less the costs of extraction.

Where ownership of the minerals is divided between life estate and remainder interests, neither the life tenant nor the remainderman, acting alone, may extract the minerals without the consent of the other. However, if the instrument creating the interest authorized one or the other to lease, or if all consent, the minerals may be extracted.

§ 2.03(b) Conveying and Reserving Oil and Gas Rights. A conveyance of a tract of land, without specific reference to the minerals, transfers title to all oil and gas rights to the grantee except those mineral rights expressly reserved.

The grant of an undivided interest "in and to all the oil, gas, and other minerals in and under the following described land", without more, will be treated as the unambiguous grant of a fully-participating mineral interest, unless additional language contained in the instrument creates doubt as to the parties' intent. The addition of a right of ingress and egress or a reference to oil and gas "in and under *and that may be produced*" from the specified land is not a necessary part of the grant, but will not render it ambiguous; such additional language is a customary part of most printed mineral deeds. (See Appendix A to this chapter for such a sample "Mineral Deed".)

If the grantor desires to reserve an interest in the minerals, the reservation should read, "excepting and reserving unto [identify the grantor by name], his heirs, and assigns, in addition to all minerals heretofore previously reserved or conveyed", then followed by the same language recommended above for a grant. To insure that this reservation is not interpreted as a mere exception from the warranty, it should be put in the granting clause, immediately following the description of the land, and should expressly contain words of reservation, not merely of exception. Likewise, to insure that the reservation is not defeated by estoppel by deed, the reservation should recite that it is an addition to previous grants and reservations, which should likewise be excepted from the warranty. (A sample mineral reservation is contained in Appendix A to this chapter.)

The foregoing language is sufficient to grant, or reserve, a fully-participating mineral interest. If the parties desire to convey less than the entire mineral ownership, there can be a division of rights as to the substances covered, the depths involved, the duration of the grant and the size of the interest. Likewise, if it is desired to grant or reserve less than all of the "incidents of ownership" of mineral rights, which include the power to develop, to execute leases, and to participate in bonus, rentals and royalties, one or more of such incidents of ownership can be eliminated from the grant.[4]

§ 2.03(c) Interpreting Conveyances of Oil and Gas Rights

§ 2.03(c)(1) In General. One of the most difficult, and recurring, problems you must deal with in checking titles is the interpretation of mineral conveyances.

Conveyances of mineral rights are frequently made by printed form deed. The printed form Mineral Deed attached as Appendix A to this chapter follows the language discussed in § 2.03(b) and does not create problems of interpretation. However, where the parties, intentionally or inadvertently, turn to a printed form of "Royalty Deed", confusion is almost certain to result. Similar problems of interpreting the nature and size of the interest being carved out arise where inept draftsmen attempt to prepare their own form of mineral conveyance, or type a reservation of some interest in the oil and gas rights into a deed conveying the land. The confusion is so widespread that in almost no instance except a conveyance of a fully-participating mineral interest under a standard form Mineral Deed can the nature or size of the interest granted or reserved be clearly resolved without a stipulation.

The two chief areas of confusion in the interpretation of a grant or reservation of mineral rights are the legal nature or "quality" of the interest, and its size or "quantum".

§ 2.03(c)(2) Legal Nature of the Interest. Problems concerning the quality of the interest granted or reserved normally involve a determination of whether the parties intended to create a fully-participating Mineral Interest, a non-participating Royalty Interest, or some form of hybrid interest possessing some, but not all, of the incidents of ownership which

---

[4] Problems concerning the size and incidents of mineral ownership are discussed in § 2.03(c). As to depth limitations and the substances covered, see Chapter Three.

accompany the fully participating "Mineral Interest".

The legal distinction between a "Mineral Interest" and a "Royalty Interest" is quite clear. A Mineral Interest possesses all of the incidents of mineral ownership including the right to explore (the "right of ingress and egress"), the right to lease (the "executive right"), and the right to share in bonus, delay rentals and royalties. It is a *right* in the oil and gas *"in place"*, and a *right to extract* the minerals. The Royalty Interest, on the other hand, is a single incident of mineral ownership: the right to share in oil and gas if, as and when produced, normally as a cost-free interest. It includes no right of ingress and egress for development of the property; no executive (leasing) right; and no right to share in bonus or delay rentals. It is a *right* to oil and gas *after capture*, and an interest in oil and gas *extracted by others*.

The distinction between the Royalty Interest and the Mineral Interest is crucial, since it determines who must join in the lease, to whom bonus and delay rentals will be paid, and how production will be shared.

In determining whether or not a given conveyance should be construed as creating a Mineral Interest or a Royalty Interest, certain construction guidelines should be followed. As a general rule, in the absence of the existence of an outstanding oil and gas lease, "royalty" is not a word of art — the grant of a "1/16 Royalty" may or may not create a non-participating Royalty Interest. Similarly, the absence of the mention of the right of ingress and egress is not determinative; it is implied unless expressly prohibited. Also, the label of the instrument is of relatively little significance.

If the direct object of the grant is an interest in "oil, gas and other minerals", or the reference is to minerals "in and under" the land, an interest in the minerals *in place* is indicated; the grant thus contains indicia of a Mineral Interest. A similar result occurs when an instrument contains a right of ingress and egress for *exploration* and *development*.[5] Conversely, if

---

[5] Ingress and egress *to receive his share of production* is not, however, inconsistent with the creation of a Royalty Interest.

the interest granted relates merely to oil and gas "produced", without more, a Royalty Interest is indicated.[6] However, the inclusion of an executive right or any interest in bonus and delay rentals is not consistent with a pure Royalty Interest.

Interpretation of the instrument may also be affected by the interest in production granted. If the interest is so large that it would be inconsistent with industry practices as to permissible royalty — i.e., "a one-half interest in oil and gas produced" — the court may tend to interpret the instrument as creating a Mineral Interest, not a Royalty Interest.[7]

The typical grant or reservation frequently blends certain language indicative of a Mineral Interest with other language indicative of a Royalty Interest, or contains inconsistent, ambiguous or unusual phraseology.[8] In attempting to construe such grants, several conclusions are possible. The interest can be interpreted as a pure Mineral Interest, or a pure Royalty Interest. The conclusion can be reached that a "hybrid" interest has been created, i.e., that one is less than a full Mineral Interest but more than a pure Royalty Interest.[9] Finally, it can be determined that the language is so unclear, or inconsistent, that the interest created cannot be determined

---

[6] However, add "in and under, and that may be produced," and you are back to a probable Mineral Interest.

[7] Until the 1970s, this meant that any interest in excess of one-eighth would be viewed with suspicion. However, with larger royalties common today, this should no longer be the case.

[8] The most common error in attempted conveyances of non-participating Royalty Interests is to include the right of ingress and egress for exploration, etc. This language — contained in most printed "Royalty Deeds" — has consistently been held by the courts to cloud the conveyance.

[9] One type of "hybrid" frequently created by the true intention of the parties is the "non-executive" mineral interest. Unfortunately, courts have often interpreted inartfully drawn grants and reservations as creating various "hybrids" which were rarely what the parties appear to have intended.

without resort to a stipulation of interest between the parties, or a ratification of the lease.

Even if the language does not appear hopelessly inconsistent, only in the clearest cases should you as an examiner unilaterally reach a determination that a specified kind of interest has been created without requiring a stipulation of the parties. This means that in most instances, unless language substantially identical to that contained in Appendix A of this chapter has been used, a stipulation or lease ratification will be required.

§ **2.03(c)(3) Size of the Interest.** The determination of the size of the interest granted or reserved is often directly related to the Mineral Interest/ Royalty Interest decision. If the interest is determined to be a Mineral Interest, the holder will receive the specified fraction *of the reserved royalty*, not the specified fraction of gross production. If, on the other hand, the interest is determined to be a Royalty Interest, the holder will receive the specified fraction of *gross production*. In other words a "1/16 Royalty" would entitle the holder to one-sixteenth of gross production if it is held to be a non-participating Royalty Interest, versus one-sixteenth of one-eighth of production (assuming a one-eighth royalty) if it is held a Mineral Interest was created.

The size determination will also be affected by the exact wording of the grant. Thus, while a grant of a "one-sixteenth royalty" or a "one-sixteenth royalty interest" would be governed by the foregoing test, a conveyance of "one-sixteenth *of the* royalty" is normally interpreted as granting or reserving only one-sixteenth of the royalty payable under the lease and not one-sixteenth of gross production. Similarly, if the Royalty Interest is so large as to make such a royalty reservation impractical — i.e., "an undivided one-half Royalty Interest" — a different rule of construction will apply.

§ **2.03(d) Other Interpretation Problems.** The grant may also attempt to limit the grantee's use of the land. As in the case of efforts to restrict the quality, quantity or duration of the grant, such intent must be clearly expressed or the grantee will acquire an unrestricted fee. In examining instruments that contain such attempted limitations, you as an examiner must thus determine whether the grantee acquired an unlimited fee, a determinable fee, a surface interest, an easement, or a fee with restrictions on the use of the land ("restrictive covenants"). This determination will affect which party, if either, has the right to extract the minerals.[10]

A recitation of the purpose of the grant in a conveyance — i.e., "for school purposes" — does not limit the grant. Even an enforceable covenant concerning the use of the land, or a condition which could cause the grant to terminate unless the land is so used, will not prevent *additional* uses. However, a grant "exclusively" for a specified use would prevent the grantee from extracting the minerals.

§ 2.04 Curing Title Defects

§ 2.04(a) In General.
Few titles are fatally defective. Valid title is vested somewhere. Your problem, as an examiner, is to determine who has the better title and how to perfect this title.

Normally, there are several curative methods which you could use to cure the title. Where a formal title opinion has been submitted, the examiner will have "required" that curative which will render title marketable. However, it is for you to determine whether or not a marketable title is in fact required for your company's present purposes; which of the alternative methods of title curative will result in a satisfactory title and, among these alternatives, to select the one which is the least expensive and the least time consuming.

Court proceedings *are* costly and time consuming. The court action may be lost, or at least may stir up trouble which could be avoided if other curative alternatives were explored. Likewise, there is often uncertainty as

---

[10] If a fee — limited or otherwise — was created, the lease should come from the grantee. If an easement or surface interest was created, the grantor should lease. In case of restrictive covenants, the minerals pass to the grantee; however, if the restrictions are severe enough, they may prevent mineral development of the land.

to whether or not the court proceeding has in fact definitely cured the title. Accordingly, while this method of curing the title in some instances must be resorted to (see § 2.04(d)), it should be avoided whenever possible.

If a defect is relatively insignificant, a proper "solution" may be to waive the requirement rather than to attempt to cure the title. This is particularly true where the defect is an ancient one. However, there is little reason to waive an easily cured defect. (Title requirements are normally easier to cure before drilling than after, since the property becomes more valuable, and the claimants more greedy, after a productive well has been drilled.)

§ 2.04(b) Curative Instruments
§ 2.04(b)(1) Curative Conveyances.
Title is frequently cured through the use of curative instruments which may range from various curative conveyances to affidavits.

Curing a defect by conveyance is the surest means of curing a title. However, use of this curative method requires an ability to identify and locate the possible claimants, plus a cooperative attitude on their part.

The *correction deed* is used to cure errors in prior conveyances, particularly errors in the description of the land or the interest conveyed. However, if the correction deed might adversely affect the grantee's interest, he must join in the instrument. Also, if the original instrument required the joinder of the spouse, the spouse must also join in the correction deed. And if the grantee's signature is required, and there is any possibility of a homestead or similar claim by the grantee's spouse, the spouse must also join in the correction deed.

The correction deed must be clearly labeled as such and as being "in lieu of" the prior conveyance. Otherwise, it may be interpreted as granting an additional interest to the grantee.

Where clouds exist on the title which do not appear to represent valid claims, the best method of title curative is to secure a *quit claim deed* from the potential claimant. Again, the spouse must join if there is any possible homestead or similar claim.

The major restriction on the use of a quit claim deed is whether or not the claimant will prove cooperative.

Where the possibility of claim is remote, and the curative is requested primarily as a method of evidencing that necessary inquiries were made or to create an estoppel, a *disclaimer* may be secured.

Again, possible homestead rights should be considered.

The typical disclaimer would not qualify as a conveyance, and thus might prove ineffective if more than a mere estoppel was required, due to the absence of words of grant. Accordingly, in more serious cases, language of quit claim and conveyance are added to the disclaimer. In such an event, the "disclaimer" becomes a quit claim deed; but problems may arise as to whether there has been misrepresentation or an inherent fraud in securing the instrument in that form without labeling it as a quit claim deed.

Where the objective is not to eliminate the claim of a possible interest holder, but to clarify the quality or size of the interest of one or more parties, a *declaration of interest*, often referred to as a stipulation of interest, is frequently used. Preferably, all affected parties should join in the execution of the stipulation. Less than all the parties can validy stipulate only if the joining parties stipulate to the least possible interest they could claim. Again, the instrument must contain words of cross conveyance, and homestead possibilities must be considered.

If some question exists as to validity of the company's lease, or if it is desired to secure coverage under the lease of some disputed interest, a *ratification* may be secured. If it is necessary to clarify to whom rentals are to be paid and the amount of rentals payable to each party, a *rental stipulation*, also referred to as a *rental division order*, should be executed.

The ratification of the oil and gas lease will frequently contain a rental stipulation provision. Likewise, the rental stipulation will normally contain words of ratification covering the lease. In either event, words of grant should be contained in the instrument, leasing the interest owned by the signing parties to the current holder of the leasehold interest on the same terms as specified in the basic lease. (Sample curative conveyances —

·taken from *The Landman's Handbook on Petroleum Land Titles*[11] — are contained in Appendix B of this chapter.)

§ **2.04(b)(2) Curative Affidavits.** In addition to curative conveyances, affidavits are frequently used as curative instruments. If a marketable title is required, an affidavit cannot be used to close a gap in the chain of title. However, an affidavit may be secured to evidence that a necessary inquiry has been made or where a less than marketable title is acceptable.

There is nothing magic about an affidavit. In determining whether or not to rely upon such curative, it must be remembered that it will be just as reliable as the person making and the person securing the affidavit.

The *affidavit of possession* is used to establish that the duty to inquire of persons in possession has been satisfied, that no one is in adverse possession, and that title may be defendable, based on adverse possession, if early defects cloud the title. Such an affidavit should be secured both from the owner of the land and from the person actually occupying the land, and it should be the result of an on-the-spot inspection.

In many instances, it may be advisable to supplement the affidavit of possession by securing either a *landowner's questionnaire* or an *inspection report* at the time the oil and gas lease is purchased. Both contain invaluable information and often will preclude the making of title requirements upon formal title examination.

Affidavits of possession should be joined in by any tenant or a *tenant's consent agreement* secured. The instrument should affirm the existence of any outstanding interests and the subordination of the tenancy to those interests and the rights of the Lessee.

The affidavit of possession should always affirm the existence of any separated mineral interests, or other interests of persons not in possession.

Where adverse possession is being relied upon to cure defects in the title, an *affidavit of adverse possession* should be secured. Such an affidavit should recite, in detail, the actual manner in which the land has been used

---

[1] Mosburg, *The Landman's Handbook on Petroleum Land Titles* (The Institute for Energy Development: 1976).

and should be substantiated by an additional affidavit from a reliable, impartial third party.

In addition to the affidavits of possession and adverse possession, affidavits of *identity, delivery, non-development, age, marital status,* and *non-homestead* are frequently secured in connection with such title problems.

Where a gap exists in the chain of title due to the death of a prior owner, and a probate of his estate is deemed too costly or time consuming, an *affidavit of death and heirship,* frequently referred to as a proof of death and heirship, is often secured. However, such an instrument is evidentiary only and does not perfect the title of the heirs or devices.

Sample affidavits of possession and adverse possession are contained in Appendix B to this chapter.

§ **2.04(c) Curative Statutes.** Legislatures often act to remove ancient defects as clouds on the chain of title after the passage of a reasonable period of time. Such curative statutes are of two types: (1) "spot" curative statutes, aimed at clearing specific defects; and (2) broad curative statutes that seek correction of all prior defects in the chain of title.

Curative statutes normally provide that upon the passage of a specified period of time following the recording of a defective instrument, or the entry of the defective judgment, the specified defect will be cured unless a notice of claim has been filed. Available curative statutes will vary from state to state. Among the defects that may be cured by such statutes are homestead defects, name discrepancies, irregularities in execution or acknowledgment, defective court proceedings and ancient mortgages. Likewise a handful of states have passed "Marketable Title Acts", which bar *all* ancient defects and limit the period of title search by barring all claims occurring prior to the recording of the specified "root of title".

Before relying on a curative statute to correct defects in the chain of title, it should be determined whether the statute merely creates a presumption of valid title or actually validates the title as against the prior defect. While most curative statutes require nothing more than the passage of the specified period following the recording of the defective instrument, etc., the statute must also be reviewed to insure that nothing more is required. The examiner likewise must determine who is protected by the curative statutes, since some only apply to "purchasers for value", and thus do not protect the present holder. In other instances, not all claims are barred by the curative statute. Rights that may survive can include rights of the United States or the state, rights of parties in possession, and severed mineral interests and future interests. You must also ascertain that no saving "notice" has been filed.

Certain curative statutes, such as the Marketable Title Acts, are so complicated that a legal interpretation may be required before deciding to rely on the statute.

§ **2.04(d) Curing Title by Judicial Action.** While judicial action is avoided as a title curative method whenever possible, it frequently will prove the only solution where the claimant is dead, cannot be located, or is non-cooperative.

The most commonly-used curative action is the *action to quiet title*. It is utilized to bar groundless claims whose invalidity are not clear of record, or to establish title which is based upon facts which do not appear of record, such as adverse possession.

The action to quiet title is an "in personam" action. The most critical area in reviewing the sufficiency of a quiet title suit is thus to insure that all necessary parties have been joined and that proper service has been secured upon them (see § **2.02(d)**).

Actions for reformation, rescission and cancellation are often available when the action to quiet title is not. If a written instrument does not accurately reflect the agreement of the parties, an action to quiet title could not correct this error, as the court must interpret the document as written. However, if it can be shown by clear and convincing evidence that the parties did have a "meeting of the minds" which is not accurately reflected in the conveyance, an *action for reformation* can be brought to reform the instrument properly to reflect the agreement actually reached. Such an action is available against parties to the conveyance and transferees with notice of the true agreement or who acquired the interest without consideration. However, the instrument cannot be reformed as against a bona fide purchaser without notice.

If there was no meeting of the minds between the parties, reformation is not available, since there is no mutual agreement to which the document can be reformed. However, if the misunderstanding was the result of mutual mistake, fraud or misrepresentation, an *action for rescission* will lie, and the document can be cancelled, with the parties returned to the same position in which they stood prior to its execution.

The action for rescission is available against the same parties against whom the action for reformation would lie but, again, is not available against a bona fide purchaser for value without notice.

The action to quiet title involves the *interpretation* of a conveyance; actions for reformation and rescission involve the *correction* or *cancellation* of a conveyance.

If the owner of an interest in the property has died, his successors in interest are normally established by a *probate* of his estate if he died leaving a will, or an *administration* of his estate if he died without a will; i.e., intestate. In addition, in some instances where the deceased has died intestate, a shortened procedure for determining his heirs may be provided by statute as a part of a quiet title action.

The primary requisite for action for probate or administration is proper compliance with statutory procedures, with particular attention to the giving of all required notices. You likewise must insure that the property in question is not omitted or misdescribed in the proceeding, although in some instances an omission of the property may not be fatal.

In those instances in which the succession appears relatively certain, and the cost of probating or administering the estate would be excessive, the company may wish to rely upon a satisfactory proof of death and heirship, previously discussed.

A copy of the deceased's will or a copy of a decree entered by a court in another state will not be binding in the absence of a probate or administration in the state in which the land is located; however, if there appears to be no question as to the identity of those persons who actually succeeded to the interest, such "business risk" evidences of the succession may be relied upon when less than a marketable title is required. Similarly, if a probate has been conducted in the state, which due to technical irregularities, would not be legally binding, you may be willing to rely upon such a decree as evidence of future succession where less than marketable title is sufficient. However, in each of these instances, it is necessary for you to satisfy yourself that the facts concerning the succession have been accurately determined.

If the interest of the deceased was as a life tenant or joint tenant, to establish a marketable title it is necessary to judicially establish the fact of his death. This can be done by the entry of a decree determining the deceased's death in an administration proceeding, or a decree admitting his will to probate or, in some states, by a special action to determine the death of the joint tenant or life tenant. On a business risk basis, a certified copy of a death certificate may be relied upon where marketable title is not required and there is no question as to the identity of the life or joint tenant with the party named in the death certificate.

§ 2.04(e) Curing Title Defects Arising from Unreleased Encumbrances.

If title to the property is clouded by an unreleased or defectively released mortgage, a *release* of the mortgage can be secured if the mortgage has been paid. If the mortgage has not been paid, a *subordination agreement*, also referred to as a waiver of priority, should be secured from the holder of the mortgage, or, if possible, a partial release as to the portion of the mortgaged premises covered by the lease. If the mortgagee cannot be located and the mortgage is ancient, curative statutes are available in certain states; in other instances the defect may be waived, particularly where the mortgage was for a small amount or a release appears of record which, although defective in execution, does indicate that the mortgage was in fact paid.

If the mortgage is substantial in amount and the mortgagee refuses to execute a release despite the mortgagor's assertion that the mortgage has been paid, an action to quiet title could be brought. If the mortgage is unsatisfied, but the mortgagee is unwilling to execute a subordination agreement, a status report should be secured to insure that the mortgage is current and to ascertain the unpaid balance.

Where title is clouded by the presence of an unreleased or a defectively released oil and gas lease, a *release* of the lease should be secured if the lessee can be located and is cooperative. If the lessee cannot be located, an *affidavit of non-development* (or, in some states, a certificate of non-development from the local conservation agency) should be secured, establishing that no activities have been conducted on the leased premises which would extend the lease beyond its primary term. Such curative can be relied upon only when the primary term has expired, and must cover all land described in the lease, not merely the part currently being leased by the company (see Chapter Three).

If the primary term has not expired, and the lessee is unwilling to execute a release despite assertions by the lessor that the lease has terminated, an action to quiet title may be the only solution. While some companies are willing to rely on an affidavit of non-payment of rentals, this is a dangerous practice. Also, before taking a "top lease" under such circumstances, care must be taken to insure that the new lease does not extend the original lease under the doctrine of "obstruction" (see Chapter Three).

Easements encumbering the leased premises normally will be permanent in nature. Normally, your only "curative" step will be to insure that such easement will not interfere with the company's intended use of the property. This will require a study of the entire instrument to ascertain the full extent of the grantee's right of use. If it does appear that the easement might interfere with the proposed development of the property, and it is not possible to relocate the easement, the drilisite must be relocated.

Unpaid tax and judgment liens may also create encumbrances against the property. Normally, taxes create a lien only after a notice of lien is filed. In such instances, the lien must be satisfied unless a sufficient period of time has passed since the filing of the notice so that the lien has become dormant. However, ad valorem and inheritance taxes often create a lien against the property without the filing of any notice, and thus require an investigation to insure that all such taxes have been paid.

§ 2.04(f) Preparation of Curative Instruments. In preparing a curative instrument, great care must be taken to avoid typing and other errors, such as misdescription of the parties or the property. The instrument will be binding only on those listed in the body; the mere signing of the instrument is not sufficient. However, a generic designation of the persons to be bound as "we" or "the undersigned" will be sufficient to bind all persons signing the instrument.

If the instrument could have the effect of conveying an interest in the property by any of those signing, the instrument must contain words of grant. Likewise, if there is any possibility of homestead or similar claims, the spouse of each party must likewise join.

If any possible claimant has since transferred all or part of his interest to a third party, such successor should join in the instrument. The instrument should specifically state that it is binding upon, and will run in favor of, heirs, successors and assigns of the parties. If, as is usually the case, a document to be executed by multiple parties has been prepared for counterpart execution, the instrument should expressly so state and likewise should state that it will be binding on all those signing, even if some of the parties named in the instrument do not sign, so long as this is the intent.

Curative instruments should ratify the basic lease with words of grant and should affirm that all delay rentals have been properly paid. If there is any question concerning the proper method of paying delay rentals, the instrument should likewise contain a rental stipulation.

Appendix A

SAMPLE MINERAL INSTRUMENTS

MINERAL DEED

Mid-Continent Royalty Owners Association
Approved Form Revised

KNOW ALL MEN BY THESE PRESENTS:

That_____

_____

of_____
Give exact Post Office Address
hereinafter called Grantor, (whether one or more) for and in consideration of the

sum of_____

_____Dollars ($_____) cash | _____

in hand paid and other good and valuable considerations, the receipt of which is hereby acknowledged, do_____hereby

grant, bargain, sell, convey, transfer, assign and deliver unto_____

_____of_____, hereinafter
Give exact Post Office Address

called Grantee (Whether one or more) an undivided_____interest in
and to all of the oil, gas and other minerals in and under and that may be produced from the following described lands

situated in_____County, State of_____, to-wit:

containing_____acres, more or less, together with the right of ingress and egress at all times for the purpose of mining, drilling, exploring, operating and developing said lands for oil, gas and other minerals, and storing, handling, transporting and marketing the same therefrom with the right to remove from said land all of Grantee's property and improvements.

This sale is made subject to any rights now existing to any lessee or assigns under any valid and subsisting oil and gas lease of record heretofore executed; it being understood and agreed that said Grantee shall have, receive, and enjoy the herein granted undivided interests in and to all bonuses, rents, royalties and other benefits which may accrue under the terms of said lease insofar as it covers the above described land from and after the date hereof, precisely as if the Grantee herein had been at the date of the making of said lease the owner of a similar undivided interest in and to the land described and Grantee one of the lessors therein.

Grantor agrees to execute such further assurances as may be requisite for the full and complete enjoyment of the rights herein granted and likewise agrees that Grantee herein shall have the right at any time to redeem for said Grantor by payment, any mortgage, taxes, or other liens on the above described land, upon default in payment by Grantor, and be subrogated to the rights of the holder thereof.

TO HAVE AND TO HOLD The above described property and easement with all and singular the rights, privileges,

and appurtenances thereunto or in any wise belonging to the said Grantee herein_____heirs, successors, personal

representatives, administrators, executors, and assigns for ever, and Grantor do_____ hereby warrant said title to

Grantee _____ heirs, executors, administrators, personal representatives, successors and assigns forever and

do____hereby agree to defend all and singular the said property unto the said Grantee herein_____heirs, successors, executors, personal representatives, and assigns against every person whomsoever claiming or to claim the same or any part thereof.

WITNESS_____hand this_____day of_____, 19____

_____

_____

66

STATE OF OKLAHOMA, County of_____ss: **Individual Acknowledgment**

Before me, the undersigned, a Notary Public in and for said County and State on this_____day of

_____, 19_____, personally appeared_____

_____

to me known to be the identical person____ who executed the within and foregoing instrument and acknowledged to me

that___ _____executed the same as_____free and voluntary act and deed for the uses and purposes therein set forth.

Given under my hand and seal of office the day and year last above written.

My commission expires _____   _____Notary Public

_____

STATE OF OKLAHOMA   } SS: CORPORATION ACKNOWLEDGMENT

COUNTY OF_____}   Oklahoma Form

Before me, the undersigned, a Notary Public in and for said County and State on this_____day of

_____, 19_____, personally appeared_____

to me known to be the identical person who subscribed the name of the maker thereof to the foregoing instrument as its

_____President and acknowledged to me that_____executed the same as his free and voluntary act and

deed and as the free and voluntary act and deed of such corporation, for the uses and purposes therein set forth.

Given under my hand and seal of office the day and year last above written.

My commission expires_____   _____Notary Public

TEXAS ACKNOWLEDGMENTS

E STATE OF TEXAS, County of_____ , ss:

FORE ME, the undersigned, a Notary Public in and for said County and State, on this day personally appeared

_____

own to me to be the person____ whose name _____subscribed to the foregoing instrument, and acknowledged to me that

he___ executed the same for the purposes and consideration therein expressed.

VEN UNDER MY HAND and the seal of this office, this_____ day of_____ , A.D., 19____

_____

E STATE OF TEXAS, County of_____ , ss:

FORE ME, the undersigned, a Notary Public in and for said County and State, on this day personally appeared

_____ wife of _____

own to me to be the person whose name is subscribed to the foregoing instrument, and having been examined by me privily

apart from her husband, and having the same fully explained to her, she, the said_____

nowledged such instrument to be her act and deed and declared that she had willingly signed the same for the purposes and

sideration therein expressed, and that she did not wish to retract it.

VEN UNDER MY HAND and the seal of this office, this_____ day of_____ , A.D., 19____

_____

STATE OF_____ } SS
COUNTY OF_____ }

ACKNOWLEDGEMENT, Applicable for lands in Oklah
Kansas. Nebraska. North and South Dakota. Arizona. Colo
Indiana. Mississippi. Oregon. Wyoming and/or New Me

BE IT REMEMBERED, That on this_____day of_____ A. D., 19____, before me, a Notary Public in

for said County and State, personally appeared_____

_____

_____to me known to be the identical person___ described in and who executed the w

and foregoing instrument and acknowledged to me that_____executed the same as_____free and voluntary act
deed for the purposes therein set forth.

IN WITNESS WHEREOF, I have hereunto set my official signature and affixed my notarial seal, the day and year
above written.

My commission expires:_____   _____ Notary P

CORPORATION ACKNOWLEDGMENT   (Oklahoma F

STATE OF_____County of_____. ss:

On this_____day of_____, A. D., 19____. before me, the undersigned, a Notary P

in and for the county and state aforesaid, personally appeared_____
to me known to be the identical person who signed the name of the maker thereof to the within and foregoing instrument a

_____President and acknowledged to me that_____executed the same as_____ free and voluntary
and deed, and as the free and voluntary act and deed of said corporation, for the uses and purposes therein set forth.

Given under my hand and seal the day and year last above written.

My commission expires:_____ Notary Pu
When instrument is executed by a corporation. the corporate name must be shown and instrument signed by its President or Vice-President
attested by its Secretary or Assistant Secretary and the Corporate Seal affixed.

NOTARY ACKNOWLEDGMENT of SIGNATURE BY MARK   (Oklahoma Fo

STATE OF_____ County of_____, ss:

Before me, _____, a Notary Public in and for said County and State on this

day of_____, 19____, personally appeared _____

_____

to me known to be the identical person ___ who executed the within and foregoing instrument by_____ mark in

presence and in the presence of_____

as witnesses and acknowledged to me that_____executed the same as_____ free and voluntary
and deed for the uses and purposes therein set forth.

In Witness Whereof, I have hereunto set my hand and official seal the day and year last above written.

My commission expires:_____   _____Notary Pu
NOTE—The signature by mark of a lessor who cannot write his name must be witnessed by two witnesses. one of whom must write lessor's n

FORM NO. 249-AF
(ORDER BY NUMBER)

MINERAL DEED
(Mid-Continent Association Form)

FROM

TO

Dated_____, 19

Lot_____Block_____Addition

Township_____, Section_____Range

No. of Acres_____County

STATE OF_____County

This instrument was filed for record on the
_____day of_____, 19____
at_____o'clock_____M., and recorded
in Book_____of_____
at page_____Fee $_____

By_____County Clerk.
_____Deputy.

RETURN TO

68

WARRANTY DEED

Statutory Form---Individual

Know All Men by These Presents:

That_____

_____

_____

of _____ County,

State of_____, part _____of the first part, in consideration of the

sum of _____ DOLLARS

in hand paid, the receipt of which is hereby acknowledged, does hereby Grant, Bargain, Sell and

Convey unto _____

of_____County, State of_____, part____

of the second part, the following described real property and premises situate in_____

County, State of_____, to-wit:

[DESCRIPTION],

excepting and reserving, however, unto [name owner], his heirs and
assigns, in addition to all minerals heretofore previously reserved
or conveyed, an undivided ____ interest in and to all of the oil,
gas, and other minerals in and under said land,

together with all the improvements thereon and the appurtenances thereunto belonging, and warrant
the title to the same, except as set forth above.

TO HAVE AND TO HOLD said described premises unto the said part_____ of the second

part,_____heirs and assigns forever, free, clear and discharged of and from all former

grants, charges, taxes, judgments, mortgages and other liens and incumbrances of whatsoever nature.

Signed and delivered this_____day of_____, 19____

_____

_____

_____

_____

INDIVIDUAL ACKNOWLEDGMENT　　　　　　　(Oklahoma Form)

STATE OF_____County of_____, ss:

Before me the undersigned, a Notary Public, in and for said County and State, on this____day of_____, 19____,

personally appeared_____

me known to be the identical person_____who executed the within and foregoing instrument and acknowledged to me that

_____executed the same as_____free and voluntary act and deed for the uses and purposes therein set forth.

Given under my hand and seal the day and year last above written.

my commission expires_____　　_____Notary Public

INDIVIDUAL ACKNOWLEDGMENT　　　　　　　(Oklahoma Form)

STATE OF_____County of_____, ss:

Before me the undersigned, a Notary Public, in and for said County and State, on this____day of_____, 19____,

personally appeared_____

me known to be the identical person_____who executed the within and foregoing instrument and acknowledged to me that

_____executed the same as_____free and voluntary act and deed for the uses and purposes therein set forth.

Given under my hand and seal the day and year last above written.

my commission expires_____　　_____Notary Public

WARRANTY DEED

FORM NO. 290-AF
(ORDER BY NUMBER)
Statutory Form Individual

FROM

TO

STATE OF_____ County } ss.

This instrument was filed for record on the

____day of_____, 19____

at____o'clock____M., and recorded

in Book____of____

at page____Fee $____

County Clerk.

By____Deputy.

RETURN TO

NOTARY ACKNOWLEDGMENT of SIGNATURE BY MARK　　　(Oklahoma Form)

STATE OF_____County of_____, ss:

Before me, _____, a Notary Public in and for said County and State on this_____

day of_____, 19____, personally appeared_____

to me known to be the identical person___ who executed the within and foregoing instrument by_____mark in my

presence and in the presence of_____

as witnesses and acknowledged to me that_____executed the same as_____free and voluntary act

and deed for the uses and purposes therein set forth.

In Witness Whereof, I have hereunto set my hand and official seal the day and year last above written.

My commission expires:_____　　_____Notary Public

NOTE—The signature by mark of a lessor who cannot write his name must be witnessed by two witnesses, one of whom must write lessor's name.

71

NOTARY ACKNOWLEDGMENT of SIGNATURE BY MARK   (Oklahoma Form)

STATE OF _____ County of _____ , ss:

Before me, _____ , a Notary Public in and for said County and State on this_____

day of_____ , 19 _____, personally appeared _____

_____

to me known to be the identical person ___ who executed the within and foregoing instrument by_____mark in my

presence and in the presence of_____

as witnesses and acknowledged to me that_____ executed the same as_____free and voluntary act
and deed for the uses and purposes therein set forth.

In Witness Whereof, I have hereunto set my hand and official seal the day and year last above written.

My commission expires:_____Notary Public

NOTE—The signature by mark of a lessor who cannot write his name must be witnessed by two witnesses, one of whom must write lessor's name.

72

Appendix B

SAMPLE CURATIVE INSTRUMENTS

Correction Deed

THIS DEED made this _____ day of _____,
19 _____ , between_____,
parties of the first part and _____,
party of the second part.

WHEREAS, a certain Mineral Deed was previously executed between the parties hereto, which deed is dated _____,
19_____ , and is recorded in Book _____ , Page _____,
of the Deed Records of _____ County, _____ ; and

WHEREAS, such deed contains certain ambiguities as to the interest granted therein;

NOW, THEREFORE, the said parties of the first part, in consideration of the sum of _____ DOLLARS to _____ duly paid, the receipt of which is hereby acknowledged, do hereby grant, bargain, sell and convey unto said party of the second part, and to his heirs and assigns, forever, an undivided _____ interest in and to all of the oil, gas and other minerals in and under the following described real estate situated in the County of _____ State of _____ , to wit:

(Description),

together with all and singular the hereditaments and appurtenances thereunto belonging.

To have and to hold the above granted premises unto the said party of the second part, his heirs and assigns, forever.

This interest is granted in lieu of, and not in addition to, the interest purported to be conveyed by the prior deed, described above, between the parties hereto.

In Witness Whereof, The said parties of the first part have hereunto set their hands the day and year first above written.

74

ACCEPTED:

_____   _____
          Second Party   First Parties
                        (Acknowledgments)

QUIT CLAIM DEED

INDIVIDUAL FORM

This Space Reserved for Filing Stamp

THIS INDENTURE, Made this_____day of_____, A. D. 19___

between_____

_____of the first part,

and_____

_____of the second part,

Witnesseth, that said part_____ of the first part, in consideration of the sum of

_____DOLLARS

to_____in hand paid, the receipt of which is hereby acknowledged, do_____hereby quitclaim, grant, bargain,

sell and convey unto the said part_____ of the second part all _____ right, title, interest, estate, and every

claim and demand, both at law and in equity, in and to all the following described property situate in

_____County, State of_____, to-wit:

together with all and singular the hereditaments and appurtenances thereunto belonging.

    To Have and to Hold the above described premises unto the said_____

_____

_____heirs and assigns forever, so that neither_____, the said_____

_____

nor any person in _____ name and behalf, shall or will hereafter claim or demand any right or title to the said premises or any part thereof; but they and everyone of them shall by these presents be excluded and forever barred.

76

In Witness Whereof, the said part_____of the first part ha_____hereunto set_____hand_____ the day and year first above written.

_____

_____

_____

_____

STATE OF OKLAHOMA   } SS: INDIVIDUAL ACKNOWLEDGMENT

COUNTY OF_____ }   Oklahoma Form

Before me, the undersigned, a Notary Public in and for said County and State on this_____day of

_____, 19_____, personally appeared _____

_____

to me known to be the identical person____ who executed the within and foregoing instrument and acknowledged to me

that_____executed the same as_____free and voluntary act and deed for the uses and purposes therein set forth.

Given under my hand and seal the day and year last above written.

My commission expires_____   _____Notary Public.

INDIVIDUAL ACKNOWLEDGMENT

(Oklahoma For

STATE OF_____ _____County of_____, ss:

Before me the undersigned, a Notary Public, in and for said County and State, on this_____day of_____, 19___

personally appeared_____

to me known to be the identical person_____who executed the within and foregoing instrument and acknowledged to me

_____executed the same as_____free and voluntary act and deed for the uses and purposes therein set forth.

Given under my hand and seal the day and year last above written.

My commission expires:_____   _____Notary Pu

INDIVIDUAL ACKNOWLEDGMENT

(Oklahoma For

STATE OF_____County of_____, ss:

Before me the undersigned, a Notary Public, in and for said County and State, on this_____day of_____, 19___

personally appeared_____

to me known to be the identical person_____who executed the within and foregoing instrument and acknowledged to me t

_____executed the same as_____free and voluntary act and deed for the uses and purposes therein set forth.

Given under my hand and seal the day and year last above written.

My commission expires:_____   _____Notary Pu

CORPORATION ACKNOWLEDGMENT

(Oklahoma For

STATE OF_____County of_____ _____, ss:

On this _____day of_____, A. D., 19____, before me, the undersigned, a Notary Pub

in and for the county and state aforesaid, personally appeared_____

to me known to be the identical person who signed the name of the maker thereof to the within and foregoing instrument as

_____President and acknowledged to me that_____executed the same as_____free and voluntary

and deed, and as the free and voluntary act and deed of said corporation, for the uses and purposes therein set forth.

Given under my hand and seal the day and year last above written.

My commission expires: _____Notary Pu

When instrument is executed by a corporation, the corporate name must be shown and instrument signed by its President or Vice-President attested by its Secretary or Assistant Secretary and the Corporate Seal affixed.

FORM NO. 280-AF
(ORDER BY NUMBER)

QUIT CLAIM DEED
(Individual Form)

FROM

TO

STATE OF _____
_____ County
ss:

This instrument was filed for record on the

_____ day of _____, 19___

at _____ o'clock _____ M., and recorded

in Book _____ of _____

at page _____ Fee $_____

County Clerk.

By _____ Deputy.

RETURN TO

78

Disclaimer

KNOW ALL MEN BY THESE PRESENTS:

That, notwithstanding any prior instrument or instruments of record covering or pertaining to the following described lands in _____ County, State of _____:

(Description)

the undersigned, _____, and _____, his wife, have and assert no right, title or interest in and to said lands or in and to the oil, gas and other minerals in and under said lands.

THAT it is the intention of the undersigned in executing this instrument to disclaim any right, title or interest in and to said lands and in and to the oil, gas and other minerals in and under said lands.

NOW, THEREFORE, for and in consideration of the sum of One Dollar ($1.00), and other good and valuable consideration, receipt of which is hereby acknowledged, the undersigned and each of them, do hereby disclaim any right, title or interest in and to the above described lands and in and to oil, gas and other minerals in and under said lands.

This instrument shall be binding upon the heirs, executors, administrators and assigns of the undersigned.

EXECUTED THIS _____ day of _____, 19 _____.

_____

_____

(Acknowledgment)

Stipulation of Interest

WHEREAS, by Warranty Deed dated _____ ,
19 _____ , recorded in Book _____ at Page _____
of the records of _____ County, _____
conveyed the following described land in said County and State, to-wit:
(Description),
unto ____ _____ , excepting and
reserving unto _____ an
undivided ¼ mineral interest in said tract; and
WHEREAS, on _____, 19 _____,
_____, joined by his wife,
_____, executed a
Quitclaim Deed to _____,
recorded in Book _____ at Page _____ of the
records of said County and State, which deed was intended solely to cure
the failure of _____
to join in the execution of the above described Warranty Deed, and was not
intended to convey any additional interest to _____,
other than that conveyed by said Warranty Deed; and
WHEREAS, some doubt and uncertainty may exist as to the intent of
the parties in executing said conveyances, and it is the desire of the
undersigned to clarify the ownership of the oil, gas and other minerals
underlying the above described land;
NOW, THEREFORE, for good and valuable consideration, the receipt
and sufficiency of which are hereby acknowledged, the undersigned hereby
stipulate and agree that the ownership of the oil, gas and other minerals in
and under the above described land is as follows:
_____ ¾ _____ ¼,

subject only to that certain Oil and Gas Lease executed by
_____ in favor of
_____ , on

80

_____ , 19_____ , and recorded in
Book _____ at Page_____ of the records
of said County and State, which lease is hereby ratified and confirmed by
each of the undersigned, and to certain term mineral conveyances of
record.

TO EFFECTUATE THE PURPOSE OF THIS AGREEMENT, each of
the parties hereto does hereby grant, bargain, sell, quitclaim and convey,
each unto the other, any interest he may own in the oil, gas and other
minerals underlying the above described land inconsistent with the
foregoing Stipulation of Interest.

This Stipulation shall be binding upon the parties hereto, their
respective heirs, devisees, personal representatives, successors and
assigns:

EXECUTED THIS _____ , day of _____ , 19 _____ .

_____

_____

(Acknowledgments)

81

RATIFICATION AND RENTAL DIVISION ORDER

KNOW ALL MEN BY THESE PRESENTS:

That, WHEREAS, that certain oil and gas lease, dated _____. from

_____. as Lessor,

to _____. as Lessee, recorded in Book _____ , Page _____

of the _____ Records of _____ County, _____. is owned by

in so far as it covers the following described land in _____ County, _____
to-wit:

NOW, THEREFORE, in consideration of the sum of One Dollar ($1.00) and other good and valuable considerations, w

and each of us, do hereby ratify, approve, confirm, and adopt the above described oil and gas lease in so far as it covers the
above described land, and do hereby lease, demise and let said land unto _____
subject to and under all of the terms and provisions of said lease, and as to said land, do hereby agree and declare that said lease
is now in full force and effect; that payment has been duly made of the entire bonus consideration and all of the delay rentals
necessary to extend said lease to the next rental paying date; and each of the undersigned agrees that any delay rentals which
may be paid under the terms of said lease with respect to the above described land may be divided as follows:

CREDIT TO	ADDRESS	AMOUNT

and that payment or tender, of the amount above set forth opposite his name, directly or to his credit in the depository bank
at the times and in the manner specified in said lease will, as to his interest in the said land, extend said lease and continue the
same in full force and effect according to its terms; provided, that if no amount is above set forth opposite his name, then payment
of the amounts above set forth to the other parties, or their successors in interest, will so extend said lease. This instrument
shall be fully binding upon, and effective as to the interest of, each of the above named persons who executes the same, without
regard to execution or lack of execution by the others or by any other person whomsoever.

82

We, and each of us, hereby release and waive all rights of dower and homestead in the above identified land, and the provisions hereof shall be binding upon the heirs, legal representatives, successors, and assigns of each of us.

WITNESS our hands and seals this_____day of _____, 19_____.

_____   _____ (SEAL)

_____   _____ (SEAL)

_____   _____ (SEAL)

_____   _____ (SEAL)

_____   _____ (SEAL)

Michigan, Indiana, Kansas, Oklahoma, New Mexico, Colorado, Wyoming, North Dakota, South Dakota, Nebraska, Arkansas, Iowa, Missouri.

ACKNOWLEDGMENT

STATE OF_____ ⎫ ss.
COUNTY OF_____ ⎭

I, _____, a Notary Public in and for said County and State, hereby certify that _____

me personally known, and known to me to be the same person ...... described in and who executed the foregoing instrument, appeared before me this day in person and acknowledged to me that ....he......... executed the same as _____ his own voluntary act and deed, for the uses, purposes and consideration therein expressed.

Given under my hand and official seal this_____day of_____, A. D., 19_____.

_____
Notary Public

My Commission Expires:_____

ACKNOWLEDGMENT

STATE OF_____ ⎫ ss.
COUNTY OF_____ ⎭

I, _____, a Notary Public in and for said County and State, hereby certify that _____

me personally known, and known to me to be the same person........ described in and who executed the foregoing instrument, appeared before me this day in person and acknowledged to me that ....he......... executed the same as _____ his own voluntary act and deed, for the uses, purposes and consideration therein expressed.

Given under my hand and official seal this_____day of_____, A. D., 19_____.

_____
Notary Public

My Commission Expires:_____

83

ACKNOWLEDGMENT

STATE OF_____ } ss.
COUNTY OF_____

I, _____, a Notary Public in and for said County and State
do hereby certify that _____

to me personally known, and known to me to be the same person........ described in and who executed the foregoing instrument
appeared before me this day in person and acknowledged to me that ....he......... executed the same as_____free and
voluntary act and deed, for the uses, purposes and consideration therein expressed.

Given under my hand and official seal this_____ ____day of_____, A. D. 19. _____.

_____
Notary Public

My Commission Expires: _____

ACKNOWLEDGMENT FOR CORPORATION

STATE OF_____ } ss.
COUNTY OF_____

I, _____, a Notary Public in and for said County and State
do hereby certify that _____
to me personally known, and known to me to be the same person who executed the foregoing instrument as _____
President of _____, a corporation, appeared before
me this day in person and, being first duly sworn, acknowledged that he is the_____President of said
corporation, that the seal affixed to said instrument is the corporate seal of said corporation, and that said instrument was signed,
sealed, and delivered in behalf of said corporation by authority of its **Board of Directors,** and further acknowledged said instrument and his execution thereof to be the free and voluntary act and deed of said corporation, and his own free and voluntary act
and deed for the uses, purposes and considerations therein expressed.

Given under my hand and official seal this_____ ____day of_____, A. D., 19_____.

_____
Notary Public

My Commission Expires: _____

STATE OF _____)
) SS.
COUNTY OF _____)

Affidavit of Possession

(To be executed by disinterested party)

The undersigned, of lawful age, being first duly sworn, deposes and says that _____ is (are) the owner(s) of the following described land in _____ County,_____ , to-wit:

The undersigned further says that _____ he is familiar with said land that _____ he knows that said owner(s) has (have) been in actual open, peaceable and undisturbed possession of all of said land and that no other person claims or occupies said land adverse to said owner(s) and that said owner(s) has (have) possession of said land as follows: (HERE INSERT WHETHER OWNER(S) ACTUALLY RESIDE(S) UPON SAID LAND OR HOLD(S) BY TENANT OR OTHERWISE AND IF BY TENANT, THE NAME AND NATURE OF SUCH TENANCY). *Owner holds possession by farming and raising livestock. The above described land is fenced.*

_____

_____
Subscribed and sworn to before me this _____ day of _____ , 19 _____.

_____
Notary Public in and for _____ County, _____
My Commission Expires: _____
(SEAL)

* * * * * * * * * *

STATE OF _____)
) SS.
COUNTY OF _____)

(To be executed by tenant)

The undersigned, of lawful age, being first duly sworn, deposes and says that he occupies the above described land as tenant of _____ , the owner(s) thereof; that he became such tenant on the _____ day of _____ ,19 _____ , and that his tenancy is for _____ years and expires on the _____ day of _____ , 19 _____ ; that he claims no title to said land other than as tenant as aforesaid and does hereby declare that his right to possession in no way interferes with the right to lease said land for oil and gas development purposes and that his possession as tenant is subject to the rights of any lessee or assignee under any oil and gas lease executed by any owner of oil and gas rights under such land.

_____

Subscribed and sworn to before me this _____ day of _____ , 19 _____ .

_____

Notary Public in and for _____ County, _____
My Commission Expires: _____
(SEAL)

* * * * * * * * * *

STATE OF _____)
) SS.
COUNTY OF _____)

(To be executed by owner(s))

The undersigned of lawful age being first duly sworn, deposes and says that _____ he is (one of) the owner(s) of said above described lands and that _ he has read the above and foregoing affidavits and certificates and personally knows that the statements made therein are true and correct and that

_____ he has owned and been in possession of the above described land for
_____ years, subject only to the rights of the owners of an undivided
_____ interest in the oil and gas rights thereunder.

| _____ | _____ |
| Address | Signature |

Subscribed and sworn to before me this _____ day of _____,
19 _____.

_____

Notary Public in and for _____ County, _____
My Commission Expires: _____
(SEAL)

Affidavit of Adverse Possession

STATE OF _____)
) SS:
COUNTY OF _____)

    Affiant _____ , of lawful age, being first duly sworn on oath states that:

    1. Affiant is the owner and is in the open, notorious, actual, exclusive, and adverse possession of the following described land situated in _____ County, State of_____ , to-wit:
(Description)

    2. Affiant holds title to the above described land under a certain Commissioner's Deed dated _____ , which was recorded on _____ , in Book _____ at Page_____ of the records of said County and State. Immediately thereafter to-wit, on _____ , Affiant went into the actual physical possession of the land by personally entering thereon and commencing the erection of an electric fence around the entire perimeter of the land as described above. Upon the completion of such fence, Affiant unloaded 12 head of angus and hereford mixed cattle upon such land and has continuously thereafter pastured cattle, horses, sheep and other livestock thereon. Affiant furthermore has visited such land on the average of three times a week, physically entering thereon and viewing the premises and inspecting the livestock. Affiant furthermore has cut timber and has cultivated certain small portions for grass pasturage by cleaning brush and weeds therefrom and planting grass and mixed seeds in those portions to improve pasturage. Such actual physical possession has continued uninterruptedly from the first day mentioned above, and during all of such time there has been no effort on the part of any third party to oust Affiant either physically, by written or oral notice to him, or by litigation.

    3. Affiant claims the full fee interest in the above described land under claim of right and by reason of adverse possession, both of the surface and

all of the mineral rights, as against all the world.
　　Further Affiant saith not.

_____

　　Subscribed and sworn to before me this _____ day of
_____ , 19 _____.
(SEAL)

_____
　　　　　　　　　　　　　Notary Public
My Commission Expires: _____
　　　　　　　　(Acknowledgment)

Form 540 6-65
Printed in U.S.A.

AFFIDAVIT OF HEIRSHIP OF

_____
Deceased

STATE OF_____ ⎫
COUNTY OF_____ ⎭ SS.

_____, of lawful age, being first duly sworn, on oath deposes and says:

That affiant was personally and well acquainted with the above named decedent during the latter's lifetime, having known deceased for_____years.

Decedent died at_____, _____County, State of_____ on or about the_____day of_____, 19_____, being_____years of age, and a resident of _____at the time of death.

That the following statements and answers to the following questions are based upon the personal knowledge of affiant and are true and correct:

1. Did decedent leave a will?_____ If so, has the will been admitted to probate_____; Give name of County and State in which such proceedings are pending, and name and address of executor.
_____
_____

(If decedent left a will, please attach a certified copy of same, together with a copy of the order of court admitting it to probate, and letters testamentary.)

2. If decedent left no will, have administration proceedings been started?_____ If so, give the name of the county and state in which said administration proceedings are pending and name and address of administrator_____
_____
_____

3. Have ancillary probate proceedings been had on decedent's estate?
If so, when?_____ Where?_____

4. If no administration proceedings have been started, are there any plans to have the estate administered?_____

5. Did decedent leave any unpaid taxes, including federal estate or state inheritance taxes or other debts?_____ If so, give as nearly as possible, the amount of such taxes or other debts, to whom owing, and whether they have since been paid_____
_____
_____

6. Was decedent surety on any bond or guarantor of any other person's indebtedness at time of death?_____ If so, give details as to principal debtor, amount, etc_____

7. Were there any suits pending or judgments rendered against decedent at time of death?_____ If so, state briefly the nature, amount involved and parties_____
_____

8. Marital Status of Decedent at Time of Death (married, single, divorced, widow, widower)_____

9. If decedent was ever married, give the following information for each marriage: (List names in order of marriage)

Name of Spouse	Date of Marriage	Living/Dead	Divorced	Date of Death or Divorce	Was there a property settlement? If "Yes"—attach copy.

90

10. If decedent had any children by any spouse, or adopted any children, give the following information:

Name of Child	Date of Birth	Address	Living/Dead	Date of Death	By Which Spouse

11. If a deceased child left descendents, give the following information

	Name of Child	Date of Birth	Address	Living Dead	Date of Death
Name of deceased child					
Name of deceased child					

12. If decedent left no children or descendents of deceased children, then please furnish the following information

a. Give names of parents of decedent

Name	Address	Living Dead	Date of Death
Father			
Mother			

b. Give names of brothers and sisters of decedent:

Name	Relation	Address	Living Dead	Date of Death

c. Give names of children of deceased brother or sister

Name of Child	Child of	Date of Birth	Address	Living Dead

13. If decedent left no heirs covered by item 12 above, then attach a full and complete affidavit of heirship of said decedent in narrative form

14. Give location or description of homestead of decedent, as of date of death _____

15. As to each tract of land or interest in land owned by the decedent at the time of his death which concerns this company, give the following information (If space provided is insufficient, please attach exhibit giving same information as to each tract)

Description	Date Acquired	From Whom?	State how Acquired (Gift, Purchase, Inheritance or Under a Will)	If Acquired by Purchase Whose Funds Were Used?

91

16. Here briefly state facts and circumstances (such as being a relative, a close friend, or attorney or agent for, decedent) which will show basis and source of information given above.

_____

_____

_____

_____
                                                          Affiant

_____
                                                           Address

Subscribed and sworn to before me this _____ day of _____ , 19 ___

My commission expires_____
                                                 Notary Public

SUPPORTING AFFIDAVIT

STATE OF_____ |
                                  | SS.
COUNTY OF _____ |

_____ , of lawful age, being first duly sworn, on oath states

That_____was personally and well acquainted with _____ during _____ lifetime

that_____has read the above affidavit by_____ and that the facts stated therein are true and correct

                                                    Affiant

Subscribed and sworn to before me this_____ day of_____ 19 ___

My commission expires _____
                                               Notary Public

92

FORM NO 28

AFFIDAVIT OF NONDEVELOPMENT
AND NONPAYMENT OF RENTAL

ATLOCK'S, INC.

STATE OF..
County of.. } ss.

.... , being first

duly sworn, deposes and says:

That .... is the present owner of the

of Section................, Township .... , Range............... in .... ,

County... , which land is described in an oil and gas mining lease executed on

...................day of .... ,..

by ....

as lessors, and ....

as lessee, recorded in Book............ , Page .... , in the office of the Register of Deeds of said county.

That since the date of said lease there has been no well drilled upon said land, nor any oil or gas produced therefrom, and that none of the rentals accruing under and by virtue of the terms of said lease have
been paid or tendered to affiant or said lessors, or to any bank for their credit, by the lessee, or his agents

or assigns, since..
and further that the lessee and his assigns had actual notice that rentals were payable to affiant under said
lease. Affiant states that he has not at any time executed any extension of said original lease, and that
the same has expired.

Affiant further states that by reason of the noncompliance with the terms of said lease by lessee and
his assigns, affiant hereby declares said lease forfeited, and will not, by acceptance of rentals, or in any
other manner, recognize the same as a valid or existing lease.

....

....

STATE OF..
County of.. } ss.

.... , being first duly

sworn deposes and says that he is .... of the.. Bank of

.... and that the records of said Bank show no rentals

have been at any time deposited in, or tendered to, said Bank under terms of the oil or gas mining lease

above described for the credit of the person who made the above affidavit or the lessor in said lease since

....

....

STATE OF... } ss. **ACKNOWLEDGMENT**
County of...

I, the undersigned, a Notary Public, in and for the said County and State, do hereby certify that

_____ personally

known to me to be the person (whether one or more) whose name is subscribed to the foregoing instrument, appeared before me this day in person and acknowledged that he (they) signed sealed and delivered the same as his (their) free and voluntary act and deed for the uses and purposes therein set forth including
release and waiver of right of homestead.

IN WITNESS WHEREOF I hereunto set my hand and seal the day and year in the instrument above written

My Commission Expires:

Residing at   _____

Notary Public in and for said County and State

WAIVER OF PRIORITY OF MORTGAGE LIEN

Whereas, the undersigned_____of_____State of

Oklahoma is the owner of a certain real estate mortgage executed by_____and his

wife, _____ to secure a loan of $_____covering the following

described land situated in_____County, State of Oklahoma, to-wit:

Which mortgage is dated the_____day of_____, 19_____ and recorded in Book_____at

page_____in the County of_____State of Oklahoma.

AND WHEREAS, on the_____ ____day of_____, 19____, _____

and his wife_____executed an Oil and Gas Lease to_____

_____for a period of_____years, covering the above described land.

NOW, THEREFORE, in consideration of One Dollar in hand paid, the receipt of which is hereby acknowledged,

the undersigned,_____does hereby except and release the working interest held under and
by virtue of said Oil and Gas lease from the lien under said mortgage, not waiving any of my rights under said mortgage
as against the royalty interest in said mortgaged land, and agrees that the Oil and Gas lease above mentioned shall have
the same validity and effect as if executed, delivered, and recorded prior to the date of execution of the said mortgage
above mentioned.

IN WITNESS WHEREOF, I have hereunto set my hand and seal, this the_____day of_____, 19_____

Attest:

_____

STATE OF OKLAHOMA }SS:
COUNTY OF_____

Before me, the undersigned, a Notary Public in and for said County and State on this_____ _____day of

_____, 19_____, personally appeared _____

to me known to be the identical person___ who executed the within and foregoing instrument and acknowledged to me
that_____executed the same as_____free and voluntary act and deed for the uses and purposes therein set forth.
Given under my hand and seal the day and year last above written.

My commission expires_____ _____Notary Public

95

STATE OF OKLAHOMA   } SS: CORPORATION ACKNOWLEDGMEI

COUNTY OF _____ }   Oklahoma Form

Before me, the undersigned, a Notary Public, in and for said County and State on this_____day

_____, 19____, personally appeared _____

to me known to be the identical person who subscribed the name of the maker thereof to the foregoing instrument as

_____President and acknowledged to me that____ _____executed the same as his free and voluntary act a

deed and as the free and voluntary act and deed of such corporation, for the uses and purposes therein set forth.

Given under my hand and seal of office the day and year last above written.

My commission expires_____   _____Notary Publ

STATE OF_____)

COUNTY OF_____ } SS   **ACKNOWLEDGMENT, Applicable for lands in Oklahoma N** Nebraska. North and South Dakota. Arizona. Colorado. Ir Mississippi. Oregon, **Wyoming.** and/or New Mexico.

BE IT REMEMBERED, That on this____day of_____ A. D., 19____, before me, a Notary Public in

for said County and State, personally appeared_____

_____to me known to be the identical person___ described in and who executed the w

and foregoing instrument and acknowledged to me that_____ executed the same as_____free and voluntary act

deed for the purposes therein set forth.

IN WITNESS WHEREOF, I have hereunto set my official signature and affixed my notarial seal. the day and year

above written.

My commission expires:_____ _____Notary F

CORPORATION ACKNOWLEDGMENT   (Oklahoma F

STATE OF_____County of_____, ss:

On this_____day of_____, A. D., 19____, before me, the undersigned, a Notary P

in and for the county and state aforesaid, personally appeared_____

to me known to be the identical person who signed the name of the maker thereof to the within and foregoing instrument a

_____President and acknowledged to me that_____executed the same as_____free and voluntar;

and deed, and as the free and voluntary act and deed of said corporation, for the uses and purposes therein set forth.

Given under my hand and seal the day and year last above written.

My commission expires:_____ _____Notary P

When instrument is executed by a corporation, the corporate name must be shown and instrument signed by its President or Vice-Presiden attested by its Secretary or Assistant Secretary and the Corporate Seal affixed.

NOTARY ACKNOWLEDGMENT of SIGNATURE BY MARK   (Oklahoma F

STATE OF_____County of_____, ss:

Before me, _____, a Notary Public in and for said County and State on this_____

day of_____, 19____, personally appeared_____

to me known to be the identical person ___ who executed the within and foregoing instrument by_____mark ir

presence and in the presence of _____

as witnesses and acknowledged to me that._____executed the same as_____free and voluntary

and deed for the uses and purposes therein set forth.

In Witness Whereof, I have hereunto set my hand and official seal the day and year last above written.

My commission expires:_____ _____Notary P

NOTE—The signature by mark of a lessor who cannot write his name must be witnessed by two witnesses, one of whom must write lessor's n

WAIVER of PRIORITY OF MORTGAGE LIEN

FROM

TO

Dated _____ 19__
Lot ____ Block ____ Addition
____, Section
Township ____ Range
County
No. of Acres ____ Terms

STATE OF _____ County } SS.

This instrument was filed for record on the ____ day of _____, 19__
at ____ o'clock ____ M., and recorded
in Book ____ of ____
at page ____ Fee $ ____
_____ County Clerk.
By _____ Deputy.

RETURN TO

TEXAS ACKNOWLEDGMENTS

STATE OF TEXAS, County of _____, ss:
RE ME, the undersigned, a Notary Public in and for said County and State, on this day personally appeared

_____
to me to be the person__ whose name_____ subscribed to the foregoing instrument, and acknowledged to me that
__ executed the same for the purposes and consideration therein expressed.

J UNDER MY HAND and the seal of this office, this____ _____ day of _____, A.D., 19____

_____

STATE OF TEXAS, County of _____, ss:
RE ME, the undersigned, a Notary Public in and for said County and State, on this day personally appeared

_____wife of_____
to me to be the person whose name is subscribed to the foregoing instrument, and having been examined by me privily
part from her husband, and having the same fully explained to her, she, the said_____
wledged such instrument to be her act and deed and declared that she had willingly signed the same for the purposes and
eration therein expressed, and that she did not wish to retract it.

N UNDER MY HAND and the seal of this office, this____ _____ day of _____, A.D., 19____

_____

97

Chapter Three

EDITOR'S COMMENTS

In "The Oil and Gas Lease", you will learn about the history and development, and the key provisions, of this critical document, which is the basis for your company's right to explore for and produce oil and gas.

The purpose of this chapter is to introduce you to the company's rights and obligations under the various "customary" lease clauses, as well as where the company stands if a particular clause is not contained, or has been stricken, from the lease. However, lease clauses are interrelated; and the lease must also be viewed as a whole. Accordingly, § **3.03** gives you such an overview of the entire Oil and Gas Lease, so that you can see how the lease clauses are related to each other and your company's overall rights under the lease.

The following points may prove of particular value to you:

- There is no "Producers 88" or any other "standard" lease form (§ **3.02(c)** and **3.03**).

- If you change a lease provision or fill in a blank without the express consent of your Lessor, you may void your lease (§ **3.02(c)**).

- What substances are and are not covered by your lease (§ **3.04(b)**)?

- How the "Mother Hubbard" clause helps lessen the damage

caused by inadvertent description or survey errors (§ **3.04(c)**).

• How the "continuous operations/cessation of production" clause both increases and reduces your company's rights (§ **3.05(d)** and **3.05(e)**).

• How an otherwise valid pooling can be destroyed, and your lease lost, by failing to record the Declaration of Pooling or conform to restrictions on unit size (§ **3.06(b)**).

• The danger of "market value" royalties (§ **3.07(b)**).

• How a shut-in well still may fail to extend your company's lease, even when the lease contains a shut-in royalty clause (§ **3.09(b)**).

• How payment of a shut-in royalty by the date specified in the shut-in clause may still prove "untimely" (§ **3.09(c)**).

— L.G.M.

Chapter Three

THE OIL AND GAS LEASE

By Lewis G. Mosburg, Jr.

§ 3.01 Introduction

It all starts with the lease. The geologist might say — correctly — that it all starts with his origination of a "prospect". However, your role as a Landman revolves around the two-page piece of paper that authorizes your company to drill a well. Your title work deals with insuring that this piece of paper has been secured from the right person, and that it covers the interest you believe it to cover. And the two contracts with which you most frequently deal — the farmout and joint operating agreements — both involve Oil and Gas Leases.

Much has been written about types of leases and lease clauses that caused considerable havoc in their day, but are no longer with us. Much has also been written about what might be contained in the "ideal" lease. However, this chapter will be concerned with the clauses you are actually going to encounter in your day-to-day land practice, rather than leases as they were, or should be.

This chapter will also deal with "fee" leases — Oil and Gas Leases executed by private landowners — rather than Federal, state and Indian leases.

101

§ 3.02 Nature and Development of the Oil and Gas Lease

§ 3.02(a) Characteristics of the Oil and Gas Lease. So what *is* an Oil and Gas Lease? The Oil and Gas Lease is an instrument under which someone who owns the mineral rights underlying a tract of land (the "Lessor") authorizes another (the "Lessee") to conduct operations for the exploration and development of the land for oil and gas purposes, at the Lessee's sole risk, in return for a cost-free interest to the Lessor in any oil and gas produced.

The Oil and Gas Lease serves multiple functions. It is a *conveyance* of a property interest from the Lessor to the Lessee[1]; an *executory contract* creating an elaborate, on-going contractual relationship between Lessor and Lessee for the life of the lease; and is also marked by a *common objective* between the Lessor and the Lessee — the extraction of minerals at a profit.

The one thing the Oil and Gas Lease is *not* is a "lease" in the "landlord-tenant" sense; it instead conveys a "defeasible", i.e., terminable, interest in the minerals in place, or a right to prospect for such minerals (see **§ 3.04(a)**).

§ 3.02(b) Evolution of the Modern Oil and Gas Lease Form. The Oil and Gas Lease form was developed as a special instrument to suit the needs of both landowner and oilman. Under this arrangement, the oilman, who possessed the financial resources and technical knowledge to develop land for oil and gas purposes, but did not own or desire to own the land, was given the right to develop the oil and gas in return for bearing the expense of development; the landowner, in return for granting the right to develop, was given a share of production in any oil and gas discovered and produced.

---

[1] Technically, except in Texas and a few other "absolute ownership" states, neither a lease nor a mineral deed conveys the minerals in place (see § **3.04(a)**). However, whatever the abstract legal classification, the practical effect of the lease is to serve as a conveyance.

Early lease forms were based on leases then being used by the salt industry (see § **1.10**). These early forms were short and contained many loopholes. Basically, immediate development was the sole consideration for the lease; and the lease accordingly required the Lessee to commence a well immediately and to prosecute its drilling with due diligence. A breach of this drilling obligation would result in a forfeiture of the lease. Furthermore, courts began to create certain "implied covenants", both for initial exploration and for further development, whether or not the lease contained such provisions.

The early Oil and Gas Leases were for a fixed term without a right of renewal or any extension in the event of production. However, such an arrangement did not prove in the interest of either Lessor or Lessee. Accordingly, as the oil industry matured, and leases were taken for future rather than immediate development, provision was made for the payment of a bonus to the Lessor, and for a "delay rental" clause under which the Lessee was given the alternative of either commencing a well or paying a specified additional consideration to the Lessor on pain of forfeiture of the lease. The lease likewise provided for a short fixed term during which exploratory operations could be conducted, followed by a "thereafter" clause under which the lease would continue in effect so long as oil or gas was being produced from the land.

§ **3.02(c) Requisites for Validity.** There is no "standard" or "customary" lease form. Thus, a contract to execute an Oil and Gas Lease must specify the terms which will be contained in that lease; a contract merely calling for the execution for a "five-year Oil and Gas Lease" is unenforceable. (The same rule applies to a contract to execute a lease on a "Producers 88" form.) However, once a written lease instrument has been entered into between the Lessor and the Lessee, the lease will be enforceable so long as it designates parties Lessor and Lessee, and contains a description of the land and interest therein being transferred to the Lessee, and so long as the general rules of conveyancing, including the requirements for a written, signed document have been satisfied.

All parties Lessor should be named in the body of the lease or at least covered by some general reference in the body, such as "et al". Courts are more liberal in permitting a subsequent insertion of the name of the Lessee in a provided blank, although this is not the best practice.

Normally, the lease need not be supported by substantial cash consideration although, in Louisiana, there must be a "serious" consideration which has in fact been paid.

Whether or not the lease must be acknowledged, witnessed or recorded will depend upon the conveyancing laws of the state. However, some states now require that the parties' mailing address be included in the instrument, at least as a condition to recording.

An undated lease is not void, but may create problems as to the calculation of the primary term, and the date for payment of delay rental and shut-in royalties.

A particularly serious problem arises if you, as a Landman, make a "material alteration", i.e., any change in the terms of the lease, without the consent of the Lessors (including the spouse if the land is homestead). (The change will be considered "material", whether beneficial or injurious to the company, if it changes the legal effect of the lease.) It is quite tempting to make such changes without contacting the Lessors if you discover you have inadvertently failed to complete the habendum or delay rental clause. However, in the majority of states, such a material alteration, even when made without fraudulent intent, will void the lease in both its altered and original form.

Initialing an agreed-to change in the lease is not a requisite for validity. However, it is evidence that the Lessors did consent to the change.

§ 3.03 Overview of a "Typical" Lease

There is no such thing as a "standard" Oil and Gas Lease. While most leases contain clauses covering many of the same general areas — pooling, commence drilling and continuous operations, shut-in wells, etc. — a particular lease may or may not contain every such "customary" clause. In any event, the exact wording of these clauses varies substantially from lease to lease, resulting in significant variations in your company's rights and obligations.

Nevertheless, it may prove difficult for you as a Landman "to see the forest for the trees". Understanding what the habendum clause says (see § 3.05) gives you only a partial idea of the duration of your lease; you must also consider how that clause is modified to increase or reduce your company's rights by the pooling, drilling-delay rental, continuous drilling, shut-in and other lease clauses.

Figures 5 and 6 show the various clauses found in a "typical" Oil and Gas Lease. (For a more readable copy of the lease itself, see Appendix A to this chapter.) Here is how those clauses interrelate to form that integrated document you know as the Oil and Gas Lease:

(1) The *date* gives the presumed effective date of the lease, which normally will establish the period of the primary term under the habendum clause, and may determine the date of payment of delay rentals and shut-in royalties.

(2) The *parties* will show who is bound by the lease, and in whose favor the lease runs.

(3) The recited *consideration* is inserted to avoid any argument that the lease is unenforceable.

(4) The *granting clause* (see § 3.04) establishes the purpose for which the land is being leased, what rights the Lessee is acquiring in the land, and what lands are subject to the lease. The lease in question also contains a *"Mother Hubbard"* clause (see § 3.04(c)), which gives the Lessee certain rights in lands adjacent or contiguous to the land specifically described.

(5) The *habendum clause* (see § 3.05), specifies how long the Lessee's rights will endure, but is subject to other provisions and limitations contained in the lease, such as the drilling and delay rental, pooling, shut-in royalty, and continuous drilling, etc., clauses.

(6) The *royalty clause* (see § 3.07) specifies what share of production or production proceeds the Lessor will receive; this may also be affected by the pooling clause. The portion of the clause referred to as the *shut-in royalty clause* (see § 3.09) also specifies the effects of a shut-in gas well.

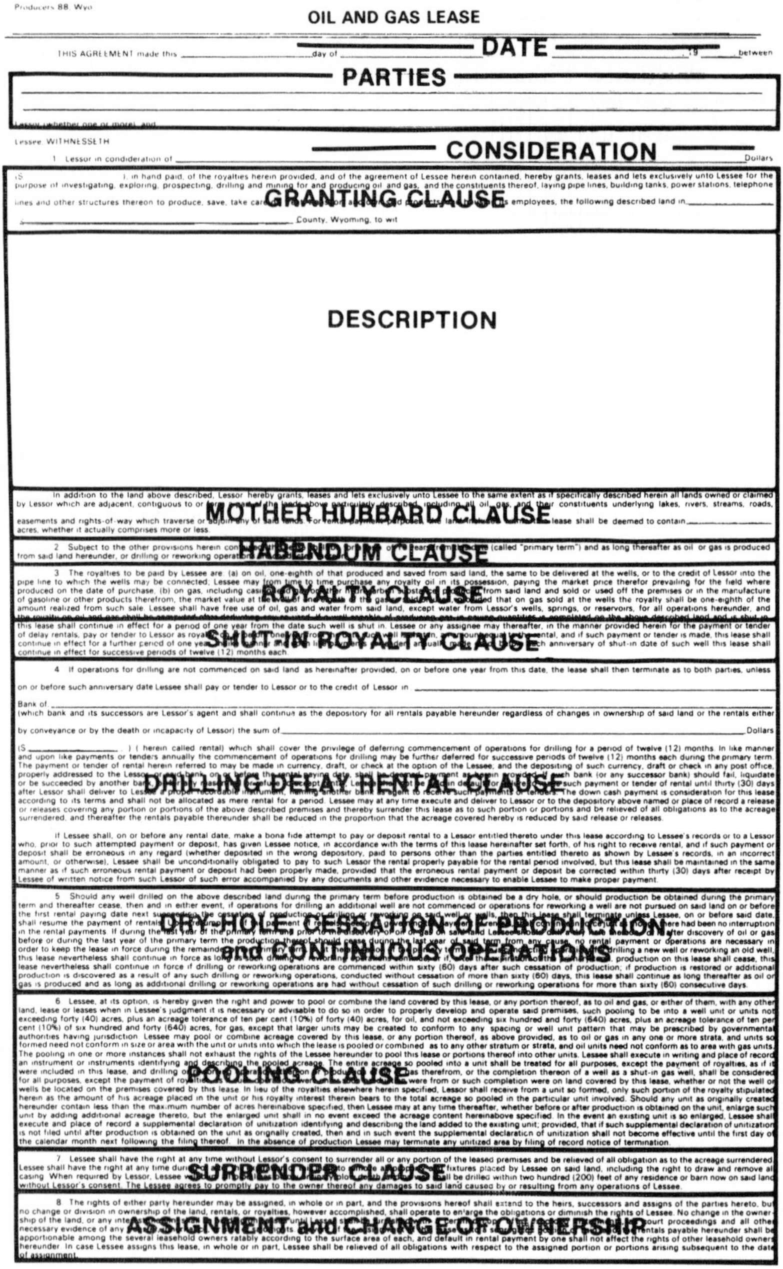

Producers 88, Wyo.

OIL AND GAS LEASE

THIS AGREEMENT made this _____ day of _____ **DATE** _____ 19___ between

PARTIES

Lessee (whether one or more), and

Lessee, WITNESSETH:

1. Lessor in consideration of _____ **CONSIDERATION** _____ Dollars

($_____), in hand paid, of the royalties herein provided, and of the agreement of Lessee herein contained, hereby grants, leases and lets exclusively unto Lessee for the purpose of investigating, exploring, prospecting, drilling and mining for and producing oil and gas, and the constituents thereof, **GRANTING CLAUSE** employees, the following described land in _____ lines and other structures thereon to produce, save, take care of _____

_____ County, Wyoming, to-wit

DESCRIPTION

In addition to the land above described, Lessor hereby grants, leases and lets exclusively unto Lessee to the same extent as if specifically described herein all lands owned or claimed by Lessor which are adjacent, contiguous to or **MOTHER HUBBARD CLAUSE** including all oil, gas, and their constituents underlying lakes, rivers, streams, roads, easements and rights-of-way which traverse or _____ lease shall be deemed to contain _____ acres, whether it actually comprises more or less.

2. Subject to the other provisions herein con **HABENDUM CLAUSE** (called "primary term") and as long thereafter as oil or gas is produced from said land hereunder, or drilling or reworking operat _____

3. The royalties to be paid by Lessee are: (a) on oil, one-eighth of that produced and saved from said land, the same to be delivered at the wells, or to the credit of Lessor into the **ROYALTY CLAUSE** that on gas sold at the wells the royalty shall be one-eighth of the _____

this lease shall continue in effect for a period of one year from the date such well is shut in. Lessee or any assignee may thereafter, in the manner provided herein for the payment or tender of delay rentals, pay or tender to Lessor as roya **SHUT-IN ROYALTY CLAUSE** anniversary of shut-in date of such well this lease shall continue in effect for successive periods of twelve (12) months each.

4. If operations for drilling are not commenced on said land as hereinafter provided, on or before one year from this date, the lease shall then terminate as to both parties, unless on or before such anniversary date Lessee shall pay or tender to Lessor or to the credit of Lessor in _____

Bank of _____ (which bank and its successors are Lessor's agent and shall continue as the depository for all rentals payable hereunder regardless of changes in ownership of said land or the rentals either by conveyance or by the death or incapacity of Lessor) the sum of _____ Dollars

($_____) herein called rental) which shall cover the privilege of deferring commencement of operations for drilling for a period of twelve (12) months. In like manner and upon like payments or tenders annually the commencement of operations for drilling may be further deferred for successive periods of twelve (12) months each during the primary term. **DRILLING DELAY RENTAL CLAUSE** the down cash payment is consideration for this lease according to its terms and shall not be allocated as mere rental for a period. Lessee may at any time execute and deliver to Lessor or to the depository above named or place of record a release or releases covering any portion or portions of the above described premises and thereby surrender this lease as to such portion or portions and be relieved of all obligations as to the acreage surrendered, and thereafter the rentals payable thereunder shall be reduced in the proportion that the acreage covered hereby is reduced by said release or releases.

If Lessee shall, on or before any rental date, make a bona fide attempt to pay or deposit rental to a Lessor entitled thereto under this lease according to Lessee's records or to a Lessor who, prior to such attempted payment or deposit, has given Lessee notice, in accordance with the terms of this lease hereinafter set forth, of his right to receive rental, and if such payment or deposit shall be erroneous in any regard (whether deposited in the wrong depository, paid to persons other than the parties entitled thereto as shown by Lessee's records, in an incorrect amount, or otherwise), Lessee shall be unconditionally obligated to pay to such Lessor the rental properly payable for the rental period involved, but this lease shall be maintained in the same manner as if such erroneous rental payment or deposit had been properly made, provided that the erroneous rental payment or deposit be corrected within thirty (30) days after receipt by Lessee of written notice from such Lessor of such error accompanied by any documents and other evidence necessary to enable Lessee to make proper payment.

5. Should any well drilled on the above described land during the primary term before production is obtained be a dry hole, or should production be obtained during the primary term and thereafter cease, then and in either event, if operations for drilling an additional well are not commenced or operations for reworking a well are not pursued on said land on or before the first rental paying date next su **DRY HOLE, CESSATION OF PRODUCTION** shall resume the payment of rental **and CONTINUOUS OPERATIONS** after discovery of oil or gas in the rental payments. If during the _____ production on this lease shall cease. This lease nevertheless shall continue in force as long _____ production is discovered as a result of any such drilling or reworking operations, conducted without cessation of production; if production is restored or additional production is discovered as a result of any such drilling or reworking operations, conducted without cessation of production, if production is restored or additional gas is produced and as long as additional drilling or reworking operations are had without cessation of such drilling or reworking operations for more than sixty (60) consecutive days.

6. Lessee, at its option, is hereby given the right and power to pool or combine the land covered by this lease, or any portion thereof, as to oil and gas, or either of them, with any other land, lease or leases when in Lessee's judgment it is necessary or advisable to do so in order to properly develop and operate said premises, such pooling to be into a well unit or units not exceeding forty (40) acres, plus an acreage tolerance of ten per cent (10%) of forty (40) acres, for oil, and not exceeding six hundred and forty (640) acres, plus an acreage tolerance of ten per cent (10%) of six hundred and forty (640) acres, plus **POOLING CLAUSE** Therefrom, or the completion thereon of a well as a shut-in gas well, shall be considered _____ for all purposes, except the payment of royalties _____ covered by this lease, whether or not the well or wells be located on the premises covered by this lease. In lieu of the royalties elsewhere herein specified, Lessor shall receive from a unit so formed, only such portion of the royalty stipulated herein as the amount of his acreage placed in the unit or his royalty interest therein bears to the total acreage so pooled in the particular unit involved. Should any unit as originally created hereunder contain less than the maximum number of acres hereinabove specified, then Lessee may at any time thereafter, whether before or after production is obtained on the unit, enlarge such unit by adding additional acreage thereto, but the enlarged unit shall in no event exceed the acreage content hereinabove specified. In the event an existing unit is so enlarged, Lessee shall execute and place of record a supplemental declaration of unitization identifying and describing the land added to the existing unit; provided, that if such supplemental declaration of unitization is not filed until after production is obtained on the unit as originally created, then and in such event the supplemental declaration of unitization shall not become effective until the first day of the calendar month next following the filing thereof. In the absence of production Lessee may terminate any unitized area by filing of record notice of termination.

7. Lessee shall have the right at any time without Lessor's consent to surrender all or any portion of the leased premises and be relieved of all obligation as to the acreage surrendered. Lessee shall have the right at any time during or after the expiration of this lease to remove all property and fixtures placed by Lessee on said land, including the right to draw and remove all casing. When required by Lessor, Lessee will **SURRENDER CLAUSE** be drilled within two hundred (200) feet of any residence or barn now on said land without Lessor's consent. The Lessee agrees to promptly pay to the owner thereof any damages to said land caused by or resulting from any operations of Lessee.

8. The rights of either party hereunder may be assigned, in whole or in part, and the provisions hereof shall extend to the heirs, successors and assigns of the parties hereto, but no change or division in ownership of the land, rentals, or royalties, however accomplished, shall operate to enlarge the obligations or diminish the rights of Lessee. No change in the ownership of the land, or any **ASSIGNMENT and CHANGE OF OWNERSHIP** rentals payable hereunder shall be _____ apportionable among the several leasehold owners ratably according to the surface area of each, and default in rental payment by one shall not affect the rights of other leasehold owners hereunder. In case Lessee assigns this lease, in whole or in part, Lessee shall be relieved of all obligations with respect to the assigned portion or portions arising subsequent to the date of assignment.

Figure 5

106

GOVERNMENT REGULATION

9. All exp[illegible] Laws, Executive Orders, Rules or Regulations, and this lease shall not be terminated in whole or in part [illegible] ented by, or if such failure is the result of any such Law, Order, Rule or Regulation, or if prevented by an act of God, of the public enemy, labor disputes, inability to obtain material, failure of transportation, or other cause beyond the control of Lessee.

FORCE MAJEURE

If, dur[illegible] on the leased premises, but Lessee is prevented from producing the same by reason of any of the causes set out in this Section, this lease [illegible] shall continue in full force and effect until Lessee is permitted to produce the oil or gas, and as long thereafter as such production continues in paying quantities or drilling or reworking operations are continued as elsewhere herein provided.

WARRANTY and PROPORTIONATE REDUCTION

10. Lessor hereby warrants and agrees to defend the title of said land and agrees that Lessee at its option may discharge any tax, mortgage or other lien upon said land, either in whole or in part, a[illegible] der toward satisfying same. Without impairment [illegible] fee simple estate, then the royalties and rentals to be paid Lessor shall be reduced proportionately.

LEGAL EFFECT

11. Lessors hereby release and waive all rights under and by virtue of the homestead exemption laws of Wyoming.

All of the [illegible] benefit of and be binding upon the parties hereto, their heirs, administrators, successors and assigns.

This agreement shall be binding on each of the above named parties who sign the same, regardless of whether it is signed by any of the other parties.

IN WITNESS WHEREOF, this instrument is executed on the date first above written.

SIGNATURES

Soc. Sec. No.
or Tax I.D. No. ________________

No. ________

OIL AND GAS LEASE

FROM

TO

Date ________, 19__

Section ____ Township ____ Range ____

________ County, Wyoming

No. of Acres ________ Term ________

STATE OF WYOMING } ss.
County of ________

This instrument was filed for record on the ________ day of ________, 19__ at ____ o'clock ____ M., and duly recorded in book ____ page ____ of the records of this office.

________ County Clerk–Register of Deeds

By ________ Deputy

Record and Mail to:

STATE OF WYOMING } ss.
County of ________

Wyoming Individual

The foregoing instrument was acknowledged before me by ________________ this ________ day of ________________, 19____

Witness my hand and official seal.

ACKNOWLEDGMENT

____
Notary Public

My commission expires ________________

STATE OF WYOMING } ss.
County of ________

Wyoming Individual

The foregoing instrument was acknowledged before me by ________________ this ________ day of ________________, 19____

Witness my hand and official seal.

____
Notary Public

My commission expires ________________

STATE OF WYOMING } ss.
County of ________

Wyoming Corporation

The foregoing instrument was acknowledged before me by ________________, President of ________________, this ________ day of ________________, 19____

Witness my hand and official seal.

____
Notary Public

My commission expires ________________

Figure 6

(7) The *drilling and delay rental clause* (see § **3.08**) permits the Lessee to defer immediate exploration on the leased premises (a former implied obligation) so long as he commences a well within a specified period, or pays periodic delay rentals. The clause also provides a method of excusing the Lessee from either obligation by releasing the lease, in whole or in part.

(8) *The dry hole, cessation of production, and continuous drilling clauses* (see § **3.05(e)**) spell out the rights of the parties if, during or after the primary term, the Lessee drills a dry hole, production ceases, or the Lessee should be in the process of conducting operations even though no oil or gas is being produced. The clause is important during the primary term to determine whether or not further drilling and/or delay rental payments are required to keep the lease alive; and, after the primary term, whether or not, and how, the lease can be kept in effect.

(9) The *pooling clause* (see § **3.06**) spells out the right of the Lessee to combine all or a portion of the leased premises with rights under other leases to form a single drilling unit, and the effects of such pooling.

(10) The *surrender clause* (see § **3.08(c)(5)**) spells out the rights of the Lessee to be relieved of his obligations by surrendering the lease. The clause also contains provisions concerning certain other rights of the Lessee and the Lessor.

(11) The *assignment clause* (see § **3.08(c)(5)**) permits a transfer by either party of his rights, in whole or in part, but prohibits any increase in the Lessee's obligations as a result of any such assignment, and relieves the Lessee of obligations as to any assigned portion of the premises. The clause also contains a *change of ownership* provision, requiring notice to the Lessee and specified documentation before he is bound by any transfers by the Lessor.

(12) *The regulatory and force majeure clauses* provide that the lease is subject to Federal and state laws, and excuse the Lessee from certain lease provisions if his failure to perform is due to circumstances beyond his control.

(13) The *warranty clause* (see § **3.10**) contains the Lessor's warranty of title and the Lessee's rights if the Lessor defaults in paying taxes and

mortgages. The clause also contains a *lesser interest* provision (see §§ **3.08(c)(5)** and **3.10**), providing for a proportionate reduction of rentals and royalties if the Lessee owns less than the entire interest in the land subject to the lease.

(14) The *legal effect clause* specifies the parties bound by the lease.

Clauses not found in this "typical" lease, but which may appear in other lease forms, are the *"notice"* and *"judicial ascertainment"* clauses, which provide the Lessee an opportunity to correct prior breaches before the lease can be declared forfeited; a *"growing crops"* and *"restoration of the premises"* clause, making more specific the rights of the Lessor as to surface damage; and an *entirety clause*, which pools the rights of the Lessors (or their successors) if ownership is not uniform throughout the lease premises. Samples of such clauses are contained in Appendix B.

In addition to the express covenants contained in the Oil and Gas Lease, there are certain *implied covenants* between the Lessor and the Lessee which courts have "written into" the lease. While the prior implied covenant to drill an exploratory well has been superseded by the drilling and delay rental clause, the Lessee is still subject to implied covenants to protect the premises against drainage; to further develop the lease premises after the drilling of an initial producing well; and to conduct his operations as a prudent operator, which includes the duty to use reasonable efforts to market the production.

§ 3.04 The Granting Clause

THIS AGREEMENT made this _____ day of _____ , 19 _____ , between _____

_____

_____

Lessor (whether one or more), and _____
Lessee, WITNESSETH:

1. Lessor in consideration of _____ Dollars ($ _____), in hand paid, of the royalties herein provided, and of the agreement of Lessee herein contained, hereby grants, leases and lets exclusively

unto Lessee for the purpose of investigating, exploring, prospecting, drilling and mining for and producing oil, and gas, and the constituents thereof, laying pipe lines, building tanks, power stations, telephone lines and other structures thereon to produce, save, take care of, treat, transport and own said products, and housing its employees, the following described land in _____ to wit:

[DESCRIPTION]

In addition to the land above described, Lessor hereby grants, leases and lets exclusively unto Lessee to the same extent as if specifically described herein all lands owned or claimed by Lessor which are adjacent, contiguous to or form a part of the lands above particularly described, including all oil, gas, and their constituents underlying lakes, rivers, streams, roads, easements and rights-of-way which traverse or adjoin any of said lands. For rental payment purposes, the land included within this lease shall be deemed to contain _____ acres, whether it actually comprises more or less.

* * *

3. ... Lessee shall have free use of oil, gas, and water from said land, except water from Lessor's wells, springs, or reservoirs, for all operations hereunder, and the royalty on oil and gas shall be computed after deducting any so used.

§ 3.04(a) Rights Granted the Lessee. As mentioned in § 3.02(a), in most states the Oil and Gas Lease is not a "lease" in the traditional sense; and, except in Louisiana, ordinary rules concerning landlord and tenant do not apply.

Insofar as the nature of the Lessee's interest is concerned, Texas holds that an Oil and Gas Lease grants the Lessee an "ownership in place" of the minerals. However, the majority of key oil-producing states hold that the Lessee acquires only a right to prospect for the oil and gas, and to reduce the minerals discovered to possession. In such "qualified ownership" states, the Lessee's interest is referred to as an "incorporeal hereditament" in the nature of a "profit a prendre". However, this theoretical legal distinction has made little practical difference in determining the extent of the Lessee's rights.

Whether in an "absolute ownership" state such as Texas, or a "qualified ownership" state, the interest of the Lessee is in the nature of a "determinable fee", since it may, but need not, endure forever.

Irrespective of the wording of the granting clause, most states hold that the rights granted the Lessee are to be determined uniformly according to the general law of the state.

The basic right granted the Lessee under the Oil and Gas Lease is the right to explore for and produce oil and gas. However, in addition to this basic right, the Oil and Gas Lease grants the Lessee certain incidental rights as to the use of the surface and the subsurface.

The rights of the Lessee to the surface are complicated by the fact that he must have sufficient surface rights to make the grant meaningful, but cannot be given exclusive rights to the surface as between himself and the surface owner.

The Lessee has an implied right of ingress and egress over the land and to make such uses of the surface as are essential or necessary to his basic right of exploration. However, these rights may only be exercised insofar as they will help carry out his basic rights under the lease, i.e., the development of the *leased premises* for oil and gas. This means that while the Lessee can construct roads, install tank batteries, etc., to facilitate the production of oil and gas from the leased premises, he cannot similarly burden the leased premises to facilitate his activities on an adjacent lease. Also, the Lessee's use of the surface cannot be negligent or unreasonable.

The Lessee's rights to incidental surface and subsurface use include the right to free use of water, gas and other substances found on the leased premises, even where these are not substances he has a right to extract under the lease for purposes of commercial sale (see § **3.04(b)**). This right exists even in the absence of a "free use" clause, and is only restricted by the requirement that the use be for the benefit of the leased premises and not be restricted by some express clause of the lease. (For instance, the lease frequently prohibits the Lessee from using water from the Lessor's wells or ponds.)

Even though the Lessee has the right to construct facilities and otherwise use the surface as an incident of his basic rights under the Oil and

Gas Lease, it is industry practice voluntarily to pay "surface damages" where demands are reasonable. And many leases contain specific provisions concerning surface damages. (For samples, see Appendix B of this chapter.)

In addition to his rights to the surface, the Lessee has a number of rights concerning subsurface usage where these rights are exercised pursuant to the primary purpose of the grant. These include the right to use or refuse to use any method of production, including repressuring and recycling, so long as the Lessee acts in good faith; to pool or unitize the land with other lands (although not to change the method of computing the Lessor's royalty); to engage in salt water disposal where incidental to operations under the lease; and to engage in various "prospecting" activities such as seismic exploration and core drilling.

The exploration right is exclusive to the Lessee; however, most of the Lessee's surface and subsurface rights are non-exclusive, so long as the use by other parties does not interfere with the Lessee's exercise of his lease rights.

§ **3.04(b) Substances Covered.** The substances which the Lessee acquires a right to extract and own are directly affected by the specific wording of the granting clause. The construction of this granting clause language is subject to the doctrine of "ejusdem generis", which holds that all items, but only those items, of the same class as those specifically named are encompassed by a subsequent general reference to other items, i.e., as in "oil, gas *and other minerals.*"

A lease whose granting clause covers only "oil and gas" obviously would not cover non-gaseous, non-hydrocarbon substances, or "hard" carbons such as coal. In addition, states are split as to whether or not substances such as casinghead gas, distillate and natural gasoline are covered under a grant of "oil and gas". Accordingly, many modern Oil and Gas Leases specifically cover such substances.

A grant of "oil and gas" also raises the question of whether the term "gas" refers to the physical state of the substances, or merely to gaseous *hydrocarbons*; i.e., would non-hydrocarbon gaseous substances such as helium, nitrogen and carbon dioxide be included within such a grant? In the lead case on helium, it was held that "gas" refers to the physical state, and is not limited to gaseous hydrocarbons. However, there is not sufficient authority on the question to consider this a closed matter.

If the granting clause covers "oil, gas, *and other minerals*", or specifically names certain minerals in addition to oil and gas, the *ejusdem generis* doctrine will be applied to determine whether or not, and which, substances other than liquid and gaseous hydrocarbons are included in the grant. In this regard, the Oil and Gas Lease grant may be construed more narrowly than similar wording in a mineral deed, since the lease contains clauses which can be interpreted as showing an intent to limit the granting clause to "minerals" extracted by the drilling of wells.

Other clauses of the lease may also affect the substances covered by the lease. Thus, while under *ejusdem generis* a grant of "oil, gas *and other minerals*" would normally not cover hard minerals (with the possible exception of coal), a royalty clause expressly providing for a royalty on coal, sulfur or lead should expand the lease coverage to include these additional minerals, and possibly other minerals of a similar type.

§ **3.04(c) Lands Covered.** While a description of the land leased is an essential element of the Oil and Gas Lease, the lease will not fail if the description will identify the land being leased under any reasonable construction. The lands leased will not be reduced because the actual acreage content exceeds the number of acres on which the bonus was computed. However, this might give grounds for a reformation of the lease on the grounds of "mutual mistake". Accordingly, many modern Oil and Gas Leases contain a "lease in gross" provision (see samples in Appendix B of this chapter), specifying that the recited acreage content will control for purposes of the lease, irrespective of actual acreage content.

Normally, an Oil and Gas Lease on land adjoining a railroad right-of-way, road, etc. will cover any mineral rights of the Lessor underlying the right-of-way. Likewise, accreted lands are normally covered by a lease of the adjacent tract. However, to avoid an inadvertent exclusion of small strips of land due to improper description of the land, an increased ownership due to adverse possession, or the like, many leases contain a so-called "Mother Hubbard" clause, which specifically covers adjacent or contiguous lands, as well as adjacent rivers, roads, easements and the like. In addition, in Texas, due to the many errors in early surveys, the "Mother Hubbard" clause frequently will cover all other lands owned by the Lessor in the survey or in adjoining surveys, so long as such lands adjoin the specified tract; this picks up small strips actually occupied by the Lessor, but inadvertently omitted from the survey.

Except in Louisiana, the courts have recognized the "Mother Hubbard" clause as a valid lease provision. However, if the Lessor owns substantial acreage in the vicinity of the specifically described tract, which technically might be considered "adjacent" under the literal wording of the "Mother Hubbard" clause, the effect of the clause has nevertheless been limited to an assumed general intent to cover only small adjacent strips.

If an Oil and Gas Lease containing a "Mother Hubbard" clause is assigned, courts are split as to whether or not the assignee will acquire acreage other than that specifically described, unless the assignment also contains a "Mother Hubbard" provision (which it seldom does).

§ 3.04(d) Depth Limitations. Frequently, a lease (or, more frequently, a lease assignment) may limit the rights granted to specified depths or formations. Such a provision can create serious problems unless it clearly identifies the rights being granted. The preferable wording of such a limitation will be to the top or base of a known geological marker, often tied to given depths on the log of a named well. The provision should never be limited solely to all rights above or below a given depth.

Often, your company may want to tie the depth assigned more closely to the depth actually drilled by an "earning well" under a farmout agreement. As to pitfalls in describing the depth assigned under such an arrangement, see Chapter Four.

§ 3.05 The Habendum Clause

2. Subject to the other provisions herein contained, this lease shall be for a term of 10 years from this date (called "primary term") and so long thereafter as oil or gas is produced from said land hereunder, or drilling or reworking operations are conducted thereon.

* * *

5. Should any well drilled on the above described land during the primary term before production is obtained be a dry hole, or should production be obtained during the primary term and thereafter cease, then and in either event, if operations for drilling an additional well are not commended or operations for reworking a well are not pursued on said land on or before the first rental paying date next succeeding the cessation of production or drilling or reworking on said well or wells, then this lease shall terminate, unless Lessee, on or before said date, shall resume the payment of rentals. Upon resumption of the payment of rentals, Section 4 governing the payment of rentals, shall continue in force just as though there had been no interruption in the rental payments. If during the last year of the primary term and prior to the discovery of oil or gas on said land Lessee should drill a dry hole thereon, or if after discovery of oil or gas before or during the last year of the primary term the production thereof should cease during the last year of said term from any cause, no rental payment or operations are necessary in order to keep the lease in force during the remainder of the primary term. If, at the expiration of the primary term, Lessee is conducting operations for drilling a new well or reworking an old well, this lease nevertheless shall continue in force as long as such drilling or reworking operations continue, or if, after the expiration of the primary term, production on this lease shall cease, this lease nevertheless shall continue in force if drilling or reworking operations are commenced within 60 days after such cessation of production; if production is restored or additional production is discovered as a result of any such drilling or reworking operations, conducted without cessation of more than 60 days, this lease shall continue as long thereafter as oil or gas is produced and as long as additional drilling or reworking operations are had without cessation of such drilling or reworking operations for more than 60 consecutive days.

§ 3.05(a) In General. The habendum clause prescribes the term of the lease — how long the rights granted will endure. However, the term of the lease is also affected by various supplemental clauses such as the drilling and delay rental clause, the shut-in clause, the dry hole and cessation of production clauses, the pooling clause and the surrender clause.

The habendum clause of the modern Oil and Gas Lease provides for a relatively short fixed term (the "primary term"), which may be extended by production (the "thereafter" clause). Such an habendum clause creates a determinable fee subject to termination by various events, such as a failure to establish production during the primary term.

The failure to establish production or meet any of the other conditions necessary to extend the lease beyond its primary term results in an automatic termination of the lease. Since the lease terminates automatically, rather than being forfeited, no notice is required the Lessor prior to termination; and, for the most part, equitable considerations are immaterial.

§ 3.05(b) Extension of the Primary Term by "Production". The typical habendum clause provides that the lease shall extend for the specified primary term "and as long thereafter as oil or gas is produced from said land...."

Courts are split as to what constitutes "production" under the habendum clause. In Oklahoma and several other states, the *discovery* of either oil or gas will satisfy the clause. However, in Louisiana, Texas, Kansas and a majority of states, "production" requires actual *extraction* of the oil or gas, and, in the case of gas, marketing. (Note that, even in Oklahoma, production must be marketed within a reasonable period of time or the lease is subject to cancellation for violation of the implied covenant to market.) In addition, in the majority of oil states, oil or gas must likewise be produced in *paying quantities* for the habendum clause to be satisfied. (Such a requirement is met if the quantity of production is sufficient to return some profit to the Lessee over operating costs, and does not require that the cost of drilling and equipping the well be taken into account.)

Even in states which require actual extraction and marketing of oil and gas to extend the lease, the mere "discovery" of oil and gas may be sufficient to extend the lease if the lease contains a shut-in clause (see § **3.09**).

Once production, however defined, is secured, a well located anywhere on the leased premises, and producing any substance covered by the lease, will, in the absence of a special provision in the lease to the contrary, extend the lease as to all of the leased premises, as to all horizons, and as to all substances. In all states except Louisiana, this rule applies even where the Lessee has subdivided the lease premises by a partial assignment of a separate portion; however, in Louisiana, the habendum clause is considered "divisible", so that production from only a part of the leased premises which has been "subdivided" by a partial assignment of a separate portion will extend the lease only as to that subdivided portion from which production is being secured. Likewise, the fact that the Lessor's interest has been assigned so that different portions of the leased premises are owned by different owners does not serve to "subdivide" the lease so as to require production from each separately owned tract; nor will the Lessee be required to drill an offset well on a non-productive portion of the leased premises to protect it against drainage from other portions of the leased premises. However, if the leased premises has not been fully developed, the lease may still be subject to partial cancellation as to the undeveloped portions for failure to comply with the implied covenant of further development.

§ 3.05(c) Extension of the Lease by Means Other than Production. In Oklahoma, Texas and several other states, drilling operations being conducted at the end of the primary term will serve to extend the lease so long as the lease contains a "commerce" type drilling and delay rental clause. However, several other states such as California, Kansas and Louisiana, have indicated contrary results. Accordingly, most modern lease forms contain express "continuous drilling" clauses specifically providing for an extension of the lease if operations are being conducted at the end of the primary term.[2]

---

[2] As to the acts necessary to "commence" a well for lease extension purposes, see § **3.08(b)**.

Except in the situations just noted, a failure to obtain production during the primary term results in an automatic termination of the lease, and equitable considerations are generally immaterial. Thus, the fact that the Lessee has expended considerable sums in unsuccessful exploratory efforts, or has been prevented from drilling due to impassable roads, storms or the unavailability of rigs or casing, is not sufficient to prevent termination of the lease. However, if the Lessee's inability to commence operations has been caused by unfair conduct on the part of the Lessor, the lease may be extended under the doctrine of "obstruction".

A finding of obstruction may result if the Lessor has prevented or interfered with the Lessee's attempts to extend the lease, either by physically interfering with the conduct of drilling operations or by commencing an attack on the Lessee's title. The attack on the title may be in the form of a lawsuit, or an unconditional demand for a release of the lease. It could even arise from the granting of a "top lease" to a new Lessee which fails to recognize the rights of a present Lessee under an existing lease.

In several states, obstruction only "tolls" the term of the lease, so that when the obstruction ceases, the Lessee merely has a period left in which to commence operations equal to the time remaining in the primary term when the obstruction commenced. However, in Oklahoma and Texas the Lessee is given a "reasonable period of time under the circumstances" in which to commence drilling operations.

§ 3.05(d) Effect of Cessation of Production.

If the lease is "producing" at the end of the primary term, but production thereafter ceases, the lease will terminate if the cessation is "permanent", but not if the cessation is merely "temporary".

An interruption of production to improve recovery, such as to increase or restore production, to drill to deeper formations during the period in which the shallow formations are still productive, or to change over to secondary recovery methods while the lease is still capable of primary production, is considered a "temporary" cessation, so long as the Lessee is acting in good faith. However, if the well permanently ceases to produce from the formation in which it is completed, in most jurisdictions the Lessee will not be permitted in the absence of a special clause in the lease to conduct a new exploratory effort or to initiate secondary recovery.

General rules as to the effects of a temporary or permanent cessation of production are affected by the "cessation of production" clause (see § 3.05(e)).

§ 3.05(e) Effect of Special Lease Provisions. The pooling clause will be discussed in § 3.06. However, at this point it should be noted that if all or any portion of the leased premises has been validly pooled with other lands, either under a voluntary pooling provision or under statutory "forced", i.e., compulsory, pooling, production from any of the pooled area will extend all leases if any portion of the lease is included within the pooled unit. As a general rule, this extension will apply both to lands lying within, and lying without, the pooled area.

The shut-in clause similarly will be discussed in greater detail in § 3.09. If the lease contains a shut-in clause, a shut-in well may likewise serve to extend the lease. However, the conditions necessary to extend a lease under the shut-in clause are dependent on a number of circumstances, including the exact wording of the particular clause. These problems will be discussed later.

Some leases contain a clause, added for the benefit of the Lessor, known as the "Pugh" clause. The general nature of such a clause is to limit the amount of acreage and/or formations that can be extended by a single well, usually to the drilling and spacing unit.[3] Normally, the "Pugh" clause applies only to a "pooling" situation. However, there is no standard form of such clause; and, again, it is necessary to refer to the exact wording of the clause in determining its effect on the general rules previously noted. (For samples, see Appendix B of this chapter.)

---

[3] In some instances, the outside acreage can be extended, at least temporarily, by the payment of delay rentals.

Most modern leases contain detailed provisions referred to as dry hole, cessation of production, and continuous drilling (operations) clauses. Such clauses prevent lease termination in the event of the drilling of a dry hole, the cessation of production, or when operations are being conducted, but production has not been established, as of the expiration of the primary term, so long as delay rental payments are resumed (if within the primary term) or other operations are commenced within a specified period.

While the dry hole, etc. clauses are normally looked at as increasing the rights of the Lessee, in certain instances the clauses may restrict rights that would otherwise be available to the Lessee. Thus, the clauses frequently require additional drilling or the resumption of delay rentals within the primary term where otherwise a dry hole would have done away with any such obligation for the balance of the primary term. Likewise, where a "temporary" cessation of production could continue for a substantial period of time without causing lease termination under general rules of law, the cessation of production clause normally requires that operations be commenced within a 60 or 90-day period following the termination of production; this may modify such "temporary" cessation rules. In any event, before relying upon such clauses you should carefully study their exact wording to determine the conditions necessary for their effectiveness.

§ 3.06 The Pooling Clause

6. Lessee, at its option, is hereby given the right and power to pool or combine, the land covered by this lease, or any portion thereof, as to oil and gas, or either of them, with any other land, lease or leases when in Lessee's judgment it is necessary or advisable to do so in order to properly develop and operate said premises, such pooling to be into a well unit or units not exceeding 40 acres, plus an acreage tolerance of 10 percent of 40 acres for oil, and not exceeding 640 acres, plus an acreage tolerance of 10 percent of 640 acres for gas, except that larger units may be created to conform to any spacing or well unit pattern that may be prescribed by governmental authorities having jurisdiction. Lessee may pool or combine acreage covered by this lease, or any portion thereof, as above provided, as to oil or gas in any

one or more strata, and units so formed need not conform in size or area with the unit or units into which the lease is pooled or combined as to any other stratum or strata, and oil units need not conform as to area with gas units. The pooling in one or more instances shall not exhaust the rights of the Lessee hereunder to pool this lease or portions thereof into other units. Lessee shall execute in writing and place of record an instrument or instruments identifying and describing the pooled acreage. The entire acreage so pooled into a unit shall be treated for all purposes, except the payment of royalties, as if it were included in this lease, and drilling or reworking operations thereon or production of oil or gas therefrom, or the completion thereon of a well as a shut-in gas well, shall be considered for all purposes, except the payment of royalties, as if such operations were on or such production were from or such completion were on land covered by this lease, whether or not the well or wells be located on the premises covered by this lease. In lieu of the royalties elsewhere herein specified, Lessor shall receive from a unit so formed, only such portion of the royalty stipulated herein as the amount of his acreage placed in the unit or his royalty interest therein bears to the total acreage so pooled in the particular unit involved. Should any unit as originally created hereunder contain less than the maximum number of acres hereinabove specified, then Lessee may at any time thereafter, whether before or after production is obtained on the unit, enlarge such unit by adding additional acreage thereto, but the enlarged unit shall in no event exceed the acreage content hereinabove specified. In the event an existing unit is so enlarged, Lessee shall execute and place of record a supplemental declaration of unitization identifying and describing the land added to the existing unit; provided that if such supplemental declaration of unitization is not filed until after production is obtained on the unit as originally created, then and in such event the supplemental declaration of unitization shall not become effective until the first day of the calendar month next following the filing thereof. In the absence of production Lessee may terminate any unitized area by filing of record notice of termination.

§ **3.06(a) Nature of the Pooling Clause.** The pooling clause is a grant of advance authority to the Lessee to consolidate all or part of the leasehold with other oil and gas rights to create a drilling unit on which usually only one well will be drilled. It is to be distinguished from "unitization", which is a consolidation of oil and gas rights covering all or substantially all of an entire field or reservoir.

The typical lease pooling clause includes: a grant of the power to pool, in whole or in part, for specified purposes, and further defines the substances and formations to be covered by the pooling; a maximum unit size; the effect of the declaration of pooling on other lease clauses, such as habendum, royalty and delay rental clauses; and the mechanics of putting the unit into effect. A typical clause might authorize the Lessee to pool all or a portion of the land covered by the lease, for oil, gas, or both, in order for proper development and operation of the premises; limit pooling units to 40 acres for oil and 640 acres for gas (with an acreage tolerance of 10 percent), except when larger units are necessary to comply with conservation orders; authorize pooling as to one or more formations, with multiple exercises of the power permitted. In some clauses, enlargement and termination of units are authorized. The effect of such pooling on other lease clauses would also be specified, and the pooling power might be exercised by the execution and recording of an instrument describing the pooled acreage.

The validity of the exercise of the pooling power under such clauses has generally been upheld. However, there is some question as to whether or not non-participating royalty owners or owners of non-executive mineral interests would be bound by the exercise of a pooling power contained in the lease which they had not joined in or ratified.

§ 3.06(b) Exercise of the Pooling Power. The extension of the lease term, or the satisfaction of drilling or delay rental requirements, by the exercise of the pooling power, will be effective only if the pooling right is timely exercised. Thus, the power must be exercised while the leases to be extended are still in effect. If the primary term of a lease has expired, or a delay rental has been missed, prior to the exercise of the pooling, the lease will have terminated; life cannot be breathed back into it by a subsequent exercise of the pooling power, even though the off-lease well was commenced prior to the drilling-delay rental, etc., date. Likewise, if the pooling power is limited to an authority of the pool "for development", pooling after production has already been obtained may not be authorized.

If the formalities for pooling have not been satisfied, such as requirements concerning the execution and recording of a written Declaration of Pooling or limitations on unit size, the exercise will not be valid.

Assuming that the pooling power has been validly exercised, production from any portion of the unit will extend all leases as to all tracts, even where only a portion of the leased premises lies within the unit. In "commence" states, the mere commencement of drilling on any portion of the pooled unit will extend all leases. However, if the lease contains a "Pugh" clause (see § **3.05(e)**), that clause may limit the extending effects of production from a pooled unit to lands lying within the unit.

The pooling clause of the lease will normally specify in detail the effect of the pooling on payment of royalty. Often, the clause will provide that all Lessors under all leases will share in unit royalties in the percentage which the acres covered by their lease bear to the acres contained within the unit. Other pooling clauses provide, however, that unit production will be *allocated* among the various leases, and that the lease royalty provisions will apply to such allocated production. The difference in such clauses may prove significant where a Lessor has assigned a separated portion of his leased premises to another party, and where, in the absence of an overriding provision in the pooling clause, there would be no "apportionment" of the royalties among the owners of the separately owned tracts. (As to apportionment of royalties, see § **3.07(c)**.)

Insofar as the payment of delay rentals is concerned, as a general rule the drilling of a well anywhere within the unit area will satisfy the drilling and delay rental clause as to all leases included within the unit; nor will any duty to pay rentals as to acreage lying outside the unit be implied. (Again, a "Pugh" clause may require resumption of rentals as to outside acreage.)

A key error in the attempted exercise of the pooling power is a failure to study the exact terms of each pooling clause contained in the various leases to be pooled to determine the steps required and the conditions surrounding a valid exercise of the pooling rights under that lease. For instance, a failure to make a timely recording of the Declaration of Pooling, or to conform to restrictions on unit size could void an attempted pooling.

Accordingly, it is encumbent upon the Landman to determine that each lease which will be included within the unit, in whole or in part, does authorize the proposed pooling, and to make sure that all conditions necessary for a valid execution of the pooling power, as specified in each of the leases, has been complied with.

§ 3.07 The Royalty Clause

3. The royalties to be paid by Lessee are: (a) on oil, one-eighth of that produced and saved from said land, the same to be delivered at the wells, or to the credit of Lessor into the pipeline to which the wells may be connected; Lessee may from time to time purchase any royalty oil in its possession, paying the market price therefore prevailing for the field where produced on the date of purchase; (b) on gas, including casinghead gas or other hydrocarbon substance, produced from said land and sold or used off the premises, or in the manufacture of gasoline or other products therefrom, the market value at the well of one-eighth of the gas so sold or used, provided that on gas sold at the wells the royalty shall be one-eighth of the amount realized from such sale. Lessee shall have free use of oil, gas and water from said land, except water from Lessor's wells, springs or reservoirs, for all operations hereunder, and the royalty on oil and gas shall be computed after deducting any so used.

§ 3.07(a) Nature of the Royalty Clause. "Royalty" is a share of production or production proceeds reserved by the Lessor as compensation for permitting the Lessee to explore and develop the leased premises. As a general rule, the royalty will relate to gross production; be free of costs of exploration, operation and production; and continue throughout the life of the production.

The typical royalty clause is subdivided into at least two subparts, one covering the royalty payable on oil, and the other covering the royalty payable on gas. Some leases contain a third clause covering the royalty payable on other substances.

The oil royalty clause is quite standardized and uniformly provides for the delivery to the Lessor *in kind* of a specified fraction of gross production, free of cost.[4] Under such a clause, title to a specified fraction of the oil remains in the Lessor.

The royalty normally applies only to oil "produced *and saved*" from the land, and usually provides for the delivery of the oil into the Lessor's tanks at the well, or to the Lessor's credit into the pipeline to which the wells are connected.

Gas royalty provisions vary to a far greater degree than do oil royalty provisions. However, the general pattern is to *pay* to the Lessor *in money* a specified fraction of either: (a) the *proceeds* received from the sale of the gas; or (b) the *"market value"* or "market price" of gas sold or used off the premises. Again, the determination is based on gross production, free of cost, but only for gas sold or used *off the premises*.

The gas royalty clause frequently provides that the price determination is to be made "at the well", and may be modified even as to "market value" or "market price" clauses by a provision that, in the event of actual sale, the sale price will govern. However, these provisions often apply to "sales at the wellhead".

Under the gas royalty clause, title to the full eight-eighths of gas production passes to the Lessee, subject only to his obligation to account to the Lessor for his share of the proceeds or market value.

Whatever the form of the royalty provision, the Lessee is under an implied obligation to market the Lessor's share of production at a fair price. Under the oil royalty clause, he is also under an express duty to deliver the oil, either into the Lessor's tanks or to the Lessor's credit to the pipeline to which the wells are connected.

Where gas is being produced, the Lessee clearly has authority to sell the full interest in production, since the lease grants him title to eight-eighths of the gas. While the Lessor need not join in the execution of the division order, it is frequently good practice to have him join to establish that the gas has been sold for fair value.

---

[4] In the past, the "customary" royalty was one-eighth. Today, however, much larger fractions, often increasing after payout, are becoming "standard" in many parts of the country.

Where oil is produced, it is normally assumed that the Lessee has an implied right to market the production for the Lessor's account to a responsible purchaser at the posted price on usual terms and conditions, unless the Lessor has provided his own pipeline connection or storage facilities. This right to market is implied even though the Lessor has actual title to his share of production under the oil royalty clause. In practice, again, the sale is normally confirmed by the execution of the division order by the Lessor.

§ 3.07(b) Calculation of the Royalty Payment. Where royalties are payable "in kind", as under the customary oil royalty provision, the Lessee must bear all costs of producing a marketable product, but is only required to bear his proportionate part of the costs of further refining, processing or marketing the product. Thus, if there is a market for the "raw" oil, the accounting to the Lessor is based on the price received for the untreated oil if this is the condition in which it is sold. If the Lessee elects to treat the oil, the Lessee accounts to the Lessor for the price actually received for the treated oil, *less the Lessor's proportionate part of the costs of treating*. If, however, treatment was necessary to prepare the oil for market, and there is no market for untreated oil, treating costs would be considered a cost of production to be borne solely by the Lessee.

If oil must be transported to the market, the Lessor must bear his proportionate share of the transportation costs.

When royalty is payable based upon "proceeds", as in many gas royalty clauses, the basis of the calculation is easy to determine where a sale is made at the wellhead to a non-affiliated purchaser. If, however, the sale is not at the wellhead and not at arms length, or the Lessee uses rather than sells the gas, the calculation becomes more complex.

Whereas under most royalty provisions the Lessor must bear his share of the costs of transporting the gas to the market, if the gas royalty is payable based upon "proceeds" and there is no provision tying such calculation to proceeds "at the wellhead", the majority of gas producing states hold that the Lessor is entitled to receive his share of the gross proceeds without any deduction for transportation costs.

If the Lessee uses rather than sells the gas, or extracts liquids prior to selling, the calculation of the Lessor's share of the royalty will be based upon the "market value" of the gas, even under a "proceeds" clause.

Where royalties are payable based upon "value", "market price" or "market value", the "market" is assumed to be at the well, even where the lease does not so provide. As in the case of royalties payable in kind, the Lessee alone must bear the costs required to make the gas of marketable quality, but need not, in the absence of a contrary provision in the lease, solely bear the costs required to improve the quality of the gas or to reach the market. Thus, if sales are made at a central point in the field, the calculation of the Lessor's royalty will be arrived at by deducting transportation costs from the field price.

Cases are split as to whether compression costs are to be treated as costs of making gas marketable, or of "improving" the quality of the gas.

If gas is processed, the Lessor is entitled to share in the proceeds and/or value of both the products and the "dry" gas remaining after processing; however, his royalty payments are likewise calculated by deducting the costs of processing.

See Figure 7 for a chart summarizing these "calculation" risks.

A special problem arises under "market value" and similar clauses where gas is being sold to an interstate pipeline at an "area" or "national" price fixed by the Federal Energy Regulatory Commission (FERC) which is less than the cost at which "new" gas or "intrastate" gas is being sold, or where the price being paid under older contracts is less than the current market price under newer contracts. Recent cases have indicated that, in such circumstances, the Lessor may be entitled to receive the highest price at which gas is currently being sold, irrespective of the price which the FERC or the older contract authorizes the producer to collect from the pipeline. Many producers find themselves faced with paying the Lessor more than total well proceeds under such "market value" royalty interpretations; and, in the case of interstate sales, the FERC may further disallow passing these royalty costs on to the

COMPUTATION OF ROYALTY

	WHO BEARS COST		
	"IN KIND"	"VALUE"	"PROCEEDS"
Lease, Drilling and Completion Costs ...	LE	LE	LE
Treating Costs:			
(a) Render Marketable	LE	LE	LE
(b) Improve Quality	LR/LE	LR/LE	LR/LE
Storage Costs	*	—	—
Compression Costs	—	*	*
Transportation Costs	LR/LE	LR/LE	*
Processing Costs	LR/LE	LR/LE	LR/LE
Taxes:			
(a) Income	LR/LE	LR/LE	LR/LE
(b) Property — O/G	LR/LE	LR/LE	LR/LE
(c) Property — Leasehold Equip. .	LE	LE	LE
(d) Production	*	*	*

* Courts divided

Figure 7

128

pipeline purchaser, or even abandoning the well!

§ **3.07(c) Whom to Pay.** Special problems may arise in determining the persons to whom the royalty, however calculated, is payable. If the Lessor owns less than the full interest in the land subject to the lease, the "lesser interest" clause provides for a proportionate reduction of the royalties. However, such clauses may be ineffective where the lease was acquired pursuant to a judicial or ministerial sale, in which event the net royalty payable to the Lessor should be provided in the lease. Likewise, problems exist as to whether the lesser interest clause applies to a separately reserved "overriding royalty", a production payment or a shut-in royalty.

If the Lessor's title fails, in whole or in part, as to one portion of the leased premises only, as a general rule royalties will be payable based upon the Lessor's ownership under the specific tract on which the well is located.

Execution of a single lease by multiple Lessors, each of whom owns a separate part of the leased premises, leads to an *"apportionment"* of royalties; each Lessor shares in production from each well on the leased premises, based on his proportionate ownership in the leased premises as a whole, irrespective of which tract the well was located on. However, if a Lessor subsequently conveys his interest in only a separated portion of the leased premises, there will be no apportionment of royalties among the various owners in the absence of a contrary provision in the conveyance or the Oil and Gas Lease; and the royalties from each well will be divided among only the persons owning interests in the tract on which the well is located. However, this result may be changed by the inclusion of an "entirety clause" in the lease, providing that if the leased premises are owned in separately owned tracts, the royalties will be apportioned, and will lead to a contrary result. (See Appendix B of this chapter for a sample entirety clause.) Likewise, in some states, courts have held that the "lesser interest" clause has the same effect.

§ **3.07(d) Effects of Acceptance of Royalties by Lessor and of Nonpayment of Royalties by Lessee.** As discussed in § **3.08(c)(7)**, the acceptance of a delay rental by a Lessor may serve as a ratification of a lease which otherwise would be considered to have terminated. However, the acceptance of royalty payments by a Lessor where the rights of the Lessee have in fact terminated is not considered a ratification of the lease.

In most states, a failure of the Lessee promptly to make royalty payments to the Lessor is not grounds for forfeiture of the lease. However, in Louisiana, an unreasonable delay in payment formerly was grounds for forfeiture, even without demand; only a good faith uncertainty as to the amount of royalties payable or a dispute as to title was considered "reasonable grounds" for delay in payment. While the Louisiana Mineral Code, adopted in 1974, has softened this harsh result in some instances, nonpayment of royalties can still have serious consequences in Louisiana.

§ 3.08 The Drilling and Delay Rental Clause

4. If operations for drilling are not commenced on said land as hereinafter provided, on or before one year from this date, the lease shall then terminate as to both parties, unless on or before such anniversary date Lessee shall pay or tender to Lessor or to the credit of Lessor in _____ Bank of _____ (which bank and its successors are Lessor's agent and shall continue as the depository for all rentals payable hereunder regardless of changes in ownership of said land or the rentals either by conveyance or by the death or incapacity of Lessor) the sum of _____ Dollars ($ _____) (herein called rental) which shall cover the privilege of deferring commencement of operations for drilling for a period of 12 months. In like manner and upon like payments or tenders annually the commencement of operations for drilling may be further deferred for successive periods of 12 months each during the primary term. The payment or tender of rental herein referred to may be made in currency, draft or check at the option of the Lessee; and the depositing of such currency, draft or check in any post office, properly addressed to the Lessor, or said bank, on or before the rental paying date, shall be deemed payment as herein provided. If such bank (or any successor bank) should fail, liquidate or be succeeded by another bank, or for any reason fail or refuse to accept rental, Lessee shall not be held in default for failure to make such payment or tender of rental until 30 days after Lessor shall deliver to Lessee a proper recordable instrument, naming another bank as agent to receive such payments or tenders. The down cash payment is

consideration for this lease according to its terms and shall not be allocated as mere rental for a period. Lessee may at any time execute and deliver to Lessor or to the depository above named or place of record a release or releases covering any portion or portions of the above described premises and thereby surrender this lease as to such portion or portions and be relieved of all obligations as to the acreage surrendered, and thereafter the rentals payable hereunder shall be reduced in the proportion that the acreage covered hereby is reduced by said release or releases.

5. Should any well drilled on the above described land during the primary term before production is obtained be a dry hole, or should production be obtained during the primary term and thereafter cease, then and in either event, if operations for drilling an additional well are not commenced or operations for reworking a well are not pursued on said land on or before the first rental paying date next succeeding the cessation of production or drilling or reworking on said well or wells, then this lease shall terminate unless Lessee, on or before said date, shall resume the payment of rentals. Upon resumption of the payment of rentals, Section 4 governing the payment of rentals, shall continue in force just as though there had been no interruption in the rental payments. If during the last year of the primary term and prior to the discovery of oil or gas on said land Lessee should drill a dry hole thereon, or if after discovery of oil or gas before or during the last year of the primary term the production thereof should cease during the last year of said term from any cause, no rental payment or operations are necessary in order to keep the lease in force during the remainder of the primary term. If, at the expiration of the primary term, Lessee is conducting operations for drilling a new well or reworking an old well, this lease nevertheless shall continue in force as long as such drilling or reworking operations continue, or, if, after the expiration of the primary term, production on this lease shall cease, this lease nevertheless shall continue in force if drilling or reworking operations are commenced within 60 days after such cessation of production; if production is restored or additional production is discovered as a result of any such drilling or reworking operations, conducted without cessation of more than 60 days, this lease shall continue as long thereafter as oil or gas is produced and as long as additional drilling or reworking operations are had without cessation of such drilling or reworking operations for more than 60 consecutive days.

* * *

7. Lessee shall have the right at any time without Lessor's consent to surrender all or any portion of the leased premises and be relieved of all obligations as to the acreage surrendered....

8. ... No change in the ownership of the land, or any interest therein, shall be binding on Lessee until Lessee shall be furnished with a certified copy of all recorded instruments, all court proceedings and all other necessary evidence of any transfer, inheritance or sale of said rights. In the event of the assignment of this lease as to a segregated portion of said land, the rentals payable hereunder shall be apportionable among the several leasehold owners ratably according to the surface area of each, and default in rental payments by one shall not affect the rights of other leasehold owners hereunder.

* * *

10. ... It is agreed that if Lessor owns an interest in said land less than the entire fee simple estate, then the royalties and rentals to be paid Lessor shall be reduced proportionately.

§ 3.08(a) Nature and Scope of the Drilling and Delay Rental Clause.

Early Oil and Gas Leases purported to give the Lessee a relatively long "exploratory" period during which he was neither required to conduct drilling activities nor to compensate the Lessor for his speculative holding of the land. Such a provision did not meet with favor in the courts, which held that in the absence of an express provision concerning the time for the commencement of exploration, the Lessee has an implied duty to drill an exploratory well upon demand by the Lessor. If the well was not commenced within a reasonable time following such a demand, the lease would be cancelled even though it was still within its fixed term.

The modern drilling and delay rental clause was developed to replace this implied obligation of immediate exploration with an express provision which permits the Lessee to defer the conduct of exploratory activities until the end of the primary term, provided he compensates the Lessor for permitting such a delay by the payment of a specified periodic delay rental.

Following some disastrous decisions that held the early form of drilling and delay rental clause obligated the Lessee to continue to pay such delay rentals for the life of the lease, even after the lease was condemned, the present clause form was developed, which provides a means for the Lessee to withdraw from the arrangement with no on-going liability for continued delay rental payments.

The modern clause eliminates the implied obligation to drill an exploratory well, and the Lessor cannot refuse to accept the rental and demand that a well be drilled.[5] It does not, however, eliminate the implied covenants to drill additional wells further to develop the premises, or to protect the lease against drainage by offsetting wells.

Most drilling and delay rental clauses are in the form of the so-called "unless" provision: the Lessee is not obligated either to drill or to pay delay rentals; however, if he does not conduct drilling within the period specified within the lease — usually, within one year of the anniversary date — the lease will terminate unless specified annual rentals are paid.

Another form of lease is the so-called "or" form, under which the Lessee is obligated to commence drilling operations or pay the specified annual delay rental; however, such "or" clauses likewise contain a "surrender" right, by which the Lessee can relieve himself of either obligation by surrendering the lease.

The "unless" clause serves as a special limitation on the habendum clause, limiting the duration of the grant. Thus, just as the failure to secure production will cause an automatic termination of the lease under the habendum clause (see § 3.05(a)), a failure either to conduct the specified drilling operations or to pay the delay rental will result in a similar automatic termination.

§ 3.08(b) Satisfaction of the Clause by Drilling. The more common type of drilling and delay rental clause provides that lease termination may be avoided by *commencing* a well on or before the specified date. In the absence of an express provision to the contrary, this does not require that the bit actually penetrate the surface of the land, but does require that the Lessee commence acts on the premises which are preparatory to drilling, such as site preparation or some type of physical activity on the premises

---

[5] A similar result occurs under "paid up" lease forms.

which is related to and would normally precede the making of a hole.[6]

Certain drilling and delay rental clauses require that a well be "completed" by the specified date in order to satisfy the clause. Normally, for the purposes of such a clause, the well will be considered "completed" when it has been drilled to a commercial horizon and prepared for treatment.

§ 3.08(c) Satisfaction by Payment of Delay Rental

§ 3.08(c)(1) In General. If the drilling and delay rental clause is to be satisfied by the payment of rentals rather than by drilling operations, it is critical that the rentals be paid strictly in the manner provided in the lease. If rentals are not so paid, the lease will automatically terminate; equitable considerations are normally immaterial.

§ 3.08(c)(2) Whom to Pay. A failure to pay the delay rentals to the proper person will result in an automatic termination of the lease. Normally, this means the person designated in the lease as "Lessor", and not the person who, under normal property law concepts, might be entitled to receive the payment. However, where there are problems as to life tenant and remainderman, or as to the holder of an executive power, securing a rental stipulation or rental division order normally is a wise precaution.[7]

If separate leases have been secured from separate Lessors, tender jointly to the credit of all Lessors is improper, even when title is in dispute. Tender jointly to husband and wife when only one is a party Lessor is likewise improper; however, if both joined in the lease and both are designated as "Lessor", joint payment, or payment to either, is permitted in most states. However, in Louisiana, the payment must be made to the credit of the spouse actually owning the interest, and joint payment is not permitted.

---

[6] A similar test is applied to "commencement" of a well for purposes of the habendum clause (see § 3.05(c)).

[7] As to the form of such a rental stipulation, see Appendix B to Chapter Two.

Where the Oil and Gas Lease designates multiple Lessors as "Lessor", tender of the gross amount of the rental to the joint account of all parties will satisfy the rental obligation. An exception again arises in Louisiana as to a joint payment to husband and wife where one spouse joined merely because of marital status and owns no interest in the land.

Separate crediting among joint Lessors, where not called for by the Oil and Gas Lease, is made at the Lessee's peril. If the payments are properly allocated, no problem arises. However, if a misallocation is made, the lease will terminate despite the Lessee's good faith. Accordingly, it is a far preferable policy, if separate payments are to be made, to secure a rental stipulation.

In the event of a subsequent change of ownership, if a sole Lessor conveys a part of his interest to another party, or if one joint Lessor conveys his interest, each owner must now be paid separately. However, if the conveyance has created uncertainty as to the respective interest of the parties, Oklahoma permits a tender to the joint account; and Texas permits the bringing of an interpleader action, with tender of the delay rentals into the court.

Even in the absence of an express provision, notice would be required to the Lessee of any change of ownership before he is required to make payment to the new owner. Normally, however, the lease will contain a specific "change of ownership" clause, which will specify the formalities required before the Lessee is bound by any subsequent conveyance by the Lessor.

Courts are split as to whether a Lessee acts at his peril in making payment to an apparent new owner in reliance on the record before receiving the notice specified in the change of ownership clause. Accordingly, the Lessee should insist upon compliance with such clause, or should secure a rental stipulation from all parties, before honoring an apparently regular conveyance which his subsequent title examination has brought to his attention.

If the Lessee has paid delay rentals in advance, and subsequent to such payment but prior to the rental date receives notice of a change of ownership, there is a serious question as to whether the grantee is bound by the prior payment to the grantor. The better view would be that the Lessee could pay reasonably in advance so long as the lease calls for payment of the delay rental on or before the specified date. However, it is preferable to specify a "grace" period in the change of ownership clause.

§ **3.08(c)(3) Manner of Payment.** The Oil and Gas Lease usually contains specific provisions concerning the manner of payment or tender of the delay rental, and usually authorizes payment or tender by check through use of the mails, to the Lessor's credit in a specified depository bank.

Where payment is made by check, the check must be one which would be honored for payment if presented; i.e., it must be signed and drawn on an account containing adequate funds.

Where the rental clause specifically provides for a mailing of the rental, the deposit of the rental in a mail chute under the control of the U.S. Postal Department in a properly addressed and properly stamped envelope is a proper tender of the delay rental; the fact that the rental is not received by the Lessor or the depository bank until after the due date, or at all, is immaterial. (Of course, you still must establish, to the satisfaction of a judge or jury, that you properly mailed the rental.)

Where deposit of the rental in a designated depository bank is authorized, the act of deposit (or of mailing where such is authorized) is the required act of tender; and it is immaterial that the bank did not credit the rental to the Lessor's account until after the due date, or refused to accept the deposit. However, the depository bank cannot waive the Lessor's right to a proper, timely payment of the delay rental unless the erroneous acceptance is acquiesced in by the Lessor.

§ **3.08(c)(4) Time of Payment.** Payment of the delay rental must be timely made to prevent the lease from terminating; and time is of the essence, whether or not the lease so states, in "unless" drilling and delay rental clauses. If the date for payment of the rental has been left blank, an amendment or rental stipulation should be secured rather than unilaterally inserting the delay rental date, which could be considered a material alteration of the lease (see § **3.02(c)**).

Normally, if the payment date falls on a Sunday, and in some states on a legal holiday, payment may be made on the next ensuing business day. However, there are isolated cases to the contrary.

Often, the dry hole, etc., clauses of the lease call for resumption of delay rental payments following the drilling of a dry hole or the cessation of production, during the primary term. Under the terms of such clauses, a different date may arise for the payment of the resumed delay rentals; and the dry hole, etc., clauses should be carefully studied in such event to determine the proper payment date.

§ 3.08(c)(5) Amount of Payment. An underpayment of the delay rental, no matter how insignificant, will vitiate the tender of rentals despite the Lessee's good faith, in the absence of a waiver of the right to payment in the proper amount by an acceptance of the delay rental. However, the amount of delay rental payable may be affected by the surrender or assignment clauses of the lease, which may provide for a reduction in the delay rental in the event part of the lease is surrendered, or provide for separate payments by assignor and assignee if the lease is assigned.

The rental relief provisions of the assignment clause only apply to an assignment of an entire *subdivided* portion of the lease premises. If the Lessee assigns an undivided interest in the lease, either assignor or assignee can maintain the lease in effect by drilling or paying the *full* delay rental; however, a payment by either of only his proportionate part of the rental will not maintain the lease in effect if the other party does not pay the balance. Likewise, if one party surrenders his lease as to his undivided interest only, there is little authority as to whether or not the lease may be maintained by paying the delay rental only as to the interest still under lease.

The amount of delay rental payable will likewise be affected by the lesser interest clause. Such clauses, in addition to providing for proportionate reduction of the delay rental where the Lessor owns less than the full estate in the leased premises, also may provide for an increase in the rental on the anniversary date following any reversion to or other increase in the Lessor's interest, and also may provide that failure to reduce delay rentals will not affect the Lessee's right to reduce royalty payments.

If the granting clause specifically leases "an undivided one-half interest", etc., in the described land, the delay rentals will not be proportionately reduced below the gross amount stated in the lease so long as the Lessor owns the interest specified. If, instead of specifying the net interest being leased, rentals are inserted calculated on a "net" basis, under the express language of the lesser interest clause there literally could be a further reduction. However, any such payment would be a dangerous practice.

§ 3.08(c)(6) Effect of Mispayment or Nonpayment of Rentals. A failure to pay the delay rental in the exact manner provided in the lease will result in an automatic termination of the lease. A good faith unilateral mistake on the part of the Lessee, where not induced by the Lessor, will not excuse a failure to pay; nor will "hardship" circumstances, such as flooding, accidents to machinery or family illness.

It is arguable that a failure to make prompt payment under the drilling and delay rental clause could be excused under the "notice", "force majeure", or "judicial ascertainment" clauses. (For examples, see Appendix B to this chapter.) However, such clauses frequently apply only to "defaults" by the Lessee. Since the "unless" clause does not *obligate* the Lessee to pay the delay rental, such clauses, where limited to excusing defaults, would not serve to excuse a failure to pay delay rentals. However, certain leases contain clauses which specifically provide that a bona fide attempt to tender the delay rental will maintain the lease in effect, thereupon unconditionally obligating the Lessee to make proper payment of the delay rental.[8]

§ 3.08(c)(7) Effect of Acceptance of Rental Payment. Acceptance of a delay rental payment after the primary term of the lease will serve to extend the lease for an additional year, even though the lease contained no

---

[8] For an example of such a clause, see Appendix B to this chapter.

provision for such an extension. An acceptance of a delay rental also serves as a waiver, for the rental period, of the right to demand drilling under the implied covenants of further development and to protect against drainage.

Where the Lessor knowingly accepts a mispaid delay rental such as a late payment, a payment in an improper manner, or one in an improper amount, he is frequently held to be estopped from denying the validity of the lease. However, the acceptance must be with knowledge that the payment was improper; and an immediate re-tender upon learning that an improper payment has been made will vitiate the acceptance.

The depository bank has no power to bind the Lessor by accepting an improper payment.

§ 3.09 The Shut-In Royalty Clause

3. ... If a well capable of producing gas in paying quantities is completed on the above described land and is shut-in, this lease shall continue in effect for a period of one year from the date such well is shut-in. Lessee or any assigns may thereafter, in the manner provided herein for the payment or tender of delay rentals, pay or tender to Lessor as royalty, on or before one year from the date such well is shut-in, an amount equal to the rental, and, if such payment or tender is made, this lease shall continue in effect for a further period of one year. In like manner and upon like payments or tenders annually made on or before each anniversary of shut-in date of such well this lease shall continue in effect for successive periods of 12 months each.

§ 3.09(a) Nature of the Shut-In Royalty Clause. As discussed in § **3.05(b)**, in the majority of states a "shut-in" well will not extend the primary term of the Oil and Gas Lease. To prevent lease termination in such "produce means extracted" states, and to placate Lessors in other states during the interim between discovery and marketing, the shut-in royalty clause was developed.

There is no "standard" shut-in royalty clause, and terms vary greatly from lease to lease. Accordingly, the specific language of the shut-in royalty clause must be studied closely to determine under what circumstances the shut-in well will extend the lease; to whom, when and how payment of any shut-in royalty should be made; and under what circumstances is the payment of the shut-in royalty necessary, either as a condition of extending the lease or as a contractual obligation.

§ 3.09(b) When Is the Clause Applicable? Since, in the majority of states, a shut-in well will not extend the habendum clause in the absence of some other lease provision, it is critical to determine: Is the shut-in clause contained in my particular lease triggered by the well my company has completed? You may assume that if you have a shut-in well, any shut-in clause contained in your lease is automatically applicable. But if it is not the *type* of well referred to in the clause, the alternative of extending the lease by the payment of a shut-in royalty is not available and the lease will still terminate in the absence of extraction and marketing.

The shut-in clause may speak in terms of a well which has been "shut-in", or of the situation where "gas is not being sold or used". In either case, it is the absence of production that makes the clause applicable.[9] Other clauses refer to wells shut-in "for lack of a market". Such clauses are dangerous, since they might offer no protection if a market were available, but the Lessee had shut-in the well to seek a better market.

Normally, the shut-in clause applies to a shut-in "gas well"; sometimes the clause is further limited to a well "capable of producing gas only". Such clauses, by their terms, would not apply to oil wells shut-in by conservation authorities to prevent the flaring of casinghead gas, and might not apply to gas wells which also produce large quantities of distillate or to wells with high gas-oil ratios. Oklahoma has held that wells incapable of being produced without a wasteful or unlawful flaring of gas are still "gas wells", and that wells which are physically capable of producing liquids (or are even oil wells with high gas-oil ratios) are still wells "capable of producing gas

---

[9] Some clauses apply when "production ceases", and arguably, would apply only to a well which has once produced, but has since ceased producing. However, Texas has held that such a clause also applies to a well which has been shut-in since completion.

140

only". However, Texas holds that a well which produces distillate along with gas will not permit lease extension under a "gas only" shut-in clause.

Some shut-in clauses require the payment of a shut-in royalty whenever a well has been "shut-in". Under such a clause, payment of shut-in royalty may be required as a contractual obligation even though the lease is still within its primary term, or there are other wells on the premises from which oil and gas are being extracted and marketed.

§ 3.09(c) Payment of the Shut-In Royalty. The wording of the shut-in royalty clause, unfortunately, is often patterned after the drilling and delay rental clause, so that a failure to comply strictly with its terms may lead to an automatic termination of the lease.

The most serious question concerning the payment of the shut-in royalty is often the *time* of payment.

In certain circumstances, payment which is properly made in accordance with the provisions of the shut-in clause may not be sufficient to maintain the lease in effect. If the Oil and Gas Lease is being maintained in effect by the operation of some other clause of the lease, including the "grace" period under the continuous operations clause, or in "discovery" states such as Oklahoma, the habendum clause, payment by the date mentioned in the shut-in clause will be sufficient. However, if the lease is being maintained only by the shut-in clause, the payment must be made on or before the time that the lease would otherwise terminate in the absence of the production or other operations. Accordingly, in states such as Texas, payment should normally be made prior to the end of the primary term or the "continuous operations" grace period, even though the shut-in royalty clause provides for a later date.

Assuming that payment by the date provided in the shut-in royalty clause is sufficient to extend the lease, the clause must still be studied carefully to determine when such payment is due under such clause. Many shut-in royalty clauses are quite explicit as to the payment date. However, some clauses will refer to the end of the "yearly period" during which gas is not sold or used. Such provisions raise questions as to whether the one-year period refers to the anniversary date of the lease, a 12-month period or the end of the calendar year. Likewise, many provisions require a determination to when the well is considered "shut-in".

As a general rule, the shut-in royalty should always be paid at the earliest possible date required to maintain the lease under the most unfavorable possible interpretation to the Lessee.

In determining to whom the shut-in royalty should be paid, the few cases which have considered the subject have treated the shut-in payment as a "royalty", to be paid to the person entitled to share in royalty payments. However, under the specific wording of a particular shut-in royalty clause, the payment could in fact be treated as bonus or delay rental.

The shut-in royalty is likewise payable to the person entitled to receive such payment as of the date that such payment should have been made under the shut-in royalty clause. In circumstances where an earlier payment has been made to prevent lease termination, subsequent transfers create difficult problems as to whether or not a duplicate payment should be made to the new owner.

The shut-in royalty clause will normally be specific as to the amount of the shut-in royalty payment. Such clauses vary from a fixed amount, to a fixed amount per well, to a fixed amount per acre, to an amount equal to the delay rental. However, it is clear that a tender of too small a shut-in royalty may result in the loss of the lease. Accordingly, it is critical that, whatever the form of the clause, the full amount be paid.

The shut-in royalty clause is sometimes unclear as to the exact manner of payment of the shut-in royalty. Frequently, such clauses provide for payment or tender "in the same manner as provided herein for the payment or tender of delay rentals". However, in the absence of such provision, payment should be made in cash directly to the Lessor, and sufficiently in advance of the necessary payment date that any miscarriage in the mails will not prevent the payment being received by the Lessor by the due date.

The problems just discussed are indicative of the confusion that may arise in trying to maintain the lease by virtue of a shut-in well. Accordingly, if any questions exist concerning the proper payment of shut-in royalty, the securing of a shut-in gas royalty stipulation clearly specifying that the shut-in well alone will maintain the lease in effect, and designating the date, parties, amount and manner of payment, coupled with ratification of the lease, may prove desirable.

§ 3.10 Other Express and Implied Lease Provisions

> 10. Lessor hereby warrants and agrees to defend the title of said land and agrees to defend the title of said land and agrees that Lessee at its option may discharge any tax, mortgage or other lien upon said land, either in whole or in part, and in event Lessee does so, it shall be subrogated to such lien with the right to enforce same and apply rentals and royalties accruing hereunder toward satisfying same. Without impairment of Lessee's rights under the warranty in event of failure of title, it is agreed that if Lessor owns an interest in said land less than the entire fee simple estate, then the royalties and rentals to be paid Lessor shall be reduced proportionately.

§ 3.10(a) Provisions Concerning Title. A number of the incidental provisions of the Oil and Gas Lease, such as the lesser interest clause, the assignment and surrender clauses, the change of ownership clause and the entirety clause, have been discussed earlier in this chapter. However, there are other clauses of the lease which are likewise of significance.

Insofar as covenants of title are concerned, it is unclear whether such covenants would be implied in the Oil and Gas Lease in the absence of an express provision. This is generally of little practical importance, however, since the Oil and Gas Lease almost invariably contains express provisions concerning title.

The typical *warranty clause* provides that the Lessor warrants title to the lease premises. Such a clause authorizes the Lessee to sue for and recover the bonus payments and previously paid rentals, or any costs incurred in curing title, even where the Lessee knew that the Lessor's title was defective. The lease likewise normally contains an express subrogation clause authorizing the Lessee to pay off mortgages, taxes and other liens, and to be subrogated to the rights of the holders of such liens, including the right to apply rentals and royalties to recoup such payments.

The main consequence of the warranty clause is to remove any doubt concerning the doctrine of estoppel by deed. Where the Oil and Gas Lease contains a warranty clause, it is generally held that the Lessor is estopped to deny that any after-acquired title inures to the Lessee's benefit, even when the lease contains a lesser interest clause (see Chapter Two). Even where the warranty clause has been stricken, any reversionary interest owned by the Lessor normally would be covered, since such interest was owned by the Lessor, and thus would be covered by the granting clause, at the time of leasing. However, in a few cases, questions have been raised as to whether or not subsequently-acquired interests not covered by the amount of the bonus are in fact covered by the lease.

The few cases on the subject have held that the inclusion of the warranty clause does not automatically have the same effect as a lesser interest clause in permitting proportionate reduction of rentals and royalties where a lesser interest clause is not included in the lease, or has been stricken. And if a Lessor insists on striking the warranty clause, you must be careful that the lesser interest clause is not stricken as well.[10]

§ 3.10(b) Effect of Implied Covenants.
You should likewise be alert to the effect of the "implied covenants" applicable to the lease (see § 3.03). These covenants can require a Lessee to drill a well, or take other action, if a "prudent operator" would take such a step.

The remedy for breach of an implied covenant may be damages, or loss of all or the undeveloped portions of the lease. Often, a court will enter an alternative decree: Comply within a specified period or the lease will be cancelled.

§ 3.11 Conclusion
The Oil and Gas Lease is a contractual relationship between the

---

[10] If the Lessor does insist on striking the lesser interest clause, be sure that rental and royalty clauses are amended to specify the amount of such payments on a "net" basis.

Lessor and Lessee. While this fact is frequently ignored when the Lessee attempts to assert its rights under the lease, it is seldom ignored when the Lessor is asserting his rights. The lease is normally construed against the Lessee and in favor of the Lessor. Accordingly, it is critical that any question concerning your company's rights be interpreted in reference to the language of the specific lease which the company has acquired, realizing that exact language may protect the Lessor but may not protect the Lessee.

Appendix A

SAMPLE OIL AND GAS LEASE

OIL AND GAS LEASE

THIS AGREEMENT made this _____ day of _____, 19 _____, between _____

_____

_____

Lessor (whether one or more), and _____
Lessee, WITNESSETH:

1. Lessor, in consideration of _____ Dollars ($ _____), in hand paid, of the royalties herein provided, and of the agreement of Lessee herein contained, hereby grants, leases and lets exclusively unto Lessee for the purpose of investigating, exploring, prospecting, drilling and mining for and producing oil, and gas, and the constituents thereof, laying pipelines, building tanks, power stations, telephone lines and other structures thereon to produce, save, take care of, treat, transport and own said products, and housing its employees, the following described land in _____ County, _____ , to wit:

[DESCRIPTION]

In addition to the land above described, Lessor hereby grants, leases and lets exclusively unto Lessee to the same extent as if specifically described herein all lands owned or claimed by Lessor which are adjacent, contiguous to or form a part of the lands above particularly described, including all oil, gas, and their constituents underlying lakes, rivers, streams, roads, easements and rights-of-way which traverse or adjoin any of said lands. For rental payment purposes, the land included within this lease shall be deemed to contain _____ acres, whether it actually comprises more or less.

2. Subject to the other provisions herein contained, this lease shall be for a term of 10 years from this date (called "primary term") and as long thereafter as oil or gas is produced from said land hereunder, or drilling or reworking operations are conducted thereon.

3. The royalties to be paid by Lessee are: (a) an oil, one-eighth of that produced and saved from said land, the same to be delivered at the wells, or to the credit of Lessor into the pipeline to which the wells may be connected; Lessee may from time to time purchase any royalty oil in its possession, paying the market price therefor prevailing for the field where produced on the date of purchase: (b) on gas, including casinghead gas or other hydrocarbon substance, produced from said land and sold or used off the premises or in the manufacture of gasoline or other products therefrom, the market value at the well of one-eighth of the gas so sold or used, provided that on gas sold at the wells the royalty shall be one-eighth of the amount realized from such sale. Lessee shall have free use of oil, gas and water from said land, except water from Lessor's wells, springs or reservoirs, for all operations hereunder, and the royalty on oil and gas shall be computed after deducting any so used. If a well capable of producing gas in

paying quantities is completed on the above described land and is shut-in, this lease shall continue in effect for a period of one year from the date such well is shut-in. Lessee or any assignee may, thereafter, in the manner provided herein for the payment or tender of delay rentals, pay or tender to Lessor as royalty, on or before one year from the date such well is shut-in, an amount equal to the rental, and, if such payment or tender is made, this lease shall continue in effect for a further period of one year. In like manner and upon like payments or tenders annually made on or before each anniversary of shut-in date of such well this lease shall continue in effect for successive periods of 12 months each.

4. If operations for drilling are not commenced on said land as hereinafter provided, on or before one year from this date, the lease shall then terminate as to both parties, unless on or before such anniversary date Lessee shall pay or tender to Lessor or to the credit of Lessor in _____ Bank of _____ (which bank and its successors are Lessor's agent and shall continue as the depository for all rentals payable hereunder regardless of changes in ownership of said land or the rentals either by conveyance or by the death or incapacity of Lessor) the sum of _____ Dollars ($ _____) (herein called rental) which shall cover the privilege of deferring commencement of operations for drilling for a period of 12 months. In like manner and upon like payments or tenders annually the commencement of operations for drilling may be further deferred for successive periods of 12 months each during the primary term. The payment or tender of rental herein referred to may be made in currency, draft or check at the option of the Lessee; and the depositing of such currency, draft or check in any post office, properly addressed to the Lessor, or said bank, on or before the rental paying date, shall be deemed payment as herein provided. If such bank (or any successor bank) should fail, liquidate or be succeeded by another bank, or for any reason fail or refuse to accept rental, Lessee shall not be held in default for failure to make such payment or tender of rental until 30 days after Lessor shall deliver to Lessee a proper recordable instrument, naming another bank as agent to receive such payments or tenders. The down cash payment is consideration for this lease according to its terms and shall not be allocated as mere rental for a period. Lessee may at any time execute and deliver to Lessor or to the depository above named or place of record a release or releases covering any portions of the above described premises and thereby surrender this lease as to such portion or portions and be relieved of all obligations as to the acreage surrendered, and thereafter the rentals payable hereunder shall be reduced in the proportion that the acreage covered hereby is reduced by said release or releases.

If Lessee shall, on or before any rental date, make a bona fide attempt to pay or deposit rental to a Lessor entitled thereto under this lease according to Lessee's records or to a Lessor who, prior to such attempted payment or deposit, has given Lessee notice, in accordance with the terms of this lease hereinafter set forth, of his right to receive rental, and if such payment or deposit shall be erroneous in any regard (whether deposited in the wrong depository, paid to persons other than the parties entitled thereto as shown by Lessee's records, in an incorrect amount, or otherwise), Lessee shall be unconditionally obligated to pay to such Lessor the rental properly payable for the rental period involved, but this lease shall be maintained in the same manner as if such erroneous rental payment or deposit had been properly made, provided that the erroneous rental payment or deposit be corrected within 30 days after receipt by Lessee of written notice from such Lessor of such error accompanied by any documents and other evidence necessary to enable Lessee to make proper payment.

5. Should any well drilled on the above described land during the primary term before production is obtained be a dry hole, or should production be obtained during the primary

term and thereafter cease, then and in either event, if operations for drilling an additional well are not commenced or operations for reworking a land well are not pursued on said land on or before the first rental paying date next succeeding the cessation of production or drilling or reworking on said well or wells, then this lease shall terminate unless Lessee, on or before said date, shall resume the payment of rentals. Upon resumption of the payment of rentals, Section 4 governing the payment of rentals shall continue in force just as though there had been no interruption in the rental payments. If during the last year of the primary term and prior to the discovery of oil or gas on said land Lessee should drill a dry hole thereon, or if after discovery of oil or gas before or during the last year of the primary term the production thereof should cease during the last year of said term from any cause, no rental payment or operations are necessary in order to keep the lease in force during the remainder of the primary term. If, at the expiration of the primary term, Lessee is conducting operations for drilling a new well or reworking an old well, this lease nevertheless shall continue in force as long as such drilling or reworking operations continue, or if, after the expiration of the primary term, production on this lease shall cease, this lease nevertheless shall continue in force if drilling or reworking operations are commenced within 60 days after such cessation of production; if production is restored or additional production is discovered as a result of any such drilling or reworking operations, conducted without cessation of more than 60 days, this lease shall continue as long thereafter as oil or gas is produced and as long as additional drilling or reworking operations are had without cessation of such drilling or reworking operations for more than 60 consecutive days.

6. Lessee, at its option is hereby given the right and power to pool or combine the land covered by this lease, or any portion thereof, as to oil and gas, or either of them, with any other land, lease or leases when in Lessee's judgment it is necessary or advisable to do so in order to properly develop and operate said premises, such pooling to be into a well unit or units not exceeding 40 acres, plus an acreage tolerance of 10 percent of 40 acres, for oil, and not exceeding 640 acres, plus an acreage tolerance of 10 percent of 640 acres, for gas, except that larger units may be created to conform to any spacing or well unit pattern that may be prescribed by governmental authorities having jurisdiction. Lessee may pool or combine acreage covered by this lease, or any portion thereof, as above provided, as to oil or gas in any one or more strata, and units so formed need not conform in size or area with the unit or units into which the lease is pooled or combined as to any other stratum or strata, and oil units need not conform as to area with gas units. The pooling in one or more instances shall not exhaust the rights of the Lessee hereunder to pool this lease or portions thereof into other units. Lessee shall execute in writing and place of record an instrument or instruments identifying and describing the pooled acreage. The entire acreage so pooled into a unit shall be treated for all purposes, except the payment of royalties, as if it were included in this lease, and drilling or reworking operations thereon or production of oil or gas therefrom, or the completion thereon of a well as a shut-in gas well shall be considered for all purposes, except the payment of royalties, as if such operations were on or such production were from or such completion were on land covered by this lease, whether or not the well or wells be located on the premises covered by this lease. In lieu of the royalties elsewhere herein specified, Lessor shall receive from a unit so formed, only such portion of the royalty stipulated herein as the amount of his acreage placed in the unit or his royalty interest therein bears to the total acreage so pooled in the particular unit involved. Should any unit as originally created hereunder contain less than the maximum number of acres hereinabove specified, then Lessee may at any time thereafter, whether before or after production is obtained on the unit, enlarge such unit by adding

150

additional acreage thereto, but the enlarged unit shall in no event exceed the acreage content hereinabove specified. In the event an existing unit is so enlarged. Lessee shall execute and place of record a supplemental declaration of unitization identifying and describing the land added to the existing unit; provided that if such supplemental declaration of unitization is not filed until after production is obtained on the unit as originally created, then and in such event the supplemental declaration of unitization shall not become effective until the first day of the calendar month next following the filing thereof. In the absence of production Lessee may terminate any unitized area by filing of record notice of termination.

7. Lessee shall have the right at any time without Lessor's consent to surrender all or any portion of the leased premises and be relieved of all obligation as to the acreage surrendered. Lessee shall have the right at any time during or after the expiration of this lease to remove all property and fixtures placed by Lessee on said land, including the right to draw and remove all casing. When required by Lessor, Lessee will bury all pipelines below ordinary plow depth, and no well shall be drilled within 200 feet of any residence or barn now on said land without Lessor's consent. The Lessee agrees to promptly pay to the owner thereof any damages to said land caused by or resulting from any operations of Lessee.

8. The rights of either party hereunder may be assigned, in whole or in part, and the provisions hereof shall extend to the heirs, successors and assigns of the parties hereto, but no change or division in ownership of the land, rentals or royalties, however accomplished, shall operate to enlarge the obligations or diminish the rights of Lessee. No change in the ownership of the land, or any interest therein, shall be binding on Lessee until Lessee shall be furnished with a certified copy of all recorded instruments, all court proceedings and all other necessary evidence of any transfer, inheritance or sale of said rights. In event of the assignment of this lease as to a segregated portion of said land, the rentals payable hereunder shall be apportionable among the several leasehold owners ratably according to the surface area of each, and default in rental payment by one shall not affect the rights of other leasehold owners hereunder. In case Lessee assigns this lease, in whole or in part, Lessee shall be relieved of all obligations with respect to the assigned portion or portions arising subsequent to the date of assignment.

9. All express or implied covenants of this lease shall be subject to all Federal and state laws, executive orders, rules or regulations, and this lease shall not be terminated, in whole or in part, nor Lessee held liable in damage, for failure to comply herewith, if compliance is prevented by, or if such failure is the result of any such law, order, rule or regulation, or if prevented by an act of God, of the public enemy, labor disputes, inability to obtain material, failure of transportation, or other cause beyond the control of Lessee.

If, during the term of this lease, oil or gas is discovered upon the leased premises, but Lessee is prevented from producing the same by reason of any of the causes set out in this section, this lease shall nevertheless be considered as producing and shall continue in full force and effect until Lessee is permitted to produce the oil or gas, and as long thereafter as such production continues in paying quantities or drilling or reworking operations are continued as elsewhere herein provided.

10. Lessor hereby warrants and agrees to defend the title of said land and agrees that Lessee at its option may discharge any tax, mortgage or other lien upon said land, either in whole or in part, and in event Lessee does so, it shall be subrogated to such lien with the right to enforce same and apply rentals and royalties accruing hereunder toward satisfying same. Without impairment of Lessee's rights under the warranty in event of failure of title, it is agreed

that if Lessor owns an interest in said land less than the entire fee simple estate, then the royalties and rentals to be paid Lessor shall be reduced proportionately.

11. Lessors hereby release and waive all rights of homestead.

All the provisions of this lease shall inure to the benefit of and be binding upon the parties hereto, their heirs, administrators, successors and assigns.

This agreement shall be binding on each of the above named parties who sign the same, regardless of whether it is signed by any of the other parties.

IN WITNESS WHEREOF, this instrument is executed on the date first above written.

_____   _____
_____   _____
_____   _____
_____   _____
_____   _____

[ACKNOWLEDGEMENT]

No. _____
OIL AND GAS LEASE
FROM

_____
_____

TO

_____
_____

Date _____ , 19 _____ . Section _____ Township _____ Range _____ ,
_____ County, Wyoming. No. of Acres _____ Term _____ .
STATE OF WYOMING
County of _____ SS.
 This instrument was filed for record on the _____ day of _____ , 19 _____ at
_____ o'clock _____ M., and duly recorded in book _____ page _____ of the records of this office.

_____
County Clerk — Register of Deeds
By _____
Deputy

Record and Mail to:

_____
_____

152

Appendix B

SAMPLE LEASE CLAUSES

GRANTING CLAUSE:

"In Gross" Provisions

"For all purposes of this lease the described premises shall be treated as comprising _____ acres, whether there be more or less."

"For the purpose of calculating the rental payments hereinafter provided for, said land is estimated to comprise _____ acres, whether it actually comprises more or less."

"This is a lease in gross and not by the acre and the bonus money paid and the rentals provided for shall be effective to cover all such lands irrespective of the number of acres contained therein; and for the purpose of calculating the payments herein provided for, the land included within the term of this lease is estimated to comprise _____ acres, whether it actually comprises more or less."

Surface and Crop Damage

"Lessee shall be responsible for all damages caused by Lessee's operations."

"Lessee shall be responsible for and agrees to make payment of all damages to lands, livestock, crops, timber and improvements caused by its operations hereunder."

"Lessee shall pay for damages caused by its operations to growing crops on said land."

"Upon termination of this lease, Lessee shall restore the premises to as near their original condition as is reasonably possible."

"Upon completion or abandonment of any well drilled on the premises, the surface of the ground appurtenant to said well shall be smoothed, and all excavations and slush pits shall be filled in by Lessee, and the surface rights of Lessors in said premises shall be impaired as little as reasonably possible."

HABENDUM CLAUSE:

"Pugh" and Similar Clauses

"Notwithstanding anything to the contrary herein contained drilling operations on or production from a pooled unit or units established under the paragraph two hereof embracing land covered hereby and other land shall maintain this lease in force only as to land included in such unit or units. The lease may be maintained in force as to the remainder of the land in any manner herein provided for, provided that if it be by rental payment, rentals shall be payable only on the number of acres not included in such unit or units. If at the end of the primary term this lease is being maintained as to a part of the land by operations on or production from a pooled unit or units embracing land covered hereby and other land lessee shall have the right to maintain the lease as to the land not included in such unit or units by rental payments exactly as if it were during the primary term provided that his right to pay rental shall terminate five years after the end of the primary term."

"Provided that at the expiration of _____ years from _____, 19 _____, the said right to explore and drill for oil and other hydrocarbon substances shall terminate, but the Lessee may thereafter retain and operate all wells then producing oil in paying quantities on the same terms as to royalty and other conditions as are herein specified, and maintain and use such structures and equipment as may be necessary in the operation of such wells."

"Any provisions above to the contrary notwithstanding, this lease shall ipso facto terminate two years after the date of the expiration of the primary term save as to the number of acres allocated by the Railroad Commission of Texas for each well from which oil or gas in paying quantities is being produced and sold in paying quantities and said lease shall also ipso facto terminate as to all formations from which oil and/or gas is not being produced and sold in paying quantities."

"At the expiration of said 20 years if oil, gas or other mineral is then being produced from said land, the Lessee shall promptly designate and define in writing the area or areas on which oil, gas and/or other minerals are then being produced, it being contemplated that Lessee shall have the right to retain under the terms of this lease the area or areas included within the geologic structures or formations proved to be productive at the end of said 20-year period; and in determining what area or areas are proven to be productive only the area or areas shall be designated from which there are being produced oil, gas and/or other minerals at the end of such 20-year period, and in determining the extent of such productive area or areas consideration shall be given to geologic information obtained from all wells, including dry holes, if any, in the vicinity and from the surface geology and from investigation with geophysical instruments or methods and/or from other methods which then may be in use in the industry; and as to said area or areas the lease shall continue so long thereafter as oil, gas or other mineral is produced therefrom."

ROYALTY CLAUSE:

Entirety Clause
"If the leased premises shall [now or] hereafter be owned severally or in separate tracts, the premises nevertheless shall be developed and operated as one lease and all royalties accruing hereunder shall be treated as an entirety and shall be divided among and paid to such separate owners in the proportion that the acreage owned by each such separate owner bears to the entire leased acreage."

Free Gas Clause
"Lessor shall have the privilege at his risk and expense of using gas from any gas well on said land for domestic use in the principal dwelling thereon out of any surplus gas not needed for operations hereunder."

"The Lessor to have gas free of charge from any gas well on the leased premises for all stoves and inside lights in the principal dwelling house on said land by making Lessor's own connections with the well at Lessor's expense."

DRILLING-DELAY RENTAL CLAUSE:

Excuse for Nonpayment

"If Lessee shall, on or before any rental date, make a bona fide attempt to pay or deposit rental to a Lessor entitled thereto under this lease according to Lessee's records or to a Lessor who prior to such attempted payment or deposit, has given Lessee notice, in accordance with the terms of this lease hereinafter set forth, of his right to receive rental, and if such payment or deposit shall be erroneous in any regard (whether deposited in the wrong depository, paid to persons other than the parties entitled thereto as shown by Lessee's records, in an incorrect amount, or otherwise), Lessee shall be unconditionally obligated to pay such Lessor the rental payable for the rental period involved, but this lease shall be maintained in the same manner as if such erroneous rental payment or deposit had been properly made, provided that the erroneous rental payment or deposit be corrected within 30 days after receipt by Lessee of written notice from such Lessor of such error accompanied by any documents and other evidence necessary to enable Lessee to make proper payment."

MISCELLANEOUS:

Notice and Judicial Ascertainment Clauses

"The breach by Lessee of any obligations arising hereunder shall not work a forfeiture or termination of this lease nor cause a termination or reversion of the estate created hereby nor be grounds for cancellation hereof in whole or in part unless Lessor shall notify Lessee in writing of the facts relied upon in claiming a breach hereof, and Lessee, if in default, shall have 60 days after receipt of such notice in which to commence the compliance with the obligations imposed by virtue of this instrument, and if Lessee shall fail to do so then Lessor shall have grounds for action in a court of law or such remedy to which he may feel entitled."

"It is agreed that this lease shall never be terminated, forfeited or cancelled for failure to perform in whole or in part any of its implied covenants, conditions or stipulations, until it shall have been first finally judicially determined that such failure exists, and after such final determination, Lessee is given a reasonable time therefrom to comply with any such covenants, conditions or stipulations."

Chapter Four

EDITOR'S COMMENTS

Ownership, Petroleum Land Titles and the Oil and Gas Lease all deal with "legal" matters — the legal system concerning land ownership; how title to property, including the minerals, is transferred; and the various legal consequences created under an Oil and Gas Lease. However, "Contracts Used in Oil and Gas Operations" involves far less of the "legal" and far more of the practical.

This may sound strange, since a contract is, by definition, a "legal" instrument. Section **4.02** of this chapter will deal with the legal aspects of contracts. However, the greater part of your Landman's role in dealing with oil and gas contracts will involve knowing what *subject* should be covered in the various types of contracts, and the various *alternatives* available to you in negotiating and drafting your "deal" in each of these subject areas.

There is no "right" or "wrong" answer in deciding how to deal with each subject area. How a given item is treated will depend on your company's objectives, the strength of your bargaining position, and how well you have fared in securing desirable concessions in other areas of the contract. The only "wrong" is in not knowing the various points you do need to negotiate, and the various ways in which these points can be handled.

Pay particular attention to the following:

- How binding is an "agreement" to leave a farmout proposal or other offer open for a specified period of time (§ **4.02(b)**).

- What are, and how to detect "fish hook" clauses (§ **4.02(c)**).

157

• What to look for in "support" letters (§ **4.03(b)**).

• How to draw depth limitations and requirements in a farmout agreement (§ **4.03(f)**).

• Methods of avoiding recent adverse tax rulings concerning farmout agreements (§ **4.03(g)**).

• The changes in terms of the AAPL Model Form Operating Agreement incorporated in its 1977 revision (§ **4.04**).

• Operations under the model form which require unanimous approval, and are not subject to "consent/non-consent" procedures (§ **4.04(f)**).

• The effects of a division order on the Lessor, the Lessee and the production purchaser (§ **4.05(b)(4)**).

• The significance of certain "abandonment" requirements of the Natural Gas Act on taking a farmout or a new lease (§ **4.05(c)(2)**).

— *L.G.M.*

Chapter Four

CONTRACTS USED
IN OIL AND GAS OPERATIONS

by Lewis G. Mosburg, Jr.

§ 4.01 Introduction

In every phase of oil and gas operations, the Landman deals with contracts. The very document which permits his company to explore for oil and gas — the Oil and Gas Lease — is itself a contract between the Lessor and the Lessee (see § **3.02(a)**). Initial exploration is conducted pursuant to farmout or joint operating agreements; and the operating agreement also controls the development of the prospect area and the operation of producing properties. Wells are drilled pursuant to drilling contracts, and production is sold pursuant to gas sales agreements or the oil sale "contract" — the division order. Finally, secondary recovery is conducted pursuant to unit operating agreements.

This chapter will cover the key things you need to know about those contracts you are going to deal with most frequently — "support" and "farmout agreements", which seek to encourage exploration and development; "joint operating agreements", which govern a cooperative exploration, development and operation of oil and gas properties; and "division orders" and "gas sale contracts", through which production is sold to the production purchaser. You will also be introduced to some of the basic principles of contract law and what to do and what to avoid when negotiating or reviewing contracts for the company.

§ 4.02 Basic Concepts

§ 4.02(a) What Is a "Contract"?

Your whole life is governed by promises. "Don't worry, I'll be sure that's taken care of today." "Sure, there's a Santa Claus." "Of course, I'll be home on time tonight." While many of these "promises" deal with your personal life, a significant number are involved in you or your company's business transactions.

Most promises depend upon basic integrity for enforcement. This applies in business as well as in personal affairs; time after time in your dealings as a Landman, you will have to rely upon someone else's good faith and in the belief that he will honor his promises.

Unfortunately, not all promises are fulfilled. We may have misplaced our trust, or the other person, despite his intentions, may simply be unable to deliver what he promised. But there are some "promises" that are so important to your future planning that you must *know* that they will be honored; or, if they are not, that steps can be taken to remedy the breach. And so some promises become enforceable by the courts of law.

It is these promises which a court will enforce — which are "legally binding" — that are referred to as *contracts*.

What raises a "promise" from the level of a "gentleman's agreement" to that mystic "contract" which the courts will enforce? To be legally binding a "contract" must be characterized by the following elements:

1. There must have been *offer* and *acceptance*, frequently referred to as "mutual assent".
2. There must have been *consideration* for each promise.
3. The subject matter of the agreement must have been "legal".
4. The parties must have possessed the "capacity to contract".

Other special elements may be required for certain kinds of contracts. For instance, some contracts must be in writing. Also, contracts in certain areas, or between certain parties, may be required to be under seal. However, the four elements set out above must be present in any "contract" which a court will enforce.

§ 4.02(b) What Are These Elements of a Contract? What do we mean by mutual assent, consideration, legal subject matter, and parties with capacity to contract? Mutual assent requires (1) an *offer* on one side, which is sufficiently definite as to its terms that the courts can determine within reason exactly what the parties were agreeing to, (2) *accepted* without qualification by the other side. This means that if I make an "offer" to you which you purport to accept, but your acceptance either changes or adds to the terms of my offer, you have in fact rejected the offer. The rejection terminates my offer. You cannot later accept it without qualification unless the offer is renewed by me. On the other hand, your "acceptance" is in fact a "counteroffer". I am then free either to accept your offer or to make a new counteroffer. This process continues until there has been a "meeting of the minds", arising out of one or the other of us accepting, without qualification, the other's counteroffer, or until it becomes obvious that we simply cannot agree.

Until there has been such a "meeting of the minds", no contract arises. Similarly, until this mutual assent occurs, the offer or counteroffer may be revoked at any time, even though the "offeror" has said that he will leave his offer open for a specified period which has not yet expired.[1]

The requirement for "consideration" is frequently assumed to require that money be paid to the promissor in return for his promise. However, there is no requirement that consideration be in the form of money. Instead, the requirement that a promise be supported by consideration requires that the promissor receive a benefit in return for his promise which he would not otherwise be legally entitled to receive. This so-called "quid pro quo" of course can be a monetary payment. However, even more frequently it is a return promise: "I promise to assign leases to you if you will promise to drill a test well to a specified depth." Or the consideration for the promise may be the performance of an act: "I promise to assign leases to you *when and if* you drill a test well to a specified depth."

In oil and gas transactions, the "promise for a promise" is the most frequent form of required consideration.

As a Landman, you will find that the contractual elements of mutual assent and consideration are the most frequent areas on which the legal enforceability of the promise may founder. However, it is equally important, though less frequently a problem, that the elements of legality of subject matter and parties with capacity to contract be present. If the subject matter of the contract involves a violation of the laws of this or another country, or would be contrary to public policy, the promises will be nonenforceable in the courts of law. Similarly, if the parties lack legal capacity to contract, i.e., are minors,

---

[1] Certain offers are accompanied by a "promise" to leave the offer open for a specified period. Like any other "promise", this is legally binding only if it meets the requirements for a contract. Normally, there will be no "consideration" to support the promise to leave the offer open; and the offer may thus be withdrawn at any time. However, if consideration has been given for the promise to leave the offer open, a subcontract in the nature of an enforceable *option* will arise.

mental incompetents or the like, the promises will be nonenforceable.[2]

As mentioned, other elements may be required for a particular type of contract to be enforceable. In oil and gas transactions, the requirement that the contract be *in writing* is a frequent additional element.

Legally speaking, a contract need not be in writing unless it falls within a particular state's "Statute of Frauds". However, since most states include contracts affecting land, or for the sale of valuable personal property, within their Statute of Frauds, most oil and gas contracts are required to be written.[3]

The Statute of Frauds does not require that the contract itself be in writing. It does require that the contract be *evidenced* by a writing, signed by the person against whom the contract is to be enforced.

It is beyond the scope of this chapter to discuss all the intricacies of the Statute of Frauds. Most of the contracts with which you deal involving oil and gas transactions will have been reduced to a signed writing, and will be enforceable under the Statute of Frauds. However, another rule significantly affects the written contract. This is the "Parol Evidence Rule" (see § **4.02(c)**), which may prevent you from varying a written contract by attempting to show agreements or communications not contained in the contract itself.

§ **4.02(c) Negotiating and Drafting the Contract.** Your major problems in dealing with contracts will not be whether or not the essential elements are present. Instead, you will have to deal with properly negotiating, drafting and interpreting a "deal" which is in fact legally binding. But does this contract represent the deal you actually intended to make?

---

[2] Legal capacity is discussed in greater detail in § **2.02(c)(3)**.

[3] The Statute of Frauds is also applied in most states to contracts not to be performed within one year or which involve guaranteeing performance by another.

The first critical step in contract negotiation is to make sure that you have reached agreement on all essential elements of the contract. In a farmout, you may have intended to retain a call on production, or to limit the depths to be earned by the farmee. But if this is not properly reflected in the contract, or if the contract, as worded, represents a different picture from the one you thought you agreed to, the Parol Evidence Rule means you will be bound by the contract you signed, not the one you negotiated.

The vital points to be covered in the various types of contracts with which you will work most frequently are dealt with later in this chapter. However, there are certain general rules that should guide you in preparing or reviewing all types of oil and gas contracts.

Are the various "promises" *covenants* or *conditions*? A condition merely sets forth the items which you must perform in order to earn your rights under the contract. If these are not performed, your rights under the contract will terminate; however, you are not legally bound to perform these "conditions". On the other hand, a "covenant" represents a firm contractual undertaking to perform a specified act. If you do not fulfill this promise, you not only lose your rights under the contract; you may also be liable in damages.

A perfect example of the *covenant versus condition* difference is the distinction between a "mandatory" and an "optional" earning well under a farmout agreement. If the well is "optional", and you run into trouble in the drilling, you can walk away from the well if, financially, this seems the best thing to do. You will not have earned your rights to a farmout of the acreage. On the other hand, you will not be required to expend several times the amount of money you originally budgeted for the well. However, if the well is "mandatory", you will be required to drill the well to the specified depth unless the contract expressly contains an "out" permitting you to stop drilling under these circumstances (see § **4.03(f)**).

When a contract is being prepared by the other party, and a draft is submitted to you, it is critical that you review every word of the contract to be sure that it properly reflects the deal which you thought you were entering into. Similarly, if your attorneys or contract section are drafting the contract, it is critical that they understand exactly what has been agreed to, and that you similarly review their draft to insure it complies with the terms negotiated.

The Parol Evidence Rule states that if the written document, on its face, appears to be complete and unambiguous, you cannot look at prior or contemporaneous actions outside the contract in interpreting its terms. This does not merely exclude oral negotiations or understandings. *Anything* outside the contract must be ignored, including prior written communications and "letters of intent". (Note that this does not exclude modifying the contract by *subsequent* agreements, although these also must comply with the necessary elements for a contract.)

Since you are bound by the contract you sign, you must be particularly careful to avoid "fish hook" clauses in contracts prepared by the other side. Often, a crafty "opponent" will draft certain provisions of the contract to appear to agree to one thing; a careful reading of the exact wording of the instrument may show that he is agreeing to something entirely different. Similarly, rights granted or obligations imposed in an earlier part of the contract may be significantly modified by provisions contained, and possibly intentionally buried, at a later point in the document.

The careful review of the contract document to avoid "fish hooks" is not aimed merely at preventing an improper attempt by the other side to mislead you. There may in fact not have been a "meeting of the minds" between you and the other party as to all the terms of the contract. The document prepared by him may represent his good faith understanding of the deal. However, even if it does not reflect what *you* thought you were agreeing to, you will still be bound by the contract you sign if you have not carefully reviewed it and secured changes of the unacceptable provisions.

The contracts with which you deal normally will be fairly lengthy documents. In some instances, it may seem more intelligent to draft a very brief letter of understanding outlining the basic terms of the deal, and leave the rest to the good faith of the parties. However, this practice frequently can have disastrous results.

There is nothing wrong with using a "letter agreement" rather than a "formal" contractual document as your oil and gas contract. So long as the letter agreement contains the necessary elements of a contract, it is just as binding as the more formal contractual document. However, to leave relatively significant terms not covered in the contractual document ignores the reason for entering into a contract rather than relying on a "gentleman's agreement". And as subsequent portions of this chapter will show, there are many areas in which you must insure that the minds have met before entering into an oil and gas transaction.

In some instances, it may seem simpler to leave certain areas which are presently difficult to define to the discretion of one or another of the parties or to subsequent agreement. "Test to the satisfaction of the assignor," "to be operated under the terms of a mutually acceptable operating agreement," or "such assignment to be on a form customary in the industry," all reflect examples of this philosophy. However, if the assignor should prove unreasonable in his testing requirements, the parties should be unable to agree upon the form of operating agreement, or it should turn out that there are significant variances in the forms of assignment "customarily used", such provisions can create real problems.

As a general rule, it is far better to attach the form of assignment, operating agreement or other document intended to be used rather than to leave terms to subsequent agreement or rely upon the vague "customary in the industry" phraseology. Similarly, while in many instances granting one party or the other discretion may be the only way of handling conditions which currently cannot be anticipated with definiteness, such discretion should be restricted to require it to be exercised in good faith, as a prudent operator, and with reasonableness.

§ 4.03 Support and Farmout Agreements

§ 4.03(a) In General. Your company may own many leases in which it has made a valuable investment, both in the form of bonus and delay rentals, and in seismic and other forms of preliminary "prospecting" activity.

Often budgetary limitations may not permit the company to develop all of such leases which appear prospectively valuable for oil and gas. In other instances, the risks involved in a "deep" test may be so great that it seems wise to spread this risk by permitting another operator to expend the dollars required to test the prospect by drilling. Finally, subsequent developments may not have condemned the acreage but may have established that the prospect, while not worthless, does not appear to be of sufficient economic potential to justify the expenditure of company funds for drilling, since better opportunities are available. In any of these instances, it may be desirable to enter into contracts under which an operator other than your company will take the risks of drilling. And if your company does not have a desired acreage position in a given area, or needs assistance in financing a test well, a support or farmout arrangement may permit it to accomplish its goals.

§ 4.03(b) Support Letters. One of the simplest forms of contracts to encourage another to take the risks of initial drilling is the bottom hole or dry hole contribution, frequently referred to as a bottom hole or dry hole "letter" or "support" letter. Such a letter will create an obligation on the part of the supporting party to pay a sum of money, or in some instances to assign certain acreage, upon the satisfaction of certain drilling obligations by the other party — either drilling the well to a specified depth ("bottom hole letter"), or drilling the well and having it prove uncommercial ("dry hole letter"). These agreements permit the supporting party to secure an evaluation of its acreage at relatively small cost and risk on its part.

A sample support letter is contained in Appendix A to this chapter.

The contents of such an agreement fall into three general areas:

1. *The Conditions of the Contribution.* This will include the location, depth and other conditions concerning the drilling of the well; the tests to be conducted and the information to be furnished the supporting party; and other rights to be granted the supporting party, such as access to the location and the right to conduct additional tests.

2. *The Supporting Party's Obligation.*

3. *Miscellaneous Provisions.*   This would include assignability of contract, notice provisions, time for acceptance, etc.

It should be noted that under most support letters, the operator does not contractually obligate itself to drill; the provisions for drilling and testing are simply conditions of the supporting party's obligation to make a dry hole or bottom hole contribution. Accordingly, if the operator determines to go "tight hole" it is free to do so; and the operator has simply lost its right to require a contribution from the supporting party.

Provisions for dry hole or bottom hole contribution are frequently included in farmout agreements as a part of a detailed arrangement to encourage the drilling of a test well. In such instances, all the customary provisions of the contribution agreement should be included in the farmout contract. In certain complicated situations, provision may be made for graduated support: one amount payable if drilling is conducted to a certain depth; another amount if the operator goes deeper; and still more if a completion attempt is made.

Where dry hole contributions are made, a question sometimes arises as to what constitutes a "dry hole". The attached form is silent

---

[4] But see paragraph four of the attached sample support letter if operations are commenced.

in this regard, though provision is made for the plugging and abandonment of the well *after obtaining the supporting party's approval*. However, if there is any question as to the reliability of the operator, it may be desirable to specify what constitutes a dry hole, particularly in the event that the operator or an assignee should later reenter the hole.

Your company may be the prospective recipient of a contribution rather than the grantor. In such instances, the contribution letter must be carefully studied to make sure that the proposed contribution is not illusory. The provisions of the agreement are conditions to your right to demand the promised contribution; if the testing or other requirements are too onerous, there is little chance that the contribution can be earned.[5] Any special requirements of the contribution letter, such as provisions to request the contribution within a specified period and various requirements for notice, should be flagged to insure they will be complied with in a timely fashion. In this regard, it should be noted that "time is of the essence" means just that; performance even a day late could lead to a loss of your company's rights.

Where the company is to receive support, it is critical to insure that the letter could not be interpreted as *obligating* you to drill and complete the test well.

§ 4.03(c) Farmout Agreements — General Concepts. A far more complicated arrangement than the contribution letter is the farmout agreement. Such agreements are contracts to assign oil and gas lease rights upon completion of a specified drilling obligation and the performance of the other obligations and conditions contained in the agreement. Here again, your company may either be making the farmout to another operator or may be anticipating earning certain

---

[5] See paragraph four of the sample; would you always be happy in agreeing to all these provisions?

169

acreage rights as a farmee.

The farmout agreement may involve a relatively "simple" arrangement in which the farmor assigns to the farmee certain leases which it does not consider of sufficient potential value personally to drill; or it may involve an extremely complicated arrangement for the testing of an extensive exploratory prospect in which, after the farmee has drilled one or more test wells to earn its right to an assignment of a portion of the acreage owned by the farmor, the farmor and farmee will then develop the balance of the prospect acreage under a joint operating agreement. In either event, however, the farmout requires a detailed contract clearly specifying the rights and obligations of both parties.

As is true of all contracts, the rights and obligations of the parties can only be determined by an examination of the specific agreement into which they will enter. General conclusions as to what a party's rights *should* be are meaningless; instead, the question is, "What does the agreement say?" You, as a Landman, may have made the best deal in the world; however, as mentioned earlier, all of your negotiating efforts will go for naught if the terms of the trade are not accurately and unambiguously reflected in the contract document.

Equally critical both in negotiating and drafting the farmout agreement is to be aware of what your company is trying to accomplish. The basic objectives of the farmout agreement may be to maintain the company's leases in effect; to secure an interest in production through the company's reserved interest; or to have the prospect area evaluated and tested. Obviously, in any farmout, all of these objectives may have some importance. However, it is necessary to know which of the objectives are of the most significance in determining what deal to negotiate and how the agreement must be drawn.

The negotiation and preparation of the agreement also will be affected by how valuable the company considers the leases which are the subject of the farmout. If the leases are of sufficient significance, the company will obviously negotiate a much harder trade and prepare a much more stringent agreement than if the leases appear to be of relatively little value.

§ 4.03(d) Contents of the Farmout Agreement — In General. The following items must be covered in any farmout agreement, whether simple or complicated:

1. The *leases* which are the subject of the agreement.

2. The *obligations of the assignee,* including the conditions for the drilling; the tests to be performed; and the information and reports to be furnished the assignor.

3. The *obligations of the assignor,* including the interests which the assignor will retain, and the conditions of the assignment.

4. Provisions for the *further operation and development* of the prospect area after the assignment.

5. Any *special provisions* applicable to this particular deal.

6. Various *miscellaneous provisions,* such as provisions for title, delay rentals, status of the parties, indemnification and assignability.

For example farmout agreements, see Appendix B to this chapter.

§ 4.03(e) Leases Subject to the Agreement. The farmout agreement must carefully identify the leases which are the subject of the agreement. If the leases are numerous, this normally will be accomplished by the attachment of an exhibit listing the leases. In some instances, if there is uncertainty as to the identity of all of the leases, a plat may be used either in lieu of or to supplement the lease list.

In addition to the identification of leases currently owned by the assignor which are to be the subject of the agreement, consideration must be given as to whether or not an "area of interest" provision is to be added to the agreement, as well as the status of leases acquired by either party prior to the commencement of the earning well. It should be remembered that the inclusion of such provisions may protect the assignor as well as the assignee in the event additional leases are acquired by the assignee.

It is to the assignor's interest that the assignee be aware of any special problems concerning the leases subject to the agreement to insure that no key leases are lost or provisions of prior contracts violated. Accordingly, a review of any unusual provisions or outstanding contracts affecting the leases subject to the agreement should be made by the assignor as well as the assignee, so that special problems can be identified.

If the rights to be earned by the assignee are to be limited, the farmout agreement, in addition to identifying the surface area and subject leases, should further define any limitations as to the depth to be earned by the assignee or the substances covered. If the rights to be assigned are to be limited as to oil or gas "wells", oil or gas "common sources of supply", or oil and gas "rights", careful draftsmanship is called for to define exactly what is meant by these terms as used in the agreement. Further, as will be discussed, specific problems arise in defining depth limitations.

§ **4.03(f) Assignee's Drilling Obligation.** The farmout agreement should spell out the date by which the earning well (or wells) are to be commenced; the location of the well; the depth to which the well should be drilled; and the date by which the well must be completed.

The specificity with which the earning well's location is spelled out may vary, depending upon the reason for farming out as well as surface conditions and spacing patterns. In certain instances, the assignee may be given a great degree of latitude; in others, a specific location (or certain alternate possibilities) may be provided, or a requirement may be added that the well be located at a "mutually acceptable" location. However, as a general rule, any provision calling for future "mutual agreement" can create major headaches, as well as clouds on the enforceability of the contract.

If your company is to be the assignee, it will prefer for the farmout agreement to provide that it will earn all formations, or at least to a specified depth, so long as it drills a test well to the specified depth, or discovers production at *any* depth. However, where the company is assignor, and if a deep test is the impetus for the farmout, the agreement may provide that the assignee will only earn to the depth drilled.

A description of the depth to be assigned by a footage description may create substantial legal problems. For legal certainty, assignment to the base or the top of a well-defined geological formation is best. However, the assignor often may wish to limit the assignment more closely to the depth drilled. Under such circumstances, the agreement often will provide for an assignment to the "stratigraphic equivalent" of the depth drilled. Since uncertainty may exist as to the exact effect of this term under certain circumstances, a further limiting to "but in no event below" a specified depth is advisable.

If your company is the assignee, it will desire to excuse failure to drill to a specified depth if the failure is beyond its control due to the encountering of impenetrable substances, the loss of circulation, the loss of the hole, etc. Under such circumstances, the agreement should at least provide that the company will have the option of drilling a substitute well or will earn to the depth actually drilled. In the absence of such a provision, the company may find that it will earn no rights, despite the fact that compliance with the depth provision was impossible.

Consideration should be given to the company's rights if drilling to the specified depth is not physically impossible but would be far more expensive and difficult than was anticipated.

The farmout agreement should cover the time by which the drilling of the test well must be commenced and the time by which the well must be completed. The degree of significance of these dates will depend on the circumstances leading to the farmout.

The agreement frequently will provide that the well must be drilled and completed "with due diligence", or some such provision, rather than an absolute cutoff date for completion.

Concerning the date for completing the well, it may be well to define the term "completed" in the agreement, usually by reference to an easily ascertainable date such as the running of a particular type of log or the rig release date. If it is required that the well be "commercial" or "producing" to earn the acreage, it may be advisable to define these terms.

On large tracts, the agreement may provide that more than one well must be drilled. The failure to drill the specified number of wells may lead to the earning of no rights or a limitation as to the acreage earned in the prospect area. And due to current Lessor practices in limiting the acreage covered by a lease, or inserting express drilling obligations or "Pugh" clauses (see § 3.05(e)), "continuous drilling obligations" requiring drilling on each potential drilling and spacing unit, or reassignment of the undrilled acreage, are regaining popularity.

As mentioned in § 4.02(c), in negotiating and drafting the farmout agreement, it is critical to determine whether the obligation to commence a well, to drill to a specified depth, to perform specified tests, and the other assignee "obligations", are merely a condition of the assignee's right to earn an assignment, or constitute "covenants". Frequently, the obligations are mere conditions. However, where the purpose of the farmout is to extend the assignor's leases, the assignor may wish to require acts on the part of the assignee which will insure that the leases are extended.

When the company is the assignee, the agreement should be carefully examined to insure that its "obligations" are mere conditions and not covenants, unless the company is willing to assume a firm contractual obligation.

Certain obligations of the assignee should be drafted as covenants and not conditions. These would include the duty to plug and abandon the test well if dry; compliance with conservation regulations; furnishing copies of tests, etc. actually performed; permiting access to the rig floor, etc.; indemnification of the assignor by the assignee; and the duty to notify the assignee before abandonment of the test well.

Provisions concerning testing and reporting should be specified by the geological or other appropriate department. Provision should also be made in the farmout agreement to insure that the assignor receives well information on all wells that are drilled by the assignee after assignment.

Where your company is the assignee, the provisions concerning testing should be carefully scrutinized to insure that none of the requested tests will be unnecessarily burdensome on the company; and the appropriate technical department should be consulted. In this regard, a provision that the well or certain formations will be tested "to the satisfaction of the assignor" could create problems if the assignor proves unreasonable. (See § **4.02(c)** for limitations on this testing discretion.)

§ 4.03(g) Rights Assigned and Reserved. A great deal of flexibility exists in negotiating a farmout agreement as to limitations on the rights to be assigned, and, conversely, the types of rights to be reserved. In addition to an overriding royalty interest or other interest in production, the assignor may reserve a carried interest, reversionary interest, a casing point election to participate, an undivided interest in the prospect acreage, or a divided "checkerboard" interest. The assignor may further retain deep rights or rights as to specified substances. The rights granted and reserved may include a combination of the above as to all or a part of the assigned acreage.

When an overriding royalty interest or similar interest in production is retained, the assignor will usually desire to retain the right to take production in kind. Where your company is the assignor, it may also attempt to reserve calls on production and processing rights and a preferential right of purchase. Similarly, where your company is the assignee, it will attempt to avoid the inclusion of these rights in the farmout agreement.

Where the company reserves certain rights at casing point, upon completion of the well, or following payout, or retains some interest in "net profits", the agreement should carefully define the meaning of each of these terms. Similarly, since the term "carried interest" is not a word of art, the agreement should carefully define the meaning of any such reserved interest.

The farmout agreement will frequently provide for certain rights of control for the assignor (such as the right to take over the test well prior to its abandonment), a right of reassignment of any earned acreage prior to abandonment, a right to take over operations, and a right to run its own tests. However, in certain states, this may create a problem as to whether or not your company, as assignor, would be liable as a "joint venturer" with the assignee in the event of damage to third parties.

Prior to 1977, the typical farmout agreement involved the earning of the full working interest in the drillsite drilling and spacing unit, subject to the reservation by the assignor of an overriding royalty interest which could be converted to an undivided working interest — usually 40 to 50 percent — upon payout. However, as to farmout acreage outside the test well drilling and spacing unit, the assignee earned only an undivided interest in the acreage, with the assignor reserving the remaining working interest.

The assignment of the full working interest in the test well acreage was required to avoid an unnecessary loss of IDC deductions by the assignee.[6] However, in 1975, the Internal Revenue Service

---

[6] See Treas. Reg. 1.612-4(a)(3); GCM 22, 730; Rev. Rul. 69-332; and Rev. Rul. 71-207. A conversion of the overriding royalty prior to payout would require capitalization of that portion of the working interest reserved by the assignor. See Rev. Rul. 70-336; Rev. Rul. 71-206. See also Chapter Six, "Basic Tax for the Landman".

indicated a potential change in policy where the farmout involved "outside" acreage.

The finalization of the IRS position in April 1977 came as a most unpleasant shock to everyone in the industry, and reflected an entirely different approach from the anticipated possible loss of IDC deductions. Instead, in Rev. Rul. 77-176, the IRS ruled that, as to the "excess" acreage, the assignee would be considered to have received ordinary income as "payment for services" in an amount equal to the fair market value of his assigned interest in the outside acreage, and the assignor would be treated as having sold his interest in the assigned excess acreage for its fair market value.

Rev. Rul. 77-176 substantially changes the economics of traditional farmout arrangements. While, subject to the requirements discussed above, the farmee may still deduct 100 percent of his IDC, and the farmor can capitalize his imputed "sale income" and add it to his tax "basis" in his retained override, the tax consequences — taxes immediately payable on a "paper" income — can wreck an otherwise promising deal.

While efforts are being made — so far without success — to secure the rescission of Rev. Rul. 77-176, the industry is concurrently exploring new farmout approaches to lessen or avoid the revenue ruling's impact. Since the valuation date of the transaction is the date of transfer of the acreage, the tax consequences can be lessened by assigning the acreage *as of the date of executing the farmout agreement*. Sublease and net profits arrangements are also being utilized. However, the most effective means of avoiding the effects of the ruling is the formation of a "tax partnership" between the assignor and assignee.

A sample "tax partnership" farmout is contained in Appendix B to this chapter.

§ **4.03(h) Provisions Concerning Assignment and Well Operation.** The farmout agreement will frequently create unnecessary problems by leaving to future negotiation the terms of the documents required to complete the farmout. If the farmee performs his obligations, an assignment of the earned acreage, also specifying the reserved rights, will be required, as will an operating agreement if the acreage is productive and any present or future working interest is retained by the assignor. Rather than referring to the subsequent execution of "mutually acceptable" or "customary" assignments or operating agreements, it is far better to attach exact forms to be used as exhibits to the farmout agreement (see § **4.02(c)**).

The assignor's obligation should clearly be limited to a duty to assign those leases which remain in effect on the date that the assignee earns its right to an assignment; it should be absolved from any duty to maintain the leases in effect or to renew leases which expire during the preassignment period. The farmout agreement should also specify whether or not any leases which are renewed will be subject to the duty to assign.

In drafting provisions concerning the assignor's reserved interest, problems of the less-than-full-interest lease must be clearly covered. It should be clearly specified whether the interest reserved is payable out of eight-eighths of production or some lesser interest; whether or not the interest assigned, and the interest reserved, are subject to their proportionate part of outstanding overriding royalty interests and similar burdens; and whether or not the interest is expressed on a gross (subject to proportionate reduction) or net basis.

Where your company is the assignee, any tendered assignment should be carefully examined to insure that it complies with the terms of the farmout agreement and is acceptable as to form and substance. Since farmout agreements now frequently provide that the assignment will be due only if a written request is made within a specified period, any such requirement should be flagged to insure that the company's rights are not lost, once earned, by a failure to make a timely request for an assignment.

§ **4.03(i) Other Provisions of the Farmout Agreement.** The farmout agreement normally will provide that the leases will be assigned without warranty of title, or, at most, with special warranty. The assignor will not be required to deliver or to furnish full abstracts of title, although it will normally furnish the assignee abstracts and title papers in its possession.

Where the assignor's obligation to drill the earning well is a covenant rather than a condition, provision for title examination prior to the drilling of a well will be important.

The farmout agreement should specify who is responsible for the payment of intervening delay rentals; how payments are to be borne between the parties (particularly where depth and similar limitations are included in the agreement); the effect of any failure to make an agreed payment of rentals; what proof of payment is to be submitted; and whether these provisions will change upon assignment.

Normally, the farmout agreement will be non-assignable prior to the earning of the assignee's rights.

The farmout agreement should clearly specify the critical time provisions, such as the time for acceptance of the farmout agreement by the assignee, and the various dates concerning the commencement and completion of the earning well. Time should be made "of the essence".

Special Federal regulations and requirements apply to farmout arrangements involving Federal leases; and should be reviewed prior to entering into a Federal lease farmout.

§ 4.04 Operating Agreements

§ **4.04(a) In General.** Where the operating interests for a prospective drillsite or a larger prospect area are co-owned, and where more than one party wishes to participate in the development of the unit or area, there is a need for an agreement to provide for the development of the co-owned interests and for the operation of the productive wells. Where a farmout agreement leaves both assignor or assignee with some operating interests in either the farmout well or surrounding acreage, it is necessary additionally to define how the jointly owned interests will be further developed and operated.

§ 4.04(b) The "Model Form" Operating Agreement (AAPL Form 610). Prior to 1956, it was often very difficult to reach agreement among the various parties jointly owning interests in a drillsite or prospect area concerning the terms of the necessary operating agreement. This was particularly a problem where several major oil companies were involved, each with its "standard" form from which there could be no variance without the approval of its management. As a result, a formal operating agreement often was not finalized until long after the well had been plugged and abandoned.

In 1956, following several years of extensive cooperative effort through the Legal Committee of the Mid-Continent Oil and Gas Association, a "model form" operating agreement was developed which became an industry standard for operations within much of the continental United States. This form was designated AAPL Form 610.

Special operating agreement forms exist for the Rocky Mountain area, to take into account the likely presence of Federal leases, and for Canada. Tailor-made agreements will often prove necessary for offshore activities or for activities in the Gulf Coast area. However, a detailed knowledge of the contents of AAPL Form 610, hereinafter referred to as the "model form", is a must for the Landman.

When initially adopted, little change was made in the model form from company to company; no one wanted to return to the prior confusion and intercompany bickering. However, as time progressed and industry practices and economics changed, the need for a revised model form became apparent.

In 1977, the revised version of the model form was adopted by the AAPL. Besides various changes in the substance of the agreement — including the addition of an optional casing point election, operator removal provisions, a simplified title examination procedure, and a flexible penalty for consent/non-consent operations — the language of the agreement has been clarified in a number of places, and the agreement completely reorganized.

In basic content, the model form covers the following areas:

1. Designation of the unit area, the agreement purpose, and the formations, leases and oil and gas interests subject to the agreement.
2. The sharing of the costs and revenues for the joint operations, including the rights of the parties as to production, the payment of royalties, and the treatment of subsequently created interests (1977 version only).
3. Provisions for the initial testing of the unit (contract) area.
4. Conduct of subsequent operations, including "consent/non-consent" operations where less than all the parties wish to participate.
5. Designation and powers of the operator, limitations on his authority, his obligations, his removal and the rights of non-operators.
6. Special rights such as preferential right of purchase, and rights as to renewal leases.
7. Miscellaneous provisions such as title, treatment of unleased interests, payment, term, access, abandonment, delay rentals, transferability, notice, insurance, and gas balancing (1977 version only).

Copies of the 1956 and 1977 versions of the model form are contained in Appendix C of this chapter.

§ 4.04(c) Subject Matter and Definitions. Initially, the model form sets forth the purpose of the agreement, the land constituting the "unit area" ("contract area" in the 1977 version), including any restrictions as to formations and depths and a designation of the leasehold and unleased interest of each party (1956 version, recitals, §1, and Ex. A; 1977 version, recitals, Art. I, and Ex. A). The initial provisions also define various terms used throughout the agreement (1956, §1; 1977, Art. I).

§ 4.04(d) Sharing of Costs and Revenues. Costs, revenues, production and equipment are borne, shared and owned in the percentages specified on Exhibit A to the form (1956, §4; 1977, Art. III.B). Each party is charged with his proportionate part of costs paid by the operator (1956, §8; 1977, Art. VII.C). The agreement thus serves as a pooling agreement for the unit area; and, unlike the division order, there is no provision for unilateral future revision or termination.

Prior to the 1977 form, AAPL Form 610 specifically prohibited charges for title examination, even when made by outside attorneys (1956, §2.A). However, the new model agreement contains an optional procedure under which the parties may be billed for fees paid outside attorneys (1977, Art. IV.A., Option No. 2).

No charge is authorized for defense of lawsuits by staff attorneys (1956, §28; 1977, Ex. C).

Special provision is made for division of costs and revenues on consent/non-consent operations (1956, §12; 1977, Art. VI), and on abandonment of a well or surrender of a lease by less than all of the parties (1956, §§16, 24; 1977, Arts. VI.E.2., VIII.A).

As to the payment and charging of royalties, etc., see 1956 version, §§14, 13; 1977 version, Articles III.B., VII.E. Note that under the 1977 agreement, where operations are conducted by less than all the parties, the consenting parties are required to pay such burdens on production (Art. VI.B.2).

Loss of leases due to title failure, or failure to pay rentals or shut-in royalties, lead to an adjustment of interest under the 1956 agreement if the "individual loss", rather than the "joint loss", form was used (§§12, 17). The 1977 version adopts an "individual loss" approach (Art. IV.B.).

Under both agreements, each party takes his own share of production (including the share of production attributable to royalty and similar interests); executes his own division order; accounts to his own royalty owners and owners of similar burdens; and receives payment directly from the pipeline for his share of production. Each party also individually bears the cost of separately disposing of his share of production (1956, § 13; 1977, Art. VI.C).

§ 4.04(e) Test Well. The model form (1956, §7; 1977, Art. VI.A.) details the terms for the initial well on the unit (contract) area. In the absence of unanimous consent, drilling must be conducted to the depth specified in that paragraph, unless granite or other practically impenetrable substances are encountered, or, under the 1977 version, "condition in the hole, which renders further drilling impractical...." "Reasonable tests" must be conducted of all formations giving indication of containing oil and gas.

The test well may be abandoned as dry only with the consent of all parties, or, under the 1977 version, by following the dry hole abandonment procedure under Article VI.E.1.

§ 4.04(f) Subsequent Operations. Drilling operations following the drilling of the test well which are not expressly provided for in the agreement must be conducted by mutual consent or pursuant to the consent/non-consent provisions (1956, §§11-12; 1977, Arts. VI.B., VIII.D). Similarly, reworking, deepening or plugging back of any well requires mutual consent or consent/non-consent procedure.

Other "projects" may be undertaken under the operator's general authority, so long as the cost does not exceed a specified figure. If the project is more costly, mutual consent is required.

Where the parties cannot agree as to the reworking, etc., of a well, or the conduct of additional operations not specifically provided for in the agreement, both agreements provide a procedure for consent/non-consent operations. This procedure is triggered by a notice of the proposed operation in the form specified in the paragraph. A 30-day response period is provided after receipt of the notice (48 hours for rework, etc., operations when a drilling rig is on location); parties who do not respond within this period are deemed to be non-consenting parties and are "out" of the operation.

The consenting parties must actually commence operations within 30 days under the 1956 form (60 days under the 1977 version) and complete them with due diligence for the consent/non-consent procedure to apply.

The consenting parties bear all the costs and risks of the operation in the ratio of their Exhibit A percentages; non-consenting parties are "deemed to relinquish" all interest in the "well, leasehold operating rights and production" until proceeds, after royalty and taxes, equal 100 percent of operating costs and surface equipment (beyond the wellhead) and a specified percentage of all other costs, less support money. After the receipt of this penalty, the original interests revert to the non-consenting parties.

There are a number of operations which are not covered by §12 or Article VI and which would require unanimous consent of the parties to conduct. These include a decision not to drill to a specified depth; the reworking of any well still capable of commercial production; and high cost projects not involving the drilling or reworking of a well. However, the 1977 version has added an elective procedure for abandonment of a dry hole (Art. VI.E.1) and an optional "casing point election" (Art. VII.D.1).

§ 4.04(g) Abandonment of Productive Wells and Surrender of Leases.
Productive wells can be abandoned by mutual consent or, if less than all parties consent, by the parties who are unwilling to abandon paying the parties who wish to abandon their part of the salvage value of the well (less costs of salvaging, plugging and abandoning). The abandoning parties must then assign their interests in the well, leasehold and equipment (limited to then productive formations, as to the drilling unit only) to the non-abandoning parties (1956, §16; 1977, Art. VI.E.2).

A similar procedure is provided under §24 of the 1956 form, and Article VIII.A of the 1977, for the surrender of leases.

§ 4.04(h) Unit Operator.
The operator designated in the model form is granted "full control of all operations... within the limits of this agreement"

(1956, §5; 1977, Art. V.A). Additional specific grants, as well as various obligations and restrictions, are contained throughout the agreement.

Under §5 of the 1956 form, the operator's liability to non-operators is limited to losses from "gross negligence" or from breaches of "the provisions of this agreement". Since the operator is required to perform in a "good and workmanlike manner", it could be argued that any failure to function in this fashion is a breach of this provision of the agreement. This has been corrected in Article V.A of the 1977 version by limiting such liability to losses resulting from "gross negligence or willful misconduct."

An operator's lien, extending to the parties' interests in leases, material, production and proceeds, is provided under both agreements, which also authorize the operator to require non-defaulting non-operators to contribute their proportionate part of amounts in default or to demand payment in advance for estimated costs for the next succeeding month (1956, §§8-9; 1977, Art. VII.B-C.). The 1977 version also grants a similar lien to the non-operators if the operator defaults.

Besides the provisions for the replacement of the operator upon his resignation or the transfer of his entire interest contained in §§19 and 21 of the 1956 agreement, the 1977 form contains a limited right of removal for "cause" (see Art. V.B).

§ 4.04(i) Other Provisions. The model form operating agreement attempts to avoid partnership liability by a declaration of non-partnership under §22 of the 1956 version, and Article VII.A and IX of the 1977 form. However, there is some question as to the effectiveness of this provision, which must also be amended if a "tax partnership" is utilized.

Both versions of the agreement grant each party a 10-day preferential right to purchase the interest of any party who wishes to sell any or all of his interest, except on merger, reorganization, or a sale on mortgage foreclosure (1956, §18, 1977, Art. VIII.G.). This provision, however, is frequently deleted from the form. Provision also is made to limit sales to all of a party's interest, or to an equal undivided interest in all leases, equipment and production in the unit area; and, if an interest is divided among four or more parties, the operator may require the appointment of a single agent for all actions, except disposition of production and receipt of production proceeds (1956, §**20**; 1977, Art. VIII.E).

A waiver of any right to partition also is contained in the 1977 form (Art. VIII.F.).

Under both agreements, each party has the right to participate in any lease acquired covering an interest covered by an expired lease so long as such lease is acquired within six months from the expiration, the 1977 form requiring response within 30 days (1956, §**23**; 1977, Art. VIII.B.). Acreage contributions received by any party are also assigned to the parties (only "Drilling Parties" under the 1977 form) in proportion to their interest, and such acreage becomes a part of the unit area (1956, §**25**; 1977, Art. VIII.C). However, there is no provision in the model form for an "area of interest"; any such desired provision must thus be added to the agreement.

Under the 1956 form, the agreement remains in effect for the life of the leases subject to the agreement, so long as operations result in production. Otherwise, the agreement terminates 90 days after abandonment of drilling operations without the conduct of new drilling (§**10**). The 1977 form provides alternate "life of lease" or "continuous drilling" options (Art. XIII).

§ **4.04(j) Accounting Procedures.** The operating agreement specifies how *proper* charges are to be borne; but in most instances, it is merely general as to *what constitutes* proper charges to the "joint account". An accounting procedure is attached to the operating agreement to specify the type of expenditures which are properly chargeable to the joint account; the proper amount which may be so charged; and protections for the non-operators against inadvertent or fraudulent mischarges.

The fact that a given expenditure is "charged to the joint account" does not necessarily mean that it is to be allocated among the parties in the same manner in which they generally share in costs and revenues. The method of allocation is instead spelled out in the operating agreement.

Early joint operating agreements contained little reference to accounting matters. Left unanswered were such questions as what expenditures were properly treated as related to the joint operations; whether any part of the operator's district expense and administrative overhead could be charged to the joint account and, if so, how such charges were to be computed; at what price should equipment, etc., be transferred to, and transferred from, the joint account; what charge, if any, should be made for the use of the operator's personnel, equipment and facilities; and provisions for financial and operational control and procedural implementation, such as audit rights, billing, payment, advance of funds, remedies for non-payment, reporting, and inventory and controllable material concepts.

The first effort to develop a standardized accounting procedure was conducted by the Unitization Committee of the Mid-Continent Oil and Gas Association. Thereafter, additional refinement was undertaken by petroleum accountants' societies of Oklahoma and Los Angeles and later by the Council of Petroleum Accountants Societies of North America, commonly referred to as COPAS.

Different accounting practices are followed in the Mid-Continent, West Coast and Canadian producing areas to reflect different technical problems encountered in each area. As a result, there are two standardized accounting procedures: The COPAS procedure with selective options to accommodate both Mid-Continent and West Coast philosophies[7], and the western Canada procedure to cover the different circumstances encountered in that area.

The COPAS accounting procedure contains definitions of various additional terms critical for an understanding of the accounting procedure. Section II details which charges may be treated as "direct charges" to the joint account, including rental and royalties, labor, material, transportation cost, cost of services, cost of repairing and replacing joint property, legal expenses, taxes, insurance premiums, and "other expenditures". Provision is made under §III of COPAS for charges for overhead, with options being

---

[7] There are currently three COPAS versions, dated 1962, 1968 and 1974, in use in the United States.

given to charge at a fixed "per well" rate or on a percentage basis, with an additional option as to whether or not technical employees are, or are not, covered. Section III permits an adjustment of such charges when charged on a "well" basis to reflect increases or decreases in specified "labor" indices and provides for amendment by mutual agreement if in practice the rates are found to be insufficient or excessive.

Provision is made for a percentage charge for "major construction" (formerly "special projects") overhead.

The accounting procedure also provides methods of valuing property acquired or disposed of for the joint account. The operator has the right to provide such equipment by purchase or "in kind" and, when not needed, he may sell, divide in kind or purchase such equipment.

Section I of the COPAS accounting procedure contains provisions for billing and payment. Payment of a bill does not prejudice the right of a non-operator to question the propriety of a charge, so long as he files written demand for adjustment within 24 months from the end of the calendar year. Non-operators are similarly given a right to conduct an audit at their own expense during this same period.

The operator is obligated to maintain detailed records of any controllable material; to conduct periodic inventories (with notice to non-operators); and to report overages and shortages resulting from lack of "reasonable diligence". In the absence of special agreement, costs of conducting such periodic inventories are not charged to the joint account.

Provision is also made for special inventories in the event of a sale or change of interest in the joint property.

A separate Canadian — PASWC — accounting procedure is used for Canadian operations.

§ 4.05 Division Orders and Gas Contracts

§ 4.05(a) In General. Once production is discovered, arrangements must be made for its sale. The "contract" for the sale of crude oil, and in many instances for the sale of casinghead gas, is a relatively simple document — the Division Order. The contract for the purchase and sale of gas well gas — the Gas Contract — is a far more complicated document. However, even the Division Order is a more involved instrument than might appear from a cursory examination; in other words, don't judge its legal affect merely from the length of the document.

§ 4.05(b) The Nature of Division and Transfer Orders.
§ 4.05(b)(1) In General. A production purchaser must be sure that the person from whom he is purchasing the oil or casinghead gas is in fact the true owner of the production, or at least has the right to sell it. If the "seller" is not the true owner — for instance if the Lessee lacked authority to sell the Lessor's share of oil, the lease had terminated or covered a smaller interest than the purchaser realized — he would be liable to the true owner for "conversion" even though he had made full payment to the seller. Similarly, if an owner of an interest in a lease subsequently assigns his interest to another, the production purchaser will be liable to the transferor if he accounts to the transferee for a larger interest than the assignment actually covered, or if the assigning instrument is invalid.

Obviously, the production purchaser needs to protect himself against these risks. The Division and Transfer Orders seek to provide this protection by requiring all persons who could possibly claim an interest in the production and its proceeds to join in an instrument stipulating exactly how these proceeds are to be divided among them. The Division Order also serves as a contract of purchase and sale of the production, with title to the production passing as a result of the contract at the time it is delivered to the purchaser.

§ 4.05(b)(2) Procedures Prior to Distribution of Proceeds. Once a production purchaser starts to take runs from a well, distribution of proceeds will be suspended while a Division Order Opinion is prepared, based upon abstracts examined by an attorney. Marketable title will normally be required by the production purchaser (see Chapter Two). Once the necessary title curative has been secured, a Division Order will be prepared based upon the "division of production" set forth in the Division Order Opinion. This Division Order will normally be prepared for "counterpart" execution — a separate, but identical, document will be forwarded to each interest owner for signature.

Title may not be cured as to all interests simultaneously. However, as various interests are cleared and a Division Order has been secured covering these interests, runs will usually be released to these owners.

Customarily, the Oil and Gas Lease does not contain any provision requiring the Lessor to execute a Division Order. However, the Lessee customarily is authorized, either expressly or by implication, to sell the Lessor's interest in production on customary terms in the industry (see § 3.07(a)); this is interpreted as including the purchaser's requirement that proceeds not be released until a Division Order has been executed by all parties. Similarly, a purchaser is not required to honor any transfer of a party's interest until both assignor and assignee have executed a Transfer Order.

The implied requirement that an interest owner or assignee execute a Division or Transfer Order assumes that the form presented contains only those terms customary in the industry. If the Division or Transfer Order contains unusual or unreasonable provisions, the interest owner is not required to execute it as a condition to this right to payment.

§ 4.05(b)(3) Contents of the Division Order. A sample Division Order is contained in Appendix D to this chapter. You will see that these Division Orders customarily contain the following provisions:

1. An identification of the *land* and *unit* (or units) affected.
2. The *type of production* being purchased: oil, gas or casinghead gas.
3. The *effective date* of the Division Order.
4. Direction as to how production proceeds are to be credited, i.e., the *division of interest*.
5. A *warranty of title* as to each party's respective interest, including a "hold harmless" provision for the protection of the purchaser.

6. *Title provisions,* including a duty on the part of the interest owner to furnish satisfactory evidence of title (including an abstract showing marketable title) or an indemnity bond (seldom used); a duty to notify the purchaser of any suit attacking the interest owner's title; and the right of the purchaser to withhold payment, without interest, pending satisfaction of the title requirements or in the event of title attack.

7. The *method of payment,* including how payment is to be computed, the method of settlement, and an authorization to deduct production taxes.

8. *Change of ownership provisions,* including a requirement that written notice and specified documentation be provided prior to honoring the transfer; an "effective date" provision (frequently 7 a.m. of the first day of the calendar month following receipt of the notice); and a provision that the "change of ownership" requirements will also apply to changes resulting from the occurrence of some contingency, such as "payout" or the termination of a production payment.

9. Various *miscellaneous provisions,* such as the right to accrue proceeds until a specified minimum amount is owed the interest holder and to require designation of a single agent for payment if the interest becomes subdivided; the address and taxpayer identification number of the interest owner; an agreement to comply with Federal, state and municipal regulations (required due to "hot oil" risks); a right to remove the purchaser's pipeline in the event of a termination of the Division Order; and provisions for witnessing of signatures and counterpart execution.

The Division Order is normally binding until "further notice". Frequently, it is prepared to run in favor of both the production purchaser and the working interest owners. (Otherwise, the Lessor may still have an action against the Lessee for failure to pay royalties properly even though he has no cause of action against the production purchaser.) And the

Division Order may also contain a provision that its terms will govern over any contrary provision in the Oil and Gas Lease.

The Transfer Order will identify the purchaser selling the interest; the person to whom the interest is being sold; the interest being assigned, and the effective hour and date of transfer; and will continue with the usual Division Order provisions, including a place for insertion of the transferee's address and taxpayer identification number.

A sample Transfer Order is also contained in Appendix D to this chapter.

§ 4.05(b)(4) Effects of Execution of a Division or Transfer Order. Once an interest holder has executed a Division or Transfer Order, the purchaser is entitled to distribute production proceeds in accordance with the provisions of the order, even if the instrument misstates the ownership. However, this will not necessarily prohibit the interest owner from suing the party who received the overpayment. Once, however, the purchaser has received notice of the revocation of the Division Order, he is no longer protected if he continues to pay in accordance with its terms.

As between the Lessor and the Lessee, the Division Order is less binding. It does not prevent the Lessor from suing to cancel his lease for failure to obtain production in paying quantities, to obtain production prior to the end of the primary term, or for violation of the implied covenants of the lease (see Chapter Three). However, in some instances, a mere joinder in a Division Order has been treated as a ratification of the lease, and may establish that the method used in distributing the royalties represents a proper accounting on behalf of the Lessee as well as the production manager.

As between individual Lessors, the Division Order has no effect whatsoever, except possibly as evidence of the construction which has been placed on the various title instruments by the interest owners. Accordingly, execution of a Division Order normally does not prevent the Lessor from suing another mineral owner in the event he feels that payments have been misallocated between them.

The provisions of the Division Order will also cover computation of the price at which the production purchaser must account to the interest owners. These provisions in an oil Division Order are relatively simple, and normally will provide for an accounting based upon the posted field price, less adjustments for treating or transportation costs, impurities, temperature and handling losses.

Where casinghead gas or natural gas is involved, more complicated problems arise since provision must be made for liquid content and how to account for it; gas *used* rather than sold; any problems concerning "market value" as contrasted with "proceeds" (see § **3.07(b)**); and a calculation where the point of delivery is other than the wellhead.

The provision concerning change of ownership protects the production purchaser against being required to account to an assignee upon receipt of actual notice of the transfer without a "grace period" or an opportunity to require satisfaction of newly-raised title requirements. However, these provisions do not apply to a successful attack against the title of someone to whom the production purchaser has been accounting. Thus, the purchaser remains liable for all prior erroneous payments, even though he has not been furnished a copy of the judgment successfully attacking the title.

The production purchaser should insure that the person executing the Division Order, if other than the owner personally, has authority to act for the interest owner. If possible homestead or community property is involved (see § **2.02(c)(7)**), both husband and wife should also join in the execution of the document.

While the Division Order generally is witnessed, it is not required as a matter of law to be acknowledged or recorded. However, if the Division Order is to serve as an irrevocable stipulation of interest by deletion of "till further notice", or a permanent ratification of an Oil and Gas Lease or gas contract, acknowledgement and recording, plus words of grant, may be required (see § **2.02(c)(4), 2.02(c)(7)**).

The production purchaser should not be expected to resolve doubtful questions in order to release production proceeds. Accordingly, until title has been rendered marketable, and all parties who could possibly claim an interest in the property have joined in a Division Order clearly agreeing as to the distribution of production proceeds, he should hold runs in suspense.

§ 4.05(c) Gas Contracts

§ **4.05(c)(1) In General.** One of the most complicated contracts with which your company must deal is the contract for the sale of natural gas, particularly gas from a gas well. Such arrangements are not simple transactions, since they involve a long term contract, a product which cannot be stored, varying periods of need on behalf of the buyer, reservation of certain portions of production and certain operating rights by the seller, and regulatory complexities under state and Federal laws. Basically, the parties must agree on what *reserves* are *subject to the contract*, the *price* to be paid for the gas purchased and sold, the buyer's *purchase obligation*, *reservations* by the seller, and the *condition in which the gas is to be delivered* to the buyer, including which party is to bear the cost of satisfying such conditions. Such items are not subject to the agreement of the parties alone, since a complex system of state and Federal regulation exists through both state conservation and public utility agencies and the Department of Energy's Federal Energy Regulatory Commission (formerly the Federal Power Commission).

§ **4.05(c)(2) Federal Regulation.** Under the Natural Gas Act, the FERC has jurisdiction over the *certification of sales for resale*, and *all transportation* of natural gas, *in interstate commerce*, including the construction, acquisition or operation of facilities for such transportation or acquisition. The FERC has the power to *review rates and contract terms* to see if they are unjust and unreasonable, specifically including a power of review of any change in rates or contract terms; and an authority to *disapprove any proposed "abandonment"* of jurisdictional sales or transportation, or the facilities involved in such sales or transportation.

Certain sales of natural gas historically have not been subject to the jurisdiction of the FERC. These include "intrastate sales" — sales of gas for ultimate consumption within the state in which the gas is produced, where the gas does not cross state lines in the course of transportation — and "direct sales" — sales directly to the ultimate user.

Another technique, the so-called sale of gas "in place", was also formerly utilized. However, many such attempted sales have been held subject to FERC jurisdiction under the Natural Gas Act.

In all attempted non-jurisdictional sales, the exemption may be lost if the gas, when transported, is co-mingled with "jurisdictional" gas. A provision for automatic termination of the contract if the sale is held to be subject to Natural Gas Act jurisdiction is ineffective. Similarly, even though the direct sale does not require certification of the sale, the *transportation* of the gas will require certification if the gas is to be taken across state lines; such certification normally will be denied if an "inferior" end use is contemplated or if the FERC is otherwise unhappy with the sale terms. (A proposed non-jurisdictional use is often enough to make the FERC "unhappy".)

The transportation certification requirement can also create problems if a producer is attempting to transport his own gas across state lines for use in a fertilizer plant, or if he attempts to transport the gas within the state in a pipeline where the producer's gas would be co-mingled with jurisdictional gas.

1978 legislation is severely affecting the "unregulated" nature of previously "non-jurisdictional" sales.

Once gas is intentionally or inadvertently committed to a jurisdictional sale or transportation, the producer cannot terminate the sale or transportation without "abandonment" approval from the FERC, even if his contract has, by its terms, expired. Such abandonment will be permitted only if the supply of natural gas is depleted, or the abandonment is otherwise in the "public convenience and necessity". The FERC will be reluctant to grant such abandonment if the reserves are not totally depleted, even though production may be uneconomical at present price levels. Under such circumstances, rather than permitting a non-jurisdictional sale or use, producers are "encouraged" to negotiate "special relief" with the jurisdictional purchaser at a price level which will still have to be approved by the Commission.

Recent decisions have indicated that a commitment of a lease to a jurisdictional Gas Contract may bind the Lessor's reversionary interest to the interstate sale, despite a termination of the committed lease. The Gas Contract may never have been recorded, and no gas may ever have been sold in interstate commerce from the particular leased premises. However, the result of these holdings makes it very difficult for you, as a Landman, to secure a lease, or take a farmout, without the possibility of being committed to a low-price or otherwise unsatisfactory market of which you aren't even aware.

Where the potential for Federal regulation exists, the Gas Contract must be prepared to take into account this implication. This will require the inclusion of various clauses which will permit an increase in the price in the event FERC approved price levels are increased, or the sale is deregulated. It will also require an elimination of those clauses which would be considered unacceptable by the FERC.

§ 4.05(c)(3) **Contents of the Gas Contract.** Aside from regulatory problems, there are four basic areas which must be covered in the Gas Contract, whether jurisdictional or non-jurisdictional. These include *commitment* of the leases, reserves and reservations; the *price* to be paid; the *quantity of gas* which the producer is required to deliver and which the pipeline is required to purchase; and the *conditions of delivery*.

Insofar as commitment and reservations are concerned, the contract must identify the leases subject to the contract; the substances covered; depth limitations; and the sellers' various reservations, including recycling rights, processing rights and the right to use gas for operations.

Pricing provisions are affected by the long-range nature of the relationship, the differences in the quality of the gas, and the conditions under which it will be delivered. The typical contract will provide for an initial "base" price; escalations, including both fixed periodic escalations and escalations based upon various "indefinite" pricing provisions, such as "favored nation" or renegotiated rates; reimbursement of certain producer costs, such as increases in taxes and royalties; and adjustments based upon the quality of the gas, such as BTU content, delivery conditions and other conditions. All of such clauses are subject to FERC review, and only specified "indefinite" provisions will be approved.

Prior to the present gas shortage, a pipeline frequently contracted for substantially more gas than it was presently prepared to take to insure an adequate supply. This led to purchases by the pipeline which were inadequate to provide a reasonable return to the producer and, in some instances, threatened the producer with lease termination. To avoid such problems, modern contracts now obligate the purchaser to purchase a specified quantity of gas, both annually and on a daily or monthly basis. This provision is usually on a "take or pay" basis — the purchaser must either take a specified volume of gas or, if it doesn't take the volume of gas, pay for it anyway. Such an arrangement may require an agreement as to the reserves subject to the contract where the "take" obligation is not based on deliverability.

While "take or pay" provisions may seem less important in the current period of gas shortage, it cannot be assumed that a given purchaser will inevitably, over the life of the contract, not find itself in at least a temporary oversupply position. Several intrastate markets have also found themselves in an overcommitted position. The producer must accordingly be alert to insure that other contract provisions do not exist which render the "take" obligation illusory.

The contract will also obligate the producer to deliver certain volumes of gas to the pipeline. While such obligations are normally dependent on the ability of the wells to deliver the specified quantity, the FERC has considered imposing obligatory drilling and reworking obligations on the producer if he is not delivering the specified quantity of gas to the pipleline.

To date, courts have not looked kindly on this extension of jurisdiction. However, if the producer has entered into a "warranty" contract, guaranteeing to deliver specified volumes, the courts may support the FERC.

One of the most critical portions of the contract is the provision concerning delivery conditions. The pipeline will normally be obligated to connect only those wells which have sufficient reserves to justify

197

construction of the necessary pipeline facilities, and further will only be required to accept gas delivered at specified pressure, and which meets specified provisions as to quality, including BTU content, liquid/solid/water vapor content, oxygen/CO_2/sulphur content, and temperature. This portion of the contract will provide for a point of delivery, either at the wellhead or a central point in the field.

Delivery conditions are normally not an obligation of the producer, but a condition of the buyer's purchase obligation. However, the producer should not assume this is so. If the producer signs a contract containing delivery conditions which all or a part of his wells cannot satisfy, he will find his reserves tied up on a long-term contract for which he will receive no pay until he has satisfied unanticipated delivery conditions. For this reason, it is highly important to determine whether or not the wells can satisfy the delivery conditions since, even when the delivery conditions are not directly "price related", the provisions as to who is required to install the facilities to bring the gas to the delivery conditions specified in the contract will directly affect the return to the producer — your company's "bottom line".

As in all contracts, the basic rule of Gas Contracts is to carefully study the entire document. This is particularly important in the Gas Contract, as pipeline attorneys have historically exhibited considerable ingenuity in developing "fish hook" provisions hidden in latter portions of a contract, which render illusory certain benefits purportedly granted to the seller in earlier provisions.

Appendix A

SAMPLE "SUPPORT" LETTER

Amoco Production Company

DRY OR BOTTOM HOLE CONTRIBUTION LETTER AGREEMENT

DATE _____

AFE _____ AREA NAME _____

_____ COUNTY OR PARISH

STATE OF _____

To

1. Herein the addressee will be referred to as "Operator" and the undersigned, Amoco Production Company, will be referred to as "Amoco"

2. Operator has indicated to Amoco Operator's intention to commence, not later than _____ the actual drilling of a test well at the location hereinafter indicated and thereafter to drill said well diligently and without unnecessary delay to the depth hereinafter indicated and complete the drilling of said well not later than _____ (   ) days after it is spudded. all at the sole cost, expense and risk of Operator, the location and depth of said well to be as follows

LOCATION

DEPTH AND PENETRATION

200

3. The area in which said well is to be drilled is believed by the parties hereto to be underlain by deposits or accumulations of oil or gas, or of both oil and gas, and it is the express purpose in drilling said well to test adequately and properly all prospective oil or gas carrying zones, as indicated by the nature of the cuttings and cores of formations penetrated, by the study of electrical logs or by other scientific methods, to determine the productive possibilities of the oil and gas reservoirs penetrated.

4. If Operator commences and drills said well as provided in paragraph 2 hereof, Operator shall:

(a) Notify Amoco in writing the date on which Operator:

    (i) Stakes the location for said well and begins preparation of location.

    (ii) Moves the material for the drilling of said well to the location.

    (iii) Commences actual drilling of said well.

    (iv) Completes said well as a producer of oil or gas in paying quantities or abandons said well as a dry hole

(b) Notify Amoco by telephone (to be confirmed by mail), telegraph, or mail, at Operator's expense.

    (i) A daily progress report of Operator's operations with respect to said well, from the date operations begin on location preparation to and including the date it is completed as a producer or plugged and abandoned as a dry hole.

    (ii) When logs are to be run, or when said well is to be tested, or any horizon is to be cored and/or drill stem tested, in sufficient time to permit a representative of Amoco to be present prior to the commencement of such operation.

    (iii) When the drilling of said well is completed and allow Amoco sufficient time for Amoco's representatives to measure the depth thereof by running a steel line measurement. If Amoco should measure the depth of said well, Amoco's measurement shall be binding on the parties hereto.

(c) Furnish, free of cost, to Amoco

    (i) Sample cuts, furnished through the facilities of a reputable Service Company, at one-foot intervals of all cores taken, and samples of drill cuttings taken at ten-foot or lesser intervals from formations encountered in said drilling operations, unless Amoco elects to take said samples itself.

      a. Intervals to be cored as required by Amoco:

201

(ii) Two field and two final, correct and complete copies if a dry hole, or three field and three final, correct and complete copies if a producer, of each of the formation surveys checked below and of any other surveys run:

☐ Conventional Electrical, Induction-Electrical or Laterolog survey, or equivalent log, from base of surface pipe to total depth (selection of log dependent on that best suited to mud conditions at logging time).

☐ Microlog or Microlaterolog, or equivalent log, with caliper of all zones in which showings of oil or gas have been encountered and the following potentially productive formations (selection of log dependent on that best suited to mud conditions at logging time).

_____

_____

☐ Gamma Ray-Sonic log from base of surface casing pipe to total depth.

_____

_____

☐ Dipmeter survey to be run as follows:

_____

_____

☐ Gamma Ray-Neutron log from base of surface pipe to total depth.
☐ Gas Analyzer to be run through following interval:

_____

_____

202

☐ Other formation surveys as set forth below:

(iii) One set of DST charts, core and fluid analysis reports and any other similar reports prepared for this well.

(iv) The plugging record of said well if it is a dry hole.

(v) Any other information requested by Amoco relative to said well or the drilling thereof.

(d) Permit Amoco, if Amoco so elects, in addition to any other privileges above mentioned:

(i) To have access to said location, well and the derrick floor at all reasonable hours for the purpose of observing Operator's operations and the progress being made by Operator.

(ii) To lower a geophone or other similar instrument into said well for the purpose of making any test desired, provided, that Amoco shall reimburse Operator for any and all costs incurred by Operator as the result of Amoco's performing any such test.

(e) Upon encountering each potentially productive oil or gas horizon, drill stem test said horizon to Amoco's satisfaction in the event Amoco's representative detects oil showings considered by said representative to be significant.

(f) If the information from any survey, considered by itself or in conjunction with other indications or evidence from cuttings, cores or showings, makes any formation encountered appear promising of being a prospective oil or gas horizon, drill stem test to Amoco's satisfaction said horizon if it was not considered by Amoco's representative to have been adequately tested at the time it was penetrated.

(g) If the information from any drill stem test, cores, cuttings, or formation surveys indicates a formation penetrated to be capable, in Amoco's opinion, of oil or gas production in commercial quantities, run a producing string and diligently attempt to complete said well as a commercial producer.

(h) Satisfy other requirements as follows:

203

5. If Operator fails to commence, drill or complete said well within the time and in the manner set out in paragraph 2 hereof, or if Operator fails to comply with one or more of the obligations herein made binding upon Operator, time being of the essence of this agreement, even though Operator's said failure may be caused, in whole or in part, by an order, rule or regulation of any present or future State or Federal authority or agency having jurisdiction in the premises, then, and in any of said events, Amoco shall be released and discharged from all obligations and liability of every kind and character hereunder.

6. If Operator commences, drills and completes said well as above provided and otherwise complies with the obligations herein made binding on Operator and if said well is a _____. Amoco, within _____ ( ) days after demand therefor by Operator, shall pay to Operator as a _____ hole contribution to said well the sum of _____

7. If said well, upon being drilled and completed as above provided, is a dry hole, Operator shall plug and abandon it within _____ ( ) days after obtaining approval of Amoco.

8. All notices, samples, reports, copies of logs and surveys and all other information and data required to be furnished by Operator to Amoco hereunder shall be delivered or sent to_____

at Amoco's Exploration Office at_____

Phone Number_____

9. This agreement shall not be assigned by Operator, in whole or in part, without the written consent and approval of Amoco; and any attempted assignment hereof or of any rights hereunder without such written consent and approval shall be void.

204

10. This agreement shall be binding upon Operator and Amoco on condition that it is accepted and approved by Operator at the place provided therefor below and _____ (   ) executed copies hereof are returned to Amoco not later than _____ (   ) days after the date hereof.

Amoco Production Company

By _____

ACCEPTED AND APPROVED

This _____ day of _____, 19 ____.

_____

_____

_____

205

Appendix B

SAMPLE FARMOUT AGREEMENTS

TRADITIONAL FARMOUT AGREEMENT

FARMOUT CONTRACT
With Operating Agreement Attached

THIS AGREEMENT, made and entered into this _____day of
_____, 1976, by and between SOME PETROLEUM
COMPANY, a corporation, authorized to do business in the
State of Anywhere, hereinafter referred to as
"Non-Operator", and EAGER TeDRILL, hereinafter referred to as
"Operator";

W I T N E S S E T H, That:

WHEREAS, Non-Operator is the owner of the Oil and gas leases
described in Exhibits "A" and "B" and covering land situated
in Some County, State of Anywhere, insofar as said leases
cover the land set out in said Exhibits (said leases and
lands therein described to the stratigraphic equivalent of
the depths, to be drilled by Operator on said lands hereinafter
sometimes being referred to as the "lease acreage"); and,

WHEREAS, Non-Operator has agreed to assign to Operator, and
Operator has agreed to accept from Non-Operator, certain interests
in the leases therein described, to the stratigraphic equivalent
of the depths to be drilled by Operator on said lands insofar
as they cover the lands therein described upon and subject
to the terms, covenants and conditions hereinafter set forth.

NOW, THEREFORE, in consideration of the premises and of
the mutual covenants and agreements hereinafter contained

208

(including all those set out in Exhibits "C" through "F"
attached hereto, said Exhibits "C" through "F" being hereby
incorporated herein by reference and made a part hereof for
all purposes), it is hereby agreed by and between the parties
hereto as follows:

1. REQUIRED TEST WELL:

Operator, not later than July 1, 1976, shall commence
the actual drilling of Required Test Well at a location on
Section 30-34N-16E, Some County, Any State, and thereafter
shall prosecute the drilling of said well diligently, without
unnecessary delay and in a workmanlike manner to a depth of
One Thousand Five Hundred (1,500) feet from the surface or
to a depth of Twenty-Five (25) feet below the top of the
Hydro Shale formation whichever is the lesser. Operator
shall complete said well as above provided not later than
thirty (30) days after commencement.

If, because of encountering impenetrable substances or
because of other conditions making further drilling impracticable, Operator shall discontinue drilling the Required Test
Well before the depth requirement therefor is satisfied,
Operator shall have the right but not the obligation to
drill a substitute well at a location selected by Operator
on said Section 30, provided the actual drilling of said
substitute well is commenced not later than thirty (30) days
after the abandonment of the original Required Test Well.
Such substitute test well shall be drilled in the manner and

209

to the depth specified for the original Required Test Well, and must be completed not later than thirty (30) days after it has been commenced. If this substitute test well is commenced, drilled and completed as herein provided, the Operator shall have complied with this contract, as to Exhibit "A" land, to the same extent as if the original Required Test Well had been commenced, drilled and completed in accordance herewith. Each reference herein to a test well shall include any substitute well therefor.

2. OPTIONAL TEST WELL:

Operator, not later than sixty (60) days after completion of the Required Test Well, has the option of commencing the actual drilling of an Optional Test Well at a location on Section 25-34N-15E, Some County, Any State, and thereafter shall prosecute the drilling of said well diligently, without unnecessary delay and in a workmanlike manner to a depth of One Thousand Five Hundred (1,500) feet from the surface or to a depth of Twenty-five (25) feet below the top of the Colorado Shale formation, whichever is the lesser. Operator shall complete said well as above provided not later than thirty (30) days after commencement.

If, because of encountering impenetrable substances or because of other conditions making further drilling impracticable, Operator shall discontinue drilling the Optional Test Well before the depth requirement therefor is satisfied, Operator shall have the right but not the obligation to drill a substitute well at a location selected by Operator

on said Section 25, provided the actual drilling of said

substitute well is commenced not later than thirty (30) days

after the abandonment of the original Optional Test Well.

Such substitute test well shall be drilled in the manner and

to the depth specified for the original Optional Test Well,

and must be completed not later than thirty (30) days after

it has been commenced. If this substitute test well is

commenced, drilled and completed as herein provided, the

Operator shall have complied with this contract, as to

Exhibit "B" land, to the same extent as if the original

Optional Test Well had been commenced, drilled and completed

in accordance herewith. Each reference herein to a test

well shall include any substitute well therefor.

3. DRILLING AND COMPLETION OF REQUIRED AND/OR OPTIONAL
    TEST WELL:

Operator shall notify Non-Operator immediately when the

location for either of said test wells is staked, when the

material for the drilling thereof is moved to the location

and when actual drilling is commenced. After actual drilling

has been commenced and continuing until Operator has completed

either of said wells as a producer, has plugged and abandoned

them as a dry hole, or has relinquished operations thereon

to Non-Operator under this Article, Operator shall furnish

to Non-Operator daily reports as to the progress of drilling,

as well as any and all other information requested by Non-

211

Operator relative to the drilling of said wells.

Non-Operator shall have the right to measure the depth of either of said test wells by running a steel line measurement; and, when said wells have reached the depth at which a decision must be reached with respect to plugging and abandoning the wells or attempting to complete them as producers, Operator shall notify Non-Operator so that Non-Operator may have an opportunity to measure the depth thereof. In the event that Non-Operator measures either of said wells as above provided, Non-Operator's measurement thereof shall be conclusive and binding on the parties hereto.

If Operator shall determine that either of the test wells is capable of producing oil or gas in commercial quantities or if the information from any drillstem test, cores, cuttings, or formation surveys indicates that either of said wells is capable, in Non-Operator's opinion, of oil or gas production in commercial quantities, Operator at its cost and expense, shall run a production string and diligently attempt to complete said well as a producer and, if successful, shall at its expense equip said well for production through and including the lease tank or separator.

If, following an attempt at completion by Operator, Operator shall determine that either of the test wells is incapable of producing oil or gas in commercial quantities and that, therefore, the said test well should be plugged and abandoned, Operator shall promptly notify Non-Operator by telephone confirmed in writing of such determination. Should Non-Operator agree with this determination (and a failure to respond in the manner and within the time indicated below shall constitute agreement), Operator shall proceed at its cost and expense to plug and abandon the said test well in accordance with applicable state laws and shall level the ground around the location and clear and clean the premises to the satisfaction of the surface owners and surface lessees. Should Non-Operator disagree with this determination, it shall have twenty-four (24) hours following the receipt of such notice, excluding non-working days in Non-Operator's office, within which to advise Operator by telephone confirmed in writing of its decision to continue operations on said well, and, if this is done, all further operations on the said test well, including, if indicated, the plugging and abandonment thereof, shall be conducted entirely by Non-Operator, at its sole cost, risk and expense. If Non-Operator shall elect to continue operations on said well under this paragraph, Operator shall be deemed to have relinquished and transferred to Non-Operator all of its right, title and interest in and to (i) the said test well, (ii) the material and equipment therein and used or acquired in connection therewith which Non-Operator retains for conducting operations hereunder, and (iii) the section of land upon which said test well was drilled. Operator in such case shall lose the right to receive an assignment

hereunder of the land described in the applicable exhibit,
provided, that, otherwise, the said test well shall be
considered to have been completed by Operator on the date
Non-Operator elected to continue operations on said well.
If Non-Operator shall elect to continue operations on said
well, as above provided, it shall reimburse Operator for the
salvage value of all material and equipment in the said test
well and used or acquired in connection therewith which are
retained by Non-Operator hereunder.

4. COST OF THE REQUIRED AND/OR OPTIONAL TEST WELL:

The entire cost, expense and risk of the drilling of
said wells shall be borne by Operator, it being understood
that the risk to be borne by the Operator shall include, but
shall not be limited to, any claim, demand, action, cause of
action, judgment, attorney's fee or expense of investigation
or litigation, for injury to, or loss or destruction of,
property or for injury to or death of any person arising out
of or in connection with the drilling of said test wells,
whether through an act or omission of a party to this Contract, or otherwise. The cost of attempting to complete the
test wells as a producer, of equipping said wells for production,
or of abandoning said wells shall be borne as provided in
Article 3, above. All risks (including those enumerated
above) incident to the completion, equipping or abandonment
of the test wells shall be borne by the party completing,

214

quipping or abandoning the same, as the case may be, under

nd pursuant to Article 3. Upon completion, and, as a

ondition precedent to Non-Operator's obligation to perform

n accordance with the provisions of this Contract, Operator

hall furnish to Non-Operator evidence satisfactory to Non-

perator establishing the payment of all bills required to

e borne by Operator in connection with said test wells.

5. GEOLOGICAL INFORMATION CONCERNING TEST WELLS:

Operator shall conduct the geological program set out

n Exhibit "C" attached hereto as a part hereof and at his

ole cost, risk and expense (except as otherwise may be

pecified in said Exhibit "C") furnish all the cores, samples,

ogs and other information and data specified in said Exhibit "C".

6. DELAY RENTALS:

In the event that, any delay rental should become due

nd payable on lease acreage covered hereby, whether before or

fter Non-Operator has executed assignments in favor of Operator,

on-Operator shall make a bona fide effort to pay such rental; and,

perator shall reimburse Non-Operator for one-half (1/2) of the

otal amount paid by Non-Operator for such rental within fifteen

15) days after receiving Non-Operator's billing therefor.

7. TITLES:

Non-Operator does not warrant the title to the lease or

ease acreage covered by this Contract, but it shall upon

equest furnish to Operator such abstracts and other title

papers as it has in its files, together with photostats of the basic lease and all intermediate assignments thereof. There shall be no obligation on the part of Non-Operator to purchase new or supplemental abstracts, nor to do any curative work in connection with the title to said lease acreage.

8. DEFAULT.

If Operator fails to comply with any of the provisions of this Contract, Non-Operator at its option may terminate this Contract; provided, that in so doing Non-Operator shall not waive, or otherwise be precluded from exercising, any other rights or remedies, at law or in equity, which it may have for the breach of the Contract by Operator or for Operator's failure to perform the Contract in whole or in part.

9. ASSIGNABILITY AND EFFECT OF CONTRACT:

Operator shall not assign this Contract in whole or in part without the written consent thereto of Non-Operator. The terms, covenants and conditions of this Contract shall be binding upon, and shall inure to the benefit of, the parties hereto and their respective heirs, successors and assigns; and said terms, covenants and conditions shall be covenants running with the land covered hereby and the leasehold estate therein and with each transfer or assignment of said land or leasehold estate.

10. PERFORMANCE:

216

10.1 _Assignments_: Upon written request by Operator made within thirty (30) days after Operator has completed the Required or Optional Test Wells as a well capable of producing oil or gas in commercial quantities, in the manner and within the time herein provided, has submitted evidence thereof and otherwise has complied with and performed all the other terms, covenants and conditions herein made binding on Operator, time being of the essence of this Contract, Non-Operator shall execute and deliver assignments to Operator as follows:

(a) Required Test Well:

(1) An assignment of all its right, title and interest in and to the drilling unit or, if no unit is established, the eighty (80) acre tract of lease acreage upon which said test well was drilled to the stratigraphic equivalent of the depth drilled in said test well but reserving to Non-Operator a one-sixteenth of eight-eighths (1/16 of 8/8) convertible overriding royalty, using therefor the form of assignment attached hereto, marked Exhibit "D" and made a part hereof.

(2) An assignment of one-half (1/2) of all its right, title and interest in and to the lease acreage described in Exhibit "A" and not covered by Paragraph (a) (1) above to the stratigraphic equivalent of the depth drilled in said test well, using therefor the form of assignment attached hereto, marked Exhibit "E" and made a part hereof.

(b) Optional Test Well:

(1) An assignment of all its right, title and interest in and to the drilling unit or, if no unit is established, the eighty (80) acre tract of lease acreage upon which said test well was drilled to the stratigraphic equivalent of the depth drilled in said test well but reserving to Non-Operator a one-sixteenth of eight-eighths (1/16 of 8/8) convertible overriding royalty, using therefor the form of assignment attached hereto, marked Exhibit "D" and made a part hereof.

(2) An assignment of one-half (1/2) of all its right, title and interest in and to the lease acreage described on Exhibit "B" and not covered by Paragraph (b) (1) above to the stratigraphic equivalent of the depth drilled in said test well, using therefor the form of assignment attached hereto, marked Exhibit "E" and made a part hereof. With respect to said form of assignments, the parties hereto approve and confirm all the terms, covenants and conditions therein set forth and agree that the assignment of said leases, as hereinabove provided, shall be subject to and in accordance with the terms, covenants and conditions contained in said Exhibits "D" and "E". Failure of Operator to request assignments as herein provided shall constitute a waiver of Operator's right to such assignment, and this Contract shall at Non-Operator's option thereupon terminate.

10.2 Recording of Assignment: Upon delivery to it thereof, Operator agrees to forthwith file for record all assignments delivered to it by Non-Operator hereunder and to furnish Non-Operator with the pertinent recording data thereof as soon as same is available.

10.3 Contributions: All contributions of cash or acreage or both toward the drilling of either of the test wells hereunder shall be owned solely by Operator.

11. NON-OPERATOR'S RIGHT TO CONVERT OVERRIDE:

The assignments provided for in Articles 10.1 (a) (1) and 10.1 (b) (1) of this Contract are subject to the right of Non-Operator, herein created, to convert its reserved one-sixteenth of eight-eights (1/16 of 8/8) overriding royalty interests to one-half (1/2) working interests at such time as the proceeds of all production from each of the applicable test wells computed separately (exclusive of royalty, overriding royalties and taxes chargeable to the working interest) equals One Hundred (100) per cent of the costs incurred by Operator in drilling, testing, completing and equipping that said well (to and including the tanks or separator), plus, during the time required to recover the foregoing, One Hundred (100) per cent of the costs incurred by Operator in operating that said well. As to each well, if and when Operator recovers the amounts aforesaid, Operator shall notify Non-Operator thereof in writing. Non-Operator shall then have the right and option, for thirty (30) days after receipt of such

219

written notice, in which to elect to convert its said overriding royalty interest to an undivided one-half (1/2) leasehold, operating and working interest in and to said lease and land and well. If Non-Operator elects to so convert its overriding royalty interest to a working interest, Non-Operator, shall, upon its request, be entitled to receive from Operator a reassignment of one-half (1/2) of the interest previously assigned to Operator pursuant to Article 10.1 (a) (1) and/or Article 10.1 (b) (1) as aforesaid, free and clear of burdens created by or through Operator. The said overriding royalty interest shall cease and terminate as of the day on which Operator has recovered the aforesaid sums as computed under the Accounting Procedure attached hereto as Exhibit "C" to Exhibit "F"; and the leasehold, operating and working interest to which said overriding royalty interest is converted shall become effective at the same time. From and after the effective date of such conversion, Non-Operator shall own an undivided one-half (1/2) leasehold, operating and working interest in and to said lease and land and well, all equipment therein and production therefrom, and thereafter said well shall be operated under the terms and provisions of the Operating Agreement attached hereto as Exhibit "F" and made a part hereof. If the acreage for a test well is pooled to form a drilling unit, the overriding royalty interest and working interest to which it may be converted shall be proportionately reduced.

As to each of said test wells, if it is completed as a

producer, Operator, within sixty (60) days thereafter, shall
furnish Non-Operator with a full statement of the costs thereof.
Until Operator has recovered the sums aforesaid from said well,
Operator shall furnish to Non-Operator, once each month, a
statement of the cost of operating each of the said wells
and a monthly statement of the quantity of oil or gas produced
therefrom during the preceding month, together with the
amount of the proceeds received from the sale thereof.

12. INSURANCE:

Operator or Operator's contractors or subcontractors
shall carry the insurance hereinafter described with companies
satisfactory to Non-Operator to cover the drilling and
completing of, and all other operations hereunder with
respect to, said test wells, as follows:

(a) Workmen's Compensation Insurance: In compliance
with the Workmen's Compensation Act of the State
of Any State with employer's liability insurance.

(b) General Liability Insurance: Limits of $100,000
for injuries or death to one person and $300,000
for injuries or death in one accident; property
damage with limits of $100,000 and $250,000 aggregate.

(c) Automobile Public Liability Insurance: Limits of
$50,000 for injuries to or death to one person;
$100,000 for injuries or death in one accident;
property damage with limits of $10,000 each accident.

13. LAWS AND REGULATIONS:

Operator shall comply with and conduct its operations
hereunder in accordance with all applicable laws, ordinances,
rules, regulations and orders of all governmental authorities

221

having jurisdiction thereof; Operator shall indemnify and hold Non-Operator harmless from any and all liability which may arise from Operator's non-compliance therewith.

14. NOTICE:

Except as herein otherwise expressly provided, any notice or other communication required or permitted here-under shall be deemed to have been properly given or delivered when delivered personally or when sent by certified mail or telegraph, with all postage or charges fully prepaid, to Non-Operator at _____, or to Operator at _____. The date of service by mail shall be the date on which such written notice or other communication is deposited in the United States Post Office, addressed as above provided. Each party hereto shall have the right to change its address for all purposes of this Article by notifying the other party hereto thereof in writing.

IN WITNESS WHEREOF, the parties hereto have executed this instrument as of the day and year first above written.

SOME PETROLEUM COMPANY

By_____
Its Attorney in Fact

_____
Eager TeDrill

(Acknowledgments)

LEASE SCHEDULE

EXHIBIT "A" PAGE ___1___   STATE OF ___Any___ COUNTY OF ___Some___

Lease No.	Lessor	Lessee	Expiration Date	Description	Recorded Book Page	ORR

Initial Test Well Block

[Describe . . . .]

223

LEASE SCHEDULE

EXHIBIT "B" PAGE ___1___   STATE OF ___Any___   COUNTY OF ___Some___

Lease No.	Lessor	Lessee	Expiration Date	Description	Recorded Book Page	ORR
Option Well Block						
[Describe ·]						

224

EXHIBIT "C"

GEOLOGICAL REQUIREMENTS

Attached to and made a part of that certain Farmout
Contract by and between __Eager TeDrill_____

as Operator, and__Some Petroleum Company_____

as Non-Operator, covering certain lands in
County of____Some_____.
State of _Any_____.

1. Formation Samples: Operator shall give Non-Operator or Non-
erator's authorized representatives access to said test well and the derrick
oor at all reasonable hours and shall furnish to Non-Operator from said well
nsecutively taken samples of cuttings as follows:

[X] at__30__ foot intervals from base of surface casing to top
of_____Hawk_____formation;

[X] at__10__foot intervals from top of____Hawk_____
_____formation to total depth;

[ ] Other_____.
_____
_____

and samples of any cores taken through the services of a
reputable sample service, unless Non-Operator elects to take
such samples itself.

2. Formation Tests: Operator shall properly test each prospective oil or gas horizon and, upon encountering such horizon in the drilling of said te well, shall notify Non-Operator when such horizon is to be tested and shall allow Non-Operator sufficient time to have a representative present when such horizon is tested. If the information from any electrical formation survey, r either before or after contract depth has been reached, and considered by its or in conjunction with other indications or evidence from cuttings, cores or showings, makes the formation appear promising of being a prospective oil or gas horizon, Operator shall properly test such horizon if it was not adequately tested at the time it was penetrated.

3. Electrical Well Formation Survey: When said well has been drilled to the contract depth, Operator shall cause to be made the well formation surveys set forth below, and shall furnish to Non-Operator____2____field print(s) and____2____final print(s) plus____1____rolled sepia(s) of the following surveys, by a surveying concern satisfactory to Non-Operator:

    [X] Induction E. S.

    [ ] Gamma Ray Sonic [ ] with caliper OR [ ] Gamma

    Ray Density, with caliper

    [ ] Lateralog

    [ ] Microlateralog through zones of interest

    [X] Other: Compensated Formation Density Log -

       Gamma Ray with Caliper

Although Operator is not obligated hereunder to cause any other electric surveys to be made, if any such other surveys are made, Operator shall furnis to Non-Operator a like number of copies of the logs of such other surveys;

226

ovided, however, that Operator is not obligated to furnish any seismic

oustical or velocity survey log.

4. <u>Mud Logging Unit</u>: Operator ☐ shall ☒ shall not be required

use a:

☐ Portable type

☐ Manned type

mud logging unit on said test well from_____to

_____.

5. <u>Mud Requirements</u>:

☐ No requirement

☒ Crude oil shall not be added to mud

☐ _____

_____.

6. <u>Geophone or Similar Instruments</u>: If Non-Operator elects so to do,
erator shall permit Non-Operator, at its sole cost, expense and risk, to
wer a geophone, or other similar instrument, into said test well for the
rpose of making any test desired; provided, that Non-Operator shall reimburse
erator for any and all costs incurred by Operator as the result of Non-
erator's performing any such test.

7. <u>Non-Operator's Office</u>: All notices, samples, reports, copies of
gs and surveys and all other information and data required to be furnished to
n-Operator by Operator under the provisions of this exhibit
all be furnished to Non-Operator at the address set out in the contract
which this exhibit is attached; provided, that daily reports
all be furnished by telephone, telegraph or teletype.

227

_____ Day Phone_____
_____ Night Phone_____

or

_____ Day Phone_____
_____ Night Phone_____

228

EXHIBIT "D"

STATE OF ANY   )
                      : ss.
COUNTY OF SOME   )

A S S I G N M E N T

KNOW ALL MEN BY THESE PRESENTS:

That, in consideration of the sum of One Dollar ($1.00) and other good and valuable consideration, the receipt and sufficiency of which are hereby acknowledged, SOME PETROLEUM COMPANY, a Delaware corporation, hereinafter referred to as "Assignor" hereby does bargain, sell, assign, transfer and convey unto EAGER TeDRILL, his heirs, successors and assigns, hereinafter referred to as "Assignee", all its right, title and interest in and to the following referenced oil and gas lease or leases, covering lands situated in Some County, State of Any, to-wit:

(to be supplied)

insofar as said lease or leases cover the following described land in said County and State, to-wit:

(to be supplied)

to the stratigraphic equivalent of the (depth drilled in the applicable test well) (said lease or leases or interest therein and land hereinabove last referred to to the depth specified sometimes being referred to herein as "lease acreage"), subject to the following terms, convenants and conditions:

1. The lease acreage covered hereby is assigned by the Assignor and accepted by the Assignee subject to the overriding royalties, production payments, net profits obligations, carried working interests and other payments out of or with respect to production which are of record and with which said lease acreage is encumbered; and the Assignee hereby assumes and agrees to pay, perform or carry, as the case may be, each of said overriding royalty royalties, production payments, net profits obligations, carried working interests and other payments out of or with respect to production, to the extent that it is or remains a burden on the lease acreage herein assigned.

2. In addition to any and all other overriding royalties, to which said lease acreage may be subject, the Assignor hereby excepts and reserves unto itself, its successors or assigns, the following overriding royalty:

(a) One-sixteenth of eight-eighths (1/16 of 8/8) of all oil, distillate, condensate and other liquid hydrocarbons produced and saved from said lease acreage under said leases, or any extension or renewal thereof, which shall be delivered free of all cost and expense, except taxes on production, at the well or wells on said lease acreage or, at the Assignor's option, to the credit of the Assignor into the pipe line to which said well or wells may be connected;

(b) One-sixteenth of eight-eighths (1/16 of 8/8) of all gas and casinghead gas produced and saved from said lease acreage under said leases, or any extension or renewal thereof, the market value (at the well) of which shall be paid to the Assignor free of all cost and expense, except taxes on production.

3. With respect to the overriding royalty herein excepted and reserved by the Assignor, the Assignor and the Assignee agree, as follows:

(a) That oil and gas used in drilling and operations on said lease acreage and in the handling of production therefrom shall be deducted before said overriding royalty is computed.

(b) That the Assignee shall furnish to the Assignor authentic itemized monthly reports of all production from said lease acreage, such reports to be mailed not later than the fifteenth day of the month following that for which the report is made.

(c) That in the event any of said leases covers less than all the oil and gas mineral rights in and to the land described in said lease and covered by this assignment, said overriding royalty as to such lease and insofar as it pertains to oil, distillate, condensate, other liquid hydrocarbons, gas and casinghead gas, shall be proportionately reduced so as to be equal to that proportion of one-sixteenth of eight-eighths (1/16 of 8/8) of said production which the interest in and to said oil and gas mineral rights in the land described in said lease and covered by this assignment bears to the full and undivided oil and gas mineral estate therein.

(d) This assignment is hereby made subject to that certain Farmout Contract (with Operating Agreement attached) between the parties hereto dated the _____ day of _____, 1976 , covering the above described and other lands and the right of Assignor set forth therein to convert the overriding royalty reserved herein to an undivided one-half (1/2) working interest under the terms and conditions provided therein.

4. As to any wells drilled on said lease acreage by the Assignee after the delivery of this assignment, the Assignee shall give the Assignor notice when the location for said well is staked, access to said wells and the derrick floor at all reasonable times and, upon request of the Assignor, shall furnish to the Assignor well samples of all cores and cuttings consecutively taken, unless the Assignor elects to take such samples; and, at the request of the Assignor, the Assignee shall furnish to the Assignor copies of any electrical well formation surveys made.

5.   In the event that the Assignee should elect to surrender, abandon or release all or any of his rights in said lease acreage, or any part thereof, the Assignee shall notify the Assignor not less than thirty (30) days in advance of such surrender, abandonment or release and, if requested so to do by the Assignor, the Assignee immediately shall reassign such rights in said lease acreage, or such part thereof, to the Assignor.

6.   This assignment is made subject to all the terms and the express and implied covenants and conditions of said leases, to the extent of the rights hereby assigned, which terms, covenants and conditions the Assignee hereby assumes and agrees to perform with respect to the lands covered hereby.   Said terms, covenants and conditions, insofar as the said lease acreage is concerned, shall be binding on the Assignee, not only in favor of the lessors and their heirs, successors and assigns, but also in favor of the Assignor and its successors and assigns.

7.   This assignment is made without warranty of any kind.

8.   All notices, reports and other communications required or permitted hereunder, or desired to be given with respect to the rights or interests herein assigned or reserved, shall be deemed to have been properly given or delivered when delivered personally or when sent by certified mail or tele-graph, with all postage or charges fully prepaid, and

addressed to the Assignor and Assignee, respectively, as
follows:

Assignor: _____

_____

_____

Assignee: _____

_____

_____

9. The terms, covenants and conditions hereof shall be
binding upon, and shall inure to the benefit of, the Assignor
and the Assignee and their respective heirs, successors or
assigns; and such terms, covenants and conditions shall be
covenants running with the lands herein described and the
lease acreage herein assigned and with each transfer or
assignment of said land or lease acreage.

10. Assignor reserves and excepts unto itself, its
successors and assigns, the option and the exclusive right
at any time, at all times and from time to time, to purchase
all oil, gas, casinghead gas and other hydrocarbons produced and
saved from said lease acreage. Payment for any oil, distillate,
condensate and other liquid hydrocarbons purchased hereunder
shall be made at the highest prevailing price for production
of similar kind and quality prevailing in the area where
produced on date of delivery. Payment for gas and casinghead
gas purchased hereunder shall be made at the highest wellhead
price under contracts for the sale of production of similar
kind and quality prevailing in the area at the time such

option and right initially is exercised; provided that, if Assignor shall contract for the resale of such gas and casinghead gas at the wellhead, such payment shall be based upon the net proceeds accruing to Assignor at the wellhead under such contract. It is understood and agreed that the right to purchase hereby reserved and excepted may be assigned by Assignor at any time, at all times and from time to time without limitation.

TO HAVE AND TO HOLD said lease acreage unto the Assignee, his heirs, successors and assigns, subject to the terms, covenants and conditions hereinabove set forth.

EXECUTED, this_____ day of _____, 1976.

(Signatures)

(Acknowledgments)

STATE OF ANY   )
                      : ss.
COUNTY OF SOME   )

A S S I G N M E N T

KNOW ALL MEN BY THESE PRESENTS:

That, in consideration of the sum of One Dollar ($1.00) and other good and valuable consideration, the receipt and sufficiency of which are hereby acknowledged, SOME PETROLEUM COMPANY, a Delaware corporation, herinafter referred to as "Assignor", hereby does bargain, sell, assign, transfer and convey unto EAGER TeDRILL, his heirs, successors or assigns, hereinafter referred to as "Assignee", one-half (1/2) of its right, title and interest in and to the following referenced oil and gas lease or leases, covering lands situated in Some County, State of Any, to-wit:

(to be supplied)

insofar as said lease or leases cover the following described land in said County and State, to-wit:

(to be supplied)

to the stratigraphic equivalent of (the depth drilled in the applicable test well) (said lease or leases or interest therein and land hereinabove last referred to to the depth specified sometimes being referred to herein as "lease acreage"), subject to the following terms, covenants and conditions:

1.   The lease acreage covered hereby is assigned by the Assignor and accepted by the Assignee subject to the overriding royalties, production payments, net profits obligations, carried working interests and other payments out of or with respect to production which are of record and with which said lease acreage is encumbered; and the Assignee hereby assumes and agrees to pay, perform or carry, as the case may be, each of said overriding royalties, production payments, net profits obligations, carried working interests and other payments out of or with respect to production, to the extent that it is or remains a burden on the lease acreage herein assigned.

2.   As to any wells drilled on said lease acreage by the Assignee after the delivery of this assignment, the Assignee shall give the Assignor access to said wells and the derrick floor at all reasonable times and, upon request of the Assignor, shall furnish to the Assignor well samples of all cores and cuttings consecutively taken, unless the Assignor elects to take such samples; and, at the request of the Assignor, the Assignee shall furnish to the Assignor copies of any electrical well formation surveys made.

3. In the event that the Assignee should elect to surrender, abandon or release all or any part of his rights in said lease acreage, or any part thereof, the Assignee shall notify the Assignor not less than thirty (30) days in advance of such surrender, abandonment or release and, if requested so to do by the Assignor, the Assignee immediately shall reassign such rights in said lease acreage, or such part thereof, to the Assignor.

4. This assignment is made subject to all the terms and the express and implied covenants and conditions of said leases, to the extent of the rights hereby assigned, which terms, covenants and conditions the Assignee hereby assumes and agrees to perform with respect to the lands covered hereby. Said terms, covenants and conditions, insofar as the said lease acreage is concerned, shall be binding on the Assignee, not only in favor of the lessors and their heirs, successors and assigns, but also in favor of the Assignor and its successors and assigns.

5. This assignment is made without warranty of any kind.

6. All notices, reports and other communications required or permitted hereunder, or desired to be given with respect to the rights or interests herein assigned or reserved shall be deemed to have been properly given or delivered when delivered personally or when sent by certified mail or

telegraph, with all postage or charges fully prepaid, and addressed to the Assignor and Assignee, respectively, as follows:

Assignor: _____
_____
_____

Assignee: _____
_____
_____

7. The terms, covenants and conditions hereof shall be binding upon, and shall inure to the benefit of, the Assignor and the Assignee and their respective heirs, successors or assigns; and such terms, covenants and conditions shall be covenants running with the lands herein described and the lease acreage herein assigned and with each transfer or assignment of said land or lease acreage.

8. Assignor reserves and excepts unto itself, its successors and assigns, the option and the exclusive right at any time, at all times and from time to time, to purchase all oil, gas, casinghead gas and other hydrocarbons produced and saved from said lease acreage. Payment for any oil, distillate, condensate and other liquid hydrocarbons purchased hereunder shall be made at the highest prevailing price for production of similar kind and quality prevailing in the area where produced on date of delivery. Payment for gas and casinghead gas purchased hereunder shall be made at the highest wellhead price under contracts for the sale of production of similar

239

kind and quality prevailing in the area at the time such
option and right initially is exercised; provided that,
if Assignor shall contract for the resale of such gas and
casinghead gas at the wellhead, such payment shall be based
upon the net proceeds accruing to Assignor at the wellhead under
such contract. It is understood and agreed that the right
to purchase hereby reserved and excepted may be assigned by
Assignor at any time, at all times and from time to time without
limitation.

TO HAVE AND TO HOLD said lease acreage unto the Assignee,
his heirs, successors and assigns, subject to the terms,
covenants and conditions hereinabove set forth.

EXECUTED, this _____ day of _____, 1976.

(Signatures)

(Acknowledgments)

240

MODEL FORM OPERATING AGREEMENT—1956
Non-Federal Lands

OPERATING AGREEMENT

DATED

———————— ————, 19 _75_,

FOR UNIT AREA IN TOWNSHIP _34 NORTH_   , RANGE 15 and 16 EAST

_____SOME_____ COUNTY, STATE OF ___ANY___

Editor's note: For sample Operating Agreements, see Appendix C to this chapter.

REVENUE RULING 77-176 FARMOUT AGREEMENT
(with Optional Tax Partnership)

FARMOUT AGREEMENT

(Present Sublease and Assignment,
Additional Acreage, Optional
Partnership Return)

FROM:

TO:

Re: Farmout Agreement

Gentlemen:

This will evidence our agreement to assign and sublease to you certain interests in oil and gas properties subject to the terms and conditions set forth herein.

1. Subject Leases

We represent that we own or have the right to acquire, but do not warrant title to, the oil and gas leases
described on Exhibit A hereto, which, to the extent such
leases cover oil and gas only in and to the following tracts,
formations, and depths, are designated the "Subject Leases":

2. Assignment and Sublease

Upon your acceptance of this Agreement, subject
to the following terms and conditions, and without warranty
of title, either expressed or implied, we hereby sublease
and assign to you all of the operating interests in the Subject Leases insofar as such Leases are included in the
Spacing Unit for any Test Well and an undivided one-half of
the operating interests in the balance of the Subject Leases.

243

3. Term

   This assignment and sublease shall remain in effect until _____ ____, 19___, (the "Initial Term")
and so long thereafter as operations for the drilling, testing or completion of a Test Well continue with due diligence.
Should any Test Well be completed as a well capable of commercial production, this assignment and lease shall continue
in effect for so long as any of the Subject Leases continue
in effect and we will, upon your request, execute and deliver
to you evidence of this assignment and sublease in recordable
form reasonably satisfactory to you, but without warranty of
title, expressed or implied. Upon the expiration of the
terms of this assignment and sublease, all rights in the Subject Leases shall revert to us, free and clear of all liens,
burdens and obligations created by and through you or your
assigns.

4. Existing Burdens

   This sublease and assignment is subject to all
existing lease burdens, overrides and payments out of production relating to the Subject Leases, which obligations
you agree to assume.

5. Test Well

   ☐ You have the right and option,

   ☐ You agree,

before the expiration of the Initial Term of this Agreement,
to commence the drilling of a Test Well at a legal location
in the following tract:

to a depth sufficient to test the _____ formation
or to a depth of _____feet, whichever is the lesser.
 Such well shall be drilled, completed, and, if not completed
as a well capable of commercial production, abandoned at
your sole cost, expense and liability, in accordance with all
applicable laws and regulations and you agree to hold us
harmless from any cost or liability associated with such
drilling, completion and abandonment.

   In the event you encounter impenetrable substances
or other drilling conditions beyond your control making
further drilling of a Test Well impractical before reaching
the target depth, you may, at your option, commence actual
drilling of a substitute Test Well within 15 days from the
time such condition or substance is encountered, in which
event the substitute Test Well shall be considered the Test

244

Well and, for purposes of this Agreement, the drilling of the Test Well shall be considered to have continued.

Additional Test Wells may be drilled under the terms and conditions set forth in Exhibit ___hereto.

6. Reservation of Override

We reserve from the sublease and assignment with respect to the Spacing Unit for any Test Well only an overriding royalty of _____of all oil, gas, casinghead gas and liquid condensates, free of all costs of development and operations and free of all taxes except applicable production and severance taxes. Such overriding royalty shall be subject to proportionate reduction to the extent the Subject Leases do not comprise the entire working interest in the Spacing Unit for the Test Well, but shall be exclusive of and in addition to all presently existing lease burdens, overrides and payments out of production.

7. Conversion of Override to Working Interest

☐ We shall have the right and option at Payout to elect to convert our overriding royalty into a

☐ At Payout, our overriding royalty shall be automatically converted into a

____% leasehold working interest in the same properties, subject to proportionate reduction, together with a like interest in all casing, surface equipment and personal property used in connection with the Test Well. "Payout" shall mean the time when the net proceeds from production from the Test Well attributable to the interest assigned and subleased to you have equalled 100% of the cost and expense of drilling, equipping, testing and operating the Test Well attributable to such interest. Such costs shall be determined in accordance with the Operating Agreement hereinafter provided. You agree to notify us promptly upon the occurance of Payout. If the conversion is elective, we shall have 60 days from the receipt of such notice to give you notice of our election to convert our override into a working interest. A failure to elect shall be deemed a waiver of our election. The election, if made, shall be effective from the time of Payout.

8. Right to Test Well Information

During the drilling of each Test Well, our representative shall at all times have access to the well and to all cores, cuttings, depths, logs, and the like. Additional requirements are set forth in Exhibit ___hereto.

9. Spacing Unit

The term "Spacing Unit" shall mean the largest spacing assigned to a well by applicable regulatory body as applicable to any formation where commercial production is encountered as determined upon completion of the well.

If no spacing is assigned the spacing shall be _____ acres in the event gas production is encountered or _____ acres if no gas production is encountered, in each case to be in the form of a square or rectangular comprised of the Subject Leases most nearly surrounding the well.

10. Delay Rentals

During the Initial Term of this Agreement, we shall make delay rental payments at the times and in the amounts which, in our opinion, are necessary to maintain the Subject Leases in force; however, we shall not be liable to you in damages or otherwise for any inadvertent error or failure with respect to such payments. We shall be entitled to reimbursement from you, without reduction by reason of any depth limitation, for the full amount of delay rentals so paid but after allocation on a surface acre to the tracts included within the Subject Leases.

11. Operating Agreement

The relationship between us as co-owners of the Subject Leases, except to the extent inconsistent herewith, shall be governed by AAPL Form 610, Model Form Operating Agreement - 1956, with the following modifications. Section 12 (Operations by Less Than All Parties) shall provide for non-consent penalties in subparagraph (a) of ____ % and ____ % and in subparagraph (b) of 500% and 500% respectively. Section 18 (Preferential Right to Purchase) shall be deleted. The Operating Agreement shall contain a "casing point" election provision incorporating the notice and penalties of Section 12. The "individual loss" form of agreement shall be used. The limitation on expenditures in Section 11 shall be completed with the amounts of $ _____ and $ _____. Exhibit "A" shall be completed to set forth the Subject Leases, our percentage interests and our addresses all as set forth above. Section 3 of the Operating Agreement relating to the drilling of a test well, shall be stricken. As between the parties hereto you shall be designated as "Operator" under the Operating Agreement. The accounting procedure to be attached as Exhibit "C" shall be the COPAS 1974 Form of Accounting Procedure with the following modifications. In Section 3, paragraph 1A (Fixed Rate Basis) shall be indicated, Item lii shall be indicated as "shall not" and paragraph 1A(1) shall be completed with the following rates:

Well Depth	Drilling Well Rate	Producing Well Rate

Exhibit "D" to the Operating Agreement (Insurance) shall be completed as set forth below:

12. Production in Kind

The right of each co-owner to take in kind or separately dispose of his proportionate share of all oil and gas produced, as set forth in Section 13 of the Operating Agreement, shall be preserved. We represent that no contract has been made for the disposition of production from the tract of land covered by this agreement.

13. No Election Out of Subchapter K (Optional)

Section.26 of the Operating Agreement referred to above shall be superceded and the following provisions shall apply to the arrangement between the parties hereto consisting of the working interests in the Subject Leases assigned and subleased by us to you and the overriding royalty and working interests in the Subject Leases retained by us.

No election shall be made for this arrangement between the parties hereto to be excluded from the provisions of Subchapter K of Chapter 1 of Subtitle A of the Internal Revenue Code or similar provisions of state income tax law.

Operator is authorized to file Federal and, if necessary, state partnership income tax returns for this arrangement and to elect on such returns to deduct intangible drilling and development costs. Operator agrees to use its best efforts in preparing and filing such returns but shall incur no liability for any non willful failure with respect to such returns. Each party agrees to furnish Operator upon request with all information readily available needed for the preparation of such returns. Operator shall consult with each other party hereto in preparing such returns and in selecting a method of accounting and shall furnish a copy of such returns to each other party. Such returns shall be filed on a calendar year basis unless all of the parties to this arrangement use the same fiscal year other than a calendar year for Federal income tax purposes or

247

agree to request permission from the Commissioner of Internal Revenue to use another taxable year and such request is granted. The parties agree that the election to deduct intangible drilling and development costs when paid or incurred shall be made for this arrangement.

To the extent permitted by law, all deductions and credits, including, but not limited to, intangible drilling and development costs, depreciation, rental expenses, and the investment qualifying for the investment tax credit where applicable, shall be allocated to the party who has been charged with the expenditure giving rise to such deductions and credits; and to the extent permitted by law, such parties shall be entitled to such deductions and credits in computing taxable income or tax liabilities to the exclusion of any other party. It is agreed the tax basis of each oil and gas property for computation of cost depletion and gain or loss on disposition or abandonment shall be allocated and reallocated when necessary based upon the capital interest under this agreement as to such property and that the capital interest for such purpose as to each property shall be considered to be owned by the parties hereto in the ratio in which the expenditures giving rise to the tax basis of such property have been borne as of the end of each year.

Should any party hereto be deemed to have realized taxable income related to this arrangement before the sale of production, the deduction or tax basis of oil and gas properties resulting therefrom shall be allocated to such party.

Income from sale of production shall be allocated in accordance with the relative interests in the production. Gains from sale of assets shall be allocated in the ratio of the excess of sale proceeds received by each party over the adjusted basis of the property sold resulting from costs borne by such party (exclusive of the basis of oil and gas property separately allocated under the preceeding paragraph). Within the limits of the foregoing allocation, gain treated as ordinary income by reason of recapture of deductions shall be allocated in the ratio in which the related deductions were shared.

Should there be a transfer of an interest under this Agreement, income and deductions attributable to such interest shall not be allocated between the transferor and transferee in a prorata manner but shall be allocated according to the date the income was accrued and the date the expense was incurred.

248

14. Binding Effect

The terms of this Agreement shall extend to and shall be binding upon the parties hereto, their respective heirs, successors, legal representatives and assigns.

15. Non-Assignability

Your rights herein shall not be assignable without our written consent until at least one Test Well has been completed as a well capable of commercial production.

16. Acceptance

If the foregoing correctly sets forth your understanding of our agreement, please endorse your acceptance by returning within ____days from the date hereof (subject to extension at our option) one copy of this agreement properly executed by you or an authorized officer in the space provided.

BY:_____

Accepted this _____day of
_____, 197_.

BY:_____

USE OF TAX PARTNERSHIP EXHIBIT

This exhibit is designed for use with typical farm-out agreements where a fractional interest in acreage beyond the drill site is earned by drilling the test well. By allowing the partnership income tax provisions to apply, it should be possible to avoid the holding of Revenue Ruling 77-176 that the value of the outside acreage must be reported as taxable income. These items should be considered:

1. The Exhibit should be specifically incorporated by reference in the original farmout agreement. The Exhibit assumes either the 1958 or 1977 edition the AAPL Model Form Operating Agreement is attached to the Farmout Agreement.

2. If an overriding royalty is retained in the test well site by the assignor, it is best for this to be automatically converted to a working interest at payout, rather than being optional, so that the assignor will clearly be a member of the tax partnership and the test well will be a part of the tax partnership.

3. The name for the arrangement may be either a combination of the names of the parties plus the date of the agreement or the prospect name.

4. The paragraph relating to production in kind is intended to avoid a risk of the arrangement being classified as an association taxable as a corporation. If the assignor insists on a call on production and is not willing to include this protective paragraph, consider use of a formal partnership under the Uniform Partnership Act or modifications to the Operating Agreement to prohibit assignment of interests except with consent of the Operator.

5. The accounting method to be adopted is left open for future determination by the Operator in consultation with other parties. In most cases the accrual method should probably be used except where the drilling contract requires a prepayment and the parties paying the drilling costs wish to deduct such costs in the year of payment.

250

6. The farmout agreement should include a "present transfer" provision (not necessarily recordable) of the interests earned or be in sublease terms in order to eliminate any risk that the transfer for purposes of the "proven property" exclusion to percentage depletion would be upon completion of the test well rather than at the time of the farmout agreement

Exhibit_____

TAX PARTNERSHIP EXHIBIT

1. Recitation
This Tax Partnership Exhibit is attached to and made a part of that certain agreement, herein called the "Farmout Agreement" dated _____(which shall be the effective date of this Exhibit) between _____ herein called the "Assignee." _____ is herein referred to as the "Operator."

2. Name
The name _____ shall be applied to this arrangement for purposes of filing income tax informational returns.

3. Supercedes Operating Agreement
This Exhibit shall supercede the first paragraph of the Section of the Operating Agreement attached to or incorporated by reference in the Farmout Agreement headed "Provision Concerning Taxation" or "Internal Revenue Code Flection" as well as any other provision of the Farmout Agreement or Exhibits thereto inconsistent with this Exhibit.

4. Subchapter K to Apply
The parties hereto agree that no election to be excluded from the application of Subchapter K, Chapter 1, Subtitle A of the Internal Revenue Code of 1954 and similar provisions of state income tax laws is to be made for the arrangement set forth in the Farmout Agreement whereby Assignor is to transfer certain interests in oil and gas leases to Assignee retaining other such interests, operating and non operating, Assignee is to pay the costs of drilling certain wells, other costs are to be shared and each to receive specified revenues as the owner of operating or non operating interests.

5. Filing of Returns
Operator is authorized to file Federal and, if necessary, state partnership income tax returns for this arrangement and to elect on such returns to deduct intangible drilling and development costs. Operator agrees to use its best efforts in preparing and filing such returns but shall incur no liability for any non willful failure with respect to such returns. Each party agrees to furnish Operator upon request with all information readily available needed for the preparation of such returns. Operator shall consult with each other party hereto in preparing such

eturns and in selecting a method of accounting and shall
urnish a copy of such returns to each other party. Such
eturns shall be filed on a calendar year basis unless all
f the parties to this arrangement use the same fiscal year
ther than a calendar year for Federal income tax purposes
r agree to request permission from the Commissioner of
nternal Revenue to use another taxable year and such
equest is granted. The partners agree that the election
o deduct intangible drilling and development costs when
aid or incurred shall be made for this arrangement.

. Allocation of Deductions and Credits
ll deductions and credits for income tax purposes, including,
ut not limited to, intangible drilling and development
osts, depreciation, rental expenses and the investment
ualifying for the investment tax credit, shall be allocated
o the party who has been charged with the expenditure giving
se to such deductions and credits and, to the extent permitted
y law, such parties shall be entitled to such deductions and
redits in computing taxable income or tax liabilities to the
xclusion of any other party. It is agreed that the tax basis
f each oil and gas property for purposes of separate computation
y each party of cost depletion and gain or loss on disposition
r abandonment shall be allocated based upon the capital
nterests under this agreement as to such property and that
he capital interests as to each property shall be considered
o be owned by the parties hereto in the ratio in which the
xpenditures giving rise to the tax basis of such property
ave been borne. Should any party hereto be deemed to
ave realized taxable income related to this arrangement
efore the sale of production, the deduction or tax basis
f oil and gas properties resulting therefrom shall be
llocated to such party.

. Allocation of Income and Gains
ncome from sale of production shall be allocated in accordance
ith the relative interests in the production. Gains from
ale of assets shall be allocated in the ratio of the excess
f sale proceeds received by each party over the adjusted
asis of the property sold resulting from costs borne by
uch party (exclusive of the basis of oil and gas property
eparately allocated under the preceeding paragraph).
ithin the limits of the foregoing allocation, gain treated
s ordinary income by reason of recapture of deductions
hall be allocated in the ratio in which the related deductions
ere shared.

. Allocation in Event of Transfer
hould there be a transfer of an interest covered by this
xhibit, whether an entire interest, a fractional undivided

interest or a defined interest, income and deductions shall
be allocated between the transferor and transferee based on
the ownership of the interest on the date income was accrued
and the date expense was incurred and not in a pro rata
manner.

9. Production in Kind
The right of each party to take in kind or separately dispose
of his proportionate share of all oil and gas produced, as
set forth in the Operating Agreement to the extent necessary
to prevent this organization from having a "joint profit
objective" as described in IT 3930, 1948-2 Internal Revenue
Cumulative Bulletin 126 and IT 3948, 1949-1 Internal Revenue
Cumulative Bulletin 161, shall be preserved and nothing in
this Exhibit or in the Farmout Agreement shall be construed
to supercede such right.

10. Tenants in Common
The rights and obligation of the parties hereto shall be
several, not joint or collective, it being the express
purpose of the parties hereto that their ownership shall
be as tenants in common. Except for income tax purposes,
nothing herein shall be construed as creating a partnership
of any kind, joint venture, association or trust, or as
imposing upon any one or more of the parties hereto any
partnership obligation.

Appendix C

AAPL MODEL FORM OPERATING AGREEMENTS

AAPL MODEL FORM OPERATING AGREEMENT — 1956

A.A.P.L. FORM 610

MODEL FORM OPERATING AGREEMENT—1956

Non-Federal Lands

OPERATING AGREEMENT

DATED

_____, 19____,

UNIT AREA IN TOWNSHIP_____, RANGE _____,

_____ COUNTY, STATE OF_____.

AMERICAN ASSOCIATION OF PETROLEUM LANDMEN
APPROVED FORM.   A.A.P.L. NO. 610
MAY BE ORDERED DIRECTLY FROM THE PUBLISHER
KRAFTBILT PRODUCTS,   BOX 800, TULSA 74101

A.A.P.L. FORM 610

TABLE OF CONTENTS

OPERATING AGREEMENT

THIS AGREEMENT, entered into this_____ day of_____, 19_____, between
_____,

hereafter designated as "Operator", and the signatory parties other than Operator.

WITNESSETH, THAT:

WHEREAS, the parties to this agreement are owners of oil and gas leases covering and, if so indicated, unleased mineral interests in the tracts of land described in Exhibit "A", and all parties have reached an agreement to explore and develop these leases and interests for oil and gas to the extent and as hereinafter provided;

NOW, THEREFORE, it is agreed as follows:

1. DEFINITIONS

As used in this agreement, the following words and terms shall have the meanings here ascribed to them.

(1) The words "party" and "parties" shall always mean a party, or parties, to this agreement.

(2) The parties to this agreement shall always be referred to as "it" or "they", whether the parties be corporate bodies, partnerships, associations, or persons real.

(3) The term "oil and gas" shall include oil, gas, casinghead gas, gas condensate, and all other liquid or gaseous hydrocarbons, unless an intent to limit the inclusiveness of this term is specifically stated.

(4) The term "oil and gas interests" shall mean unleased fee and mineral interests in tracts of land lying within the Unit Area which are owned by parties to this agreement.

(5) The term "Unit Area" shall refer to and include all of the lands, oil and gas leasehold interests and oil and gas interests intended to be developed and operated for oil and gas purposes under this agreement. Such lands, oil and gas leasehold interests and oil and gas interests are described in Exhibit "A".

(6) The term "drilling unit" shall mean the area fixed for the drilling of one well by order or rule of any state or federal body having authority. If a drilling unit is not fixed by any such rule or order, a drilling unit shall be the drilling unit as established by the pattern of drilling in the Unit Area or as fixed by express agreement of the parties.

(7) All exhibits attached to this agreement are made a part of the contract as fully as though copied in full in the contract.

(8) The words "equipment" and "materials" as used here are synonymous and shall mean and include all oil field supplies and personal property acquired for use in the Unit Area.

2. TITLE EXAMINATION, LOSS OF LEASES AND OIL AND GAS INTERESTS

A. Title Examination:

Each party other than Operator shall promptly submit to Operator abstracts certified from beginning to recent date, together with all title papers in its possession covering leases and oil and gas interests which it is subjecting to this contract. All of these abstracts and title records shall be examined for the benefit of all parties by Operator's attorneys.

Operator shall promptly submit abstracts certified from beginning to recent date, together with all title papers in its possession covering leases and oil and gas interests which it is subjecting to this agreement, to _____ for examination by the latter's attorney for the benefit of all parties.

All title examinations shall be made without charge. Each examining attorney shall prepare a complete title report on each separate tract based upon the abstract record and title papers submitted to him. Each title report shall contain a list of fee owners and their interests, shall state the attorney's opinion concerning validity of their interests, and shall contain an enumeration and description of title defects, if any, a report upon mortgages, taxes, pending suits, and judgments, and unreleased oil and gas leases, and a list of requirements, if any, upon which the examiner's approval of title to the lease or oil and gas interest is contingent. The title report shall also contain a specific description of the oil and gas lease being subjected to this contract, with a statement of its form, term (which will be satisfactory if it has a primary term expiring not sooner than_____), amount of royalty, status of delay rental payments, and unusual drilling

"Joint Loss"

obligations and of excess royalty, oil payments, and other special burdens. A copy of each title opinion, and of each supplemental opinion, and of all final opinions, shall be sent promptly to each party. The opinion of the examining attorney concerning the validity of the title to each oil and gas interest and each lease, and the amount of interest covered thereby shall be binding and conclusive on the parties, but the acceptability of leases as to primary term, royalty provisions, drilling obligations, and special burdens, shall be a matter for approval and acceptance by an authorized representative of each party.

All title examinations shall be made, and title reports submitted, within a period of_____days after the submission of abstracts and title papers. Each party shall, in good faith, try to satisfy the requirements of the examining attorneys concerning its leases and interests, and each shall have a period of_____ days from receipt of title report for this purpose. If the title to any lease, or oil and gas interest, is finally rejected by the examining attorney, all parties shall then be asked to state in writing whether they will waive the title defects and accept the leases or interests, or whether they will stand on the attorney's opinion. If one or more parties refuse to waive title defects, this agreement shall, in that case, be terminated and abandoned, and all abstracts and title papers shall be returned to their senders. If all titles are approved by the examining attorneys, or are accepted by all parties, and if all leases are accepted as to primary terms, royalty provisions, drilling obligations and special burdens, all subsequent provisions of this agreement shall become operative immediately, and the parties shall proceed to their performance as they are hereinafter stated.

B. Failure of Title:

After all titles are approved or accepted, any defects of title that may develop shall be the joint responsibility of all parties and, if a title loss occurs, it shall be the loss of all parties, with each bearing its proportionate part of the loss and of any liabilities incurred in the loss. If such a loss occurs, there shall be no change in, or adjustment of, the interests of the parties in the remaining portion of the Unit Area.

C. Loss of Leases For Other Than Title Failure:

If any lease or interest subject to this agreement be lost through failure to develop or because express or implied covenants have not been performed, or if any lease be permitted to expire at the end of its primary term and not be renewed or extended, the loss shall not be considered a failure of title and all such losses shall be joint losses and shall be borne by all parties in proportion to their interests and there shall be no readjustment of interests in the remaining portion of the Unit Area.

3. UNLEASED OIL AND GAS INTERESTS

If any party owns an unleased oil and gas interest in the Unit Area, that interest shall be treated for the purpose of this agreement as if it were a leased interest under the form of oil and gas lease attached as Exhibit "B" and for the primary term therein stated. As to such interests, the owner shall receive royalty on production as prescribed in the form of oil and gas lease attached hereto as Exhibit "B". Such party shall, however, be subject to all of the provisions of this agreement relating to lessees, to the extent that it owns the lessee interest.

4. INTERESTS OF PARTIES

Exhibit "A" lists all of the parties, and their respective percentage or fractional interests under this agreement. Unless changed by other provisions, all costs and liabilities incurred in operations under this contract shall be borne and paid, and all equipment and material acquired in operations on the Unit Area shall be owned, by the parties as their interests are given in Exhibit "A". All production of oil and gas from the Unit Area, subject to the payment of lessor's royalties, shall also be owned by the parties in the same manner.

If the interest of any party in any oil and gas lease covered by this agreement is subject to an overriding royalty, production payment, or other charge over and above the usual one-eigthh (⅛) royalty, such party shall assume and alone bear all such excess obligations and shall account for them to the owners thereof out of its share of the working interest production of the Unit Area.

5. OPERATOR OF UNIT

_____shall be the Operator of the Unit Area, and shall conduct and direct and have full control of all operations on the Unit Area as permitted and required by, and within the limits of, this agreement. It shall conduct all such operations in a good and workmanlike manner, but it shall have no liability as Operator to the other parties for losses sustained, or liabilities incurred, except such as may result from gross negligence or from breach of the provisions of this agreement.

6. EMPLOYEES

The number of employees and their selection, and the hours of labor and the compensation for services performed, shall be determined by Operator. All employees shall be the employees of Operator.

7. TEST WELL

On or before the_____ day of_____, 19_____, Operator shall commence the drilling of a well for oil and gas in the following location:

and shall thereafter continue the drilling of the well with due diligence to

unless granite or other practically impenetrable substance is encountered at a lesser depth or unless all parties agree to complete the well at a lesser depth.

Operator shall make reasonable tests of all formations encountered during drilling which give indication of containing oil and gas in quantities sufficient to test, unless this agreement shall be limited in its application to a specific formation or formations, in which event Operator shall be required to test only the formation or formations to which this agreement may apply.

If in Operator's judgment the well will not produce oil or gas in paying quantities, and it wishes to plug and abandon the test as a dry hole, it shall first secure the consent of all parties to the plugging, and the well shall then be plugged and abandoned as promptly as possible.

8. COSTS AND EXPENSES

Except as herein otherwise specifically provided, Operator shall promptly pay and discharge all costs and expenses incurred in the development and operation of the Unit Area pursuant to this agreement and shall charge each of the parties hereto with their respective proportionate shares upon the cost and expense basis provided in the Accounting Procedure attached hereto and marked Exhibit "C". If any provision of Exhibit "C" should be inconsistent with any provision contained in the body of this agreement, the provisions in the body of this agreement shall prevail.

Operator, at its election, shall have the right from time to time to demand and receive from the other parties payment in advance of their respective shares of the estimated amount of the costs to be incurred in operations hereunder during the next succeeding month, which right may be exercised only by submission to each such party of an itemized statement of such estimated costs, together with an invoice for its share thereof. Each such statement and invoice for the payment in advance of estimated costs shall be submitted on or before the 20th day of the next preceding month. Each party shall pay to Operator its proportionate share of such estimate within fifteen (15) days after such estimate and invoice is received. If any party fails to pay its share of said estimate within said time, the amount due shall bear interest at the rate of six percent (6%) per annum until paid. Proper adjustment shall be made monthly between advances and actual cost, to the end that each party shall bear and pay its proportionate share of actual costs incurred, and no more.

Revised 1967

9. OPERATOR'S LIEN

Operator is given a first and preferred lien on the interest of each party covered by this contract, and in each party's interest in oil and gas produced and the proceeds thereof, and upon each party's interest in material and equipment, to secure the payment of all sums due from each such party to Operator.

In the event any party fails to pay any amount owing by it to Operator as its share of such costs and expense or such advance estimate within the time limited for payment thereof, Operator, without prejudice to other existing remedies, is authorized, at its election, to collect from the purchaser or purchasers of oil or gas, the proceeds accruing to the working interest or interests in the Unit Area of the delinquent party up to the amount owing by such party, and each purchaser of oil or gas is authorized to rely upon Operator's statement as to the amount owing by such party.

In the event of the neglect or failure of any non-operating party to promptly pay its proportionate part of the cost and expense of development and operation when due, the other non-operating parties and Operator, within thirty (30) days after the rendition of statements therefor by Operator, shall proportionately contribute to the payment of such delinquent indebtedness and the non-operating parties so contributing shall be entitled to the same lien rights as are granted to Operator in this section. Upon the payment by such delinquent or defaulting party to Operator of any amount or amounts on such delinquent indebtedness, or upon any recovery on behalf of the non-operating parties under the lien conferred above, the amount or amounts so paid or recovered shall be distributed and paid by Operator to the other non-operating parties and Operator proportionately in accordance with the contributions theretofore made by them.

10. TERM OF AGREEMENT

This agreement shall remain in full force and effect for as long as any of the oil and gas leases subjected to this agreement remain or are continued in force as to any part of the Unit Area, whether by production, extension, renewal or otherwise; provided, however, that in the event the first well drilled hereunder results in a dry hole and no other well is producing oil or gas in paying quantities from the Unit Area, then at the end of ninety (90) days after abandonment of the first test well, this agreement shall terminate unless one or more of the parties are then engaged in drilling a well or wells pursuant to Section 12 hereof, or all parties have agreed to drill an additional well or wells under this agreement, in which event this agreement shall continue in force until such well or wells shall have been drilled and completed. If production results therefrom this agreement shall continue in force thereafter as if said first test well had been productive in paying quantities, but if production in paying quantities does not result therefrom this agreement shall terminate at the end of ninety (90) days after abandonment of such well or wells. It is agreed, however, that the termination of this agreement shall not relieve any party hereto from any liability which has accrued or attached prior to the date of such termination.

11. LIMITATION ON EXPENDITURES

Without the consent of all parties: (a) No well shall be drilled on the Unit Area except any well expressly provided for in this agreement and except any well drilled pursuant to the provisions of Section 12 of this agreement, it being understood that the consent to the drilling of a well shall include consent to all necessary expenditures in the drilling, testing, completing, and equipping of the well, including necessary tankage; (b) No well shall be reworked, plugged back or deepened except a well reworked, plugged back or deepened pursuant to the provisions of Section 12 of this agreement, it being understood that the consent to the reworking, plugging back or deepening of a well shall include consent to all necessary expenditures in conducting such operations and completing and equipping of said well to produce, including necessary tankage; (c) Operator shall not undertake any single project reasonably estimated to require an expenditure in excess of_____Dollars ($_____) except in connection with a well the drilling, reworking, deepening, or plugging back of which has been previously authorized by or pursuant to this agreement; provided, however, that in case of explosion, fire, flood, or other sudden emergency, whether of the same or different nature, Operator may take such steps and incur such expenses as in its opinion are required to deal with the emergency and to safeguard life and property, but Operator shall, as promptly as possible, report the emergency to the other parties. Operator shall, upon request, furnish copies of its "Authority for Expenditures" for any single project costing in excess of $_____.

12. OPERATIONS BY LESS THAN ALL PARTIES

If all the parties cannot mutually agree upon the drilling of any well on the Unit Area other than the test well provided for in Section 7, or upon the reworking, deepening or plugging back of a dry hole drilled at the joint expense of all parties or a well jointly owned by all the parties and not then producing in paying quantities on the Unit Area, any party or parties wishing to drill, rework, deepen or plug back such a well may give the other parties written notice of the proposed operation, specifying the work to be performed, the location, proposed depth, objective formation and the estimated cost of the operation. The parties receiving such a notice shall have thirty (30) days (except as to reworking, plugging back or drilling deeper, where a drilling rig is on location, the period shall be limited to forty-eight (48) hours exclusive of Saturday or Sunday) after receipt of the notice within which to notify the parties wishing to do the work whether they elect to participate in the cost of the proposed operation. Failure of a party receiving such a notice to so reply to it within the period above fixed shall constitute an election by that party not to participate in the cost of the proposed operation.

If any party receiving such a notice elects not to participate in the proposed operation (such party or parties being hereafter referred to as "Non-Consenting Party"), then in order to be entitled to the benefits of this section, the party or parties giving the notice and such other parties as shall elect to participate in the operation (all such parties being hereafter referred to as the "Consenting Parties") shall, within thirty (30) days after the expiration of the notice period of thirty (30) days (or as promptly as possible after the expiration of the 48-hour period where the drilling rig is on location, as the case may be) actually commence work on the proposed operation and complete it with due diligence.

The entire cost and risk of conducting such operations shall be borne by the Consenting Parties in the proportions that their respective interests as shown in Exhibit "A" bear to the total interests of all Consenting Parties. Consenting Parties shall keep the leasehold estates involved in such operations free and clear of all liens and encumbrances of every kind created by or arising from the operations of the Consenting Parties. If such an operation results in a dry hole, the Consenting Parties shall plug and abandon the well at their sole cost, risk and expense. If any well drilled, reworked, deepened or plugged back under the provisions of this section results in a producer of oil and/or gas in paying quantities, the Consenting Parties shall complete and equip the well to produce at their sole cost and risk, and the well shall then be turned over to Operator and shall be operated by it at the expense and for the account of the Consenting Parties. Upon commencement of operations for the drilling, reworking, deepening or plugging back of any such well by Consenting Parties in accordance with the provisions of this section, each Non-Consenting Party shall be deemed to have relinquished to Consenting Parties, and the Consenting Parties shall own and be entitled to receive, in proportion to their respective interests, all of such Non-Consenting Party's interest in the well, its leasehold operating rights, and share of production therefrom until the proceeds or market value thereof (after deducting production taxes, royalty, overriding royalty and other interests payable out of or measured by the production from such well accruing with respect to such interest until it reverts) shall equal the total of the following:

(A) 100% of each such Non-Consenting Party's share of the cost of any newly acquired surface equipment beyond the wellhead connections (including, but not limited to, stock tanks, separators, treaters, pumping equipment and piping), plus 100% of each such Non-Consenting Party's share of the cost of operation of the well commencing with first production and continuing until each such Non-Consenting Party's relinquished interest shall revert to it under other provisions of this section, it being agreed that each Non-Consenting Party's share of such costs and equipment will be that interest which would have been chargeable to each Non-Consenting Party had it participated in the well from the beginning of the operation; and

(B) 200% of that portion of the costs and expenses of drilling, reworking, deepening or plugging back, testing and completing, after deducting any cash contributions received under Section 25, and 200% of that portion of the cost of newly acquired equipment in the well (to and including the wellhead connections), which would have been chargeable to such Non-Consenting Party if it had participated therein.

In the case of any reworking, plugging back or deeper drilling operation, the Consenting Parties shall be permitted to use, free of cost, all casing, tubing and other equipment in the well, but the ownership of all such equipment shall remain unchanged; and upon abandonment of a well after such reworking, plugging back or deeper drilling, the Consenting Parties shall account for all such equipment to the owners thereof, with each party receiving its proportionate part in kind or in value.

Within sixty (60) days after the completion of any operation under this section, the party conducting the operations for the Consenting Parties shall furnish each Non-Consenting Party with an inventory of the equipment in and connected to the well, and an itemized statement of the cost of drilling, deepening, plugging back, testing, completing, and equipping the well for production; or, at its option, the operating party, in lieu of an itemized statement of such costs of operation, may submit a detailed statement of monthly billings. Each month thereafter, during the time the Consenting Parties are being reimbursed as provided above, the Consenting Parties shall furnish the Non-Consenting Parties with an itemized statement of all costs and liabilities incurred in the operation of the well, together with a statement of the quantity of oil and gas produced from it and the amount of proceeds realized from the sale of the well's working interest production during the preceding month. Any amount realized from the sale or other disposition of equipment newly acquired in connection with any such operation which would have been owned by a Non-Consenting Party had it participated therein shall be credited against the total unreturned costs of the work done and of the equipment purchased, in determining when the interest of such Non-Consenting Party shall revert to it as above provided; if there is a credit balance it shall be paid to such Non-Consenting Party.

If and when the Consenting Parties recover from a Non-Consenting Party's relinquished interest the amounts provided for above, the relinquished interests of such Non-Consenting Party shall automatically revert to it and from and after such reversion such Non-Consenting Party shall own the same interest in such well, the operating rights and working interest therein, the material and equipment in or pertaining thereto, and the production therefrom as such Non-Consenting Party would have owned had it participated in the drilling, reworking, deepening or plugging back of said well. Thereafter, such Non-Consenting Party shall be charged with and shall pay its proportionate part of the further costs of the operation of said well in accordance with the terms of this agreement and the accounting procedure schedule, Exhibit "C", attached hereto.

Notwithstanding the provisions of this Section 12, it is agreed that without the mutual consent of all parties, no wells shall be completed in or produced from a source of supply from which a well located elsewhere on the Unit Area is producing, unless such well conforms to the then-existing well spacing pattern for such source of supply.

The provisions of this section shall have no application whatsoever to the drilling of the initial test well on the Unit Area, but shall apply to the reworking, deepening, or plugging back of the initial test well after it has been drilled to the depth specified in Section 7, if it is, or thereafter shall prove to be, a dry hole or non-commercial well, and to all other wells drilled, reworked, deepened, or plugged back, or proposed to be drilled, reworked, deepened, or plugged back, upon the Unit Area subsequent to the drilling of the initial test well.

13. RIGHT TO TAKE PRODUCTION IN KIND

Each party shall take in kind or separately dispose of its proportionate share of all oil and gas produced from the Unit Area, exclusive of production which may be used in development and producing operations and in preparing and treating oil for marketing purposes and production unavoidably lost. Each party shall pay or deliver, or cause to be paid or delivered, all royalties, overriding royalties, or other payments due on its share of such production, and shall hold the other parties free from any liability therefor. Any extra expenditure incurred in the taking in kind or separate disposition by any party of its proportionate share of the production shall be borne by such party.

Each party shall execute all division orders and contracts of sale pertaining to its interest in production from the Unit Area, and shall be entitled to receive payment direct from the purchaser or purchasers thereof for its share of all production.

In the event any party shall fail to make the arrangements necessary to take in kind or separately dispose of its proportionate share of the oil and gas produced from the Unit Area, Operator shall have the right, subject to revocation at will by the party owning it, but not the obligation, to purchase such oil and gas or sell it to others for the time being, at not less than the market price prevailing in the area, which shall in no event be less than the price which Operator receives for its portion of the oil and gas produced from the Unit Area. Any such purchase or sale by Operator shall be subject always to the right of the owner of the production to exercise at any time its right to take in kind, or separately dispose of, its share of all oil and gas not previously delivered to a purchaser. Notwithstanding the foregoing, Operator shall not make a sale into interstate commerce of any other party's share of gas production without first giving such other party sixty (60) days notice of such intended sale.

14. ACCESS TO UNIT AREA

Each party shall have access to the Unit Area at all reasonable times, at its sole risk, to inspect or observe operations, and shall have access at reasonable times to information pertaining to the development or operation thereof, including Operator's books and records relating thereto. Operator shall, upon request, furnish each of the other parties with copies of all drilling reports, well logs, tank tables, daily gauge and run tickets and reports of stock on hand at the first of each month, and shall make available samples of any cores or cuttings taken from any well drilled on the Unit Area.

15. DRILLING CONTRACTS

All wells drilled on the Unit Area shall be drilled on a competitive contract basis at the usual rates prevailing in the area. Operator, if it so desires, may employ its own tools and equipment in the drilling of wells, but its charges therefor shall not exceed the prevailing rates in the field, and the rate of such charges shall be agreed upon by the parties in writing before drilling operations are commenced, and such work shall be performed by Operator under the same terms and conditions as shall be customary and usual in the field in contracts of independent contractors who are doing work of a similar nature.

16. ABANDONMENT OF WELLS

No well, other than any well which has been drilled or reworked pursuant to Section 12 hereof for which the Consenting Parties have not been fully reimbursed as therein provided, which has been completed as a producer shall be plugged and abandoned without the consent of all parties; provided, however, if all parties do not agree to the abandonment of any well, those wishing to continue its operation shall tender to each of the other parties its proportionate share of the value of the well's salvable material and equipment, determined in accordance with the provisions of Exhibit "C", less the estimated cost of salvaging and the estimated cost of plugging and abandoning. Each abandoning party shall then assign to the non-abandoning parties, without warranty, express or implied, as to title or as to quantity, quality, or fitness for use of the equipment and material, all of its interest in the well and its equipment, together with its interest in the leasehold estate as to, but only as to, the interval or intervals of the formation or formations then open to production. The assignments so limited shall encompass the "drilling unit" upon which the well is located. The payments by, and the assignments to, the assignees shall be in a ratio based upon the relationship of their respective percentages of participation in the Unit Area to the aggregate of the percentages of participation in the Unit Area of all assignees. There shall be no readjustment of interest in the remaining portion of the Unit Area.

After the assignment, the assignors shall have no further responsibility, liability, or interest in the operation of or production from the well in the interval or intervals then open. Upon request of the assignees, Operator shall continue to operate the assigned well for the account of the non-abandoning parties at the rates and charges contemplated by this agreement, plus any additional cost and charges which may arise as the result of the separate ownership of the assigned well.

17. DELAY RENTALS AND SHUT-IN WELL PAYMENTS

Each party shall pay all delay rentals and shut-in well payments which may be required under the terms of its lease or leases and submit evidence of each payment to the other parties at least ten (10) days prior to the payment date. The paying party shall be reimbursed by Operator for 100% of any such delay rental payment and 100% of any such shut-in well payment. The amount of such reimbursement shall be charged by Operator to the joint account of the parties and treated in all respects the same as costs incurred in the development and operation of the Unit Area. Each party responsible for such payments shall diligently attempt to make proper payment, but shall not be held liable to the other parties in damages for the loss of any lease or interest therein if, through mistake or oversight, any rental or shut-in well payment is not paid or is erroneously paid. The loss of any lease or interest therein which results from a failure to pay or an erroneous payment of rental or shut-in well payment shall be a joint loss and there shall be no readjustment of interests in the remaining portion of the Unit Area. If any party secures a new lease covering the terminated interest, such acquisiton shall be subject to the provisions of Section 23 of this agreement.

Operator shall promptly notify each other party hereto of the date on which any gas well located on the Unit Area is shut in and the reason therefor.

18. PREFERENTIAL RIGHT TO PURCHASE

Should any party desire to sell all or any part of its interests under this contract, or its rights and interests in the Unit Area, it shall promptly give written notice to the other parties, with full information concerning its proposed sale, which shall include the name and address of the prospective purchaser (who must be ready, willing and able to purchase), the purchase price, and all other terms of the offer. The other parties shall then have an optional prior right, for a period of ten (10) days after receipt of the notice, to purchase on the same terms and conditions the interest which the other party proposes to sell; and, if this optional right is exercised, the purchasing parties shall share the purchased interest in the proportions that the interest of each bears to the total interest of all purchasing parties. However, there shall be no preferential right to purchase in those cases where any party wishes to mortgage its interests, or to dispose of its interests by merger, reorganization, consolidation, or sale of all of its assets, or a sale or transfer of its interests to a subsidiary or parent company, or subsidiary of a parent company, or to any company in which any one party owns a majority of the stock.

19. SELECTION OF NEW OPERATOR

Should a sale be made by Operator of its rights and interests, the other parties shall have the right within sixty (60) days after the date of such sale, by majority vote in interest, to select a new Operator. If a new Operator is not so selected, the transferee of the present Operator shall assume the duties of and act as Operator. In either case, the retiring Operator shall continue to serve as Operator, and discharge its duties in that capacity under this agreement, until its successor Operator is selected and begins to function, but the present Operator shall not be obligated to continue the performance of its duties for more than 120 days after the sale of its rights and interests has been completed.

"Joint Loss"

20. MAINTENANCE OF UNIT OWNERSHIP

For the purpose of maintaining uniformity of ownership in the oil and gas leasehold interests covered by this contract, and notwithstanding any other provisions to the contrary, no party shall sell, encumber, transfer or make other disposition of its interest in the leases embraced within the Unit Area and in wells, equipment and production unless such disposition covers either:

(1) the entire interest of the party in all leases and equipment and production; or

(2) an equal undivided interest in all leases and equipment and production in the Unit Area.

Every such sale, encumbrance, transfer or other disposition made by any party shall be made expressly subject to this agreement, and shall be made without prejudice to the rights of the other parties.

If at any time the interest of any party is divided among and owned by four or more co-owners, Operator may, at its discretion, require such co-owners to appoint a single trustee or agent with full authority to receive notices, approve expenditures, receive billings for and approve and pay such party's share of the joint expenses, and to deal generally with, and with power to bind, the co-owners of such party's interests within the scope of the operations embraced in this contract; however, all such co-owners shall enter into and execute all contracts or agreements for the disposition of their respective shares of the oil and gas produced from the Unit Area and they shall have the right to receive, separately, payment of the sale proceeds thereof.

21. RESIGNATION OF OPERATOR

Operator may resign from its duties and obligations as Operator at any time upon written notice of not less than ninety (90) days given to all other parties. In this case, all parties to this contract shall select by majority vote in interest, not in numbers, a new Operator who shall assume the responsibilities and duties, and have the rights, prescribed for Operator by this agreement. The retiring Operator shall deliver to its successor all records and information necessary to the discharge by the new Operator of its duties and obligations.

22. LIABILITY OF PARTIES

The liability of the parties shall be several, not joint or collective. Each party shall be responsible only for its obligations, and shall be liable only for its proportionate share of the costs of developing and operating the Unit Area. Accordingly, the lien granted by each party to Operator in Section 9 is given to secure only the debts of each severally. It is not the intention of the parties to create, nor shall this agreement be construed as creating, a mining or other partnership or association, or to render them liable as partners.

23. RENEWAL OR EXTENSION OF LEASES

If any party secures a renewal of any oil and gas lease subject to this contract, each and all of the other parties shall be notified promptly, and shall have the right to participate in the ownership of the renewal lease by paying to the party who acquired it their several proper proportionate shares of the acquisition cost, which shall be in proportion to the interests held at that time by the parties in the Unit Area.

If some, but less than all, of the parties elect to participate in the purchase of a renewal lease, it shall be owned by the parties who elect to participate therein, in a ratio based upon the relationship of their respective percentage of participation in the unit area to the aggregate of the percentages of participation in the unit area of all parties participating in the purchase of such renewal lease. Any renewal lease in which less than all the parties elect to participate shall not be subject to this agreement.

Each party who participates in the purchase of a renewal lease shall be given an assignment of its proportionate interest therein by the acquiring party.

The provisions of this section shall apply to renewal leases whether they are for the entire interest covered by the expiring lease or cover only a portion of its area or an interest therein. Any renewal lease taken before the expiration of its predecessor lease, or taken or contracted for within six (6) months after the expiration of the existing lease shall be subject to this provision; but any lease taken or contracted for more than six (6) months after the expiration of an existing lease shall not be deemed a renewal lease and shall not be subject to the provisions of this section.

The provisions in this section shall apply also and in like manner to extensions of oil and gas leases.

24. SURRENDER OF LEASES

The leases covered by this agreement, in so far as they embrace acreage in the Unit Area, shall not be surrendered in whole or in part unless all parties consent.

However, should any party desire to surrender its interest in any lease or in any portion thereof, and other parties not agree or consent, the party desiring to surrender shall assign, without express or implied warranty of title, all of its interest in such lease, or portion thereof, and any well, material and equipment which may be located thereon and any rights in production thereafter secured, to the parties not desiring to surrender it. Upon such assignment, the assigning party shall be relieved from all obligations thereafter accruing, but not theretofore accrued, with respect to the acreage assigned and the operation of any well thereon, and the assigning party shall have no further interest in the lease assigned and its equipment and production. The parties assignee shall pay to the party assignor the reasonable salvage value of the latter's interest in any wells and equipment on the assigned acreage, determined in accordance with the provisions of Exhibit "C", less the estimated cost of salvaging and the estimated cost of plugging and abandoning. If the assignment is in favor of more than one party, the assigned interest shall be shared by the parties assignee in the proportions that the interest of each bears to the interest of all parties assignee.

Any assignment or surrender made under this provision shall not reduce or change the assignors' or surrendering parties' interest, as it was immediately before the assignment, in the balance of the Unit Area; and the acreage assigned or surrendered, and subsequent operations thereon, shall not thereafter be subject to the terms and provisions of this agreement.

25. ACREAGE OR CASH CONTRIBUTIONS

If any party receives while this agreement is in force a contribution of cash toward the drilling of a well or any other operation on the Unit Area, such contribution shall be paid to the party who conducted the drilling or other operation and shall be applied by it against the cost of such drilling or other operation. If the contribution be in the form of acreage, the party to whom the contribution is made shall promptly execute an assignment of the acreage, without warranty of title, to all parties to this agreement in proportion to their interests in the Unit Area at that time, and such acreage shall become a part of the Unit Area and be governed by all the provisions of this contract. Each party shall promptly notify all other parties of all acreage or money contributions it may obtain in support of any well or any other operation on the Unit Area.

26. PROVISION CONCERNING TAXATION

Each of the parties hereto elects, under the authority of Section 761(a) of the Internal Revenue Code of 1954, to be excluded from the application of all of the provisions of Subchapter K of Chapter 1 of Subtitle A of the Internal Revenue Code of 1954. If the income tax laws of the state or states in which the property covered hereby is located contain, or may hereafter contain, provisions similar to those contained in the Subchapter of the Internal Revenue Code of 1954 above referred to under which a similar election is permitted, each of the parties agrees that such election shall be exercised. Each party authorizes and directs the Operator to execute such an election or elections on its behalf and to file the election with the proper governmental office or agency. If requested by the Operator so to do, each party agrees to execute and join in such an election.

Operator shall render for ad valorem taxation all property subject to this agreement which by law should be rendered for such taxes, and it shall pay all such taxes assessed thereon before they become delinquent. Operator shall bill all other parties for their proportionate share of all tax payments in the manner provided in Exhibit "C".

If any tax assessment is considered unreasonable by Operator, it may at its discretion protest such valuation within the time and manner prescribed by law, and prosecute the protest to a final determination, unless all parties agree to abandon the protest prior to final determination. When any such protested valuation shall have been finally determined, Operator shall pay the assessment for the joint account, together with interest and penalty accrued, and the total cost shall then be assessed against the parties, and be paid by them, as provided in Exhibit "C".

27. INSURANCE

At all times while operations are conducted hereunder, Operator shall comply with the Workmen's Compensation Law of the State where the operations are being conducted. Operator shall also carry or provide insurance for the benefit of the joint account of the parties as may be outlined in Exhibit "D" attached to and made a part hereof. Operator shall require all contractors engaged in work on or for the Unit Area to comply with the Workmen's Compensation Law of the State where the operations are being conducted and to maintain such other insurance as Operator may require.

In the event Automobile Public Liability Insurance is specified in said Exhibit "D", or subsequently receives the approval of the parties, no direct charge shall be made by Operator for premiums paid for such insurance for operator's fully owned automotive equipment.

28. CLAIMS AND LAWSUITS

If any party to this contract is sued on an alleged cause of action arising out of operations on the Unit Area, or on an alleged cause of action involving title to any lease or oil and gas interest subjected to this contract, it shall give prompt written notice of the suit to the Operator and all other parties.

The defense of lawsuits shall be under the general direction of a committee of lawyers representing the parties, with Operator's attorney as Chairman. Suits may be settled during litigation only with the joint consent of all parties. No charge shall be made for services performed by the staff attorneys for any of the parties, but otherwise all expenses incurred in the defense of suits, together with the amount paid to discharge any final judgment, shall be considered costs of operation and shall be charged to and paid by all parties in proportion to their then interests in the Unit Area. Attorneys, other than staff attorneys for the parties, shall be employed in lawsuits involving Unit Area operations only with the consent of all parties; if outside counsel is employed, their fees and expenses shall be considered Unit Area expense and shall be paid by Operator and charged to all of the parties in proportion to their then interests in the Unit Area. The provisions of this paragraph shall not be applied in any instance where the loss which may result from the suit is treated as an individual loss rather than a joint loss under prior provisions of this agreement, and all such suits shall be handled by and be the sole responsibility of the party or parties concerned.

Damage claims caused by and arising out of operations on the Unit Area, conducted for the joint account of all parties, shall be handled by Operator and its attorneys, the settlement of claims of this kind shall be within the discretion of Operator so long as the amount paid in settlement of any one claim does not exceed one thousand ($1000.00) dollars and, if settled, the sums paid in settlement shall be charged as expense to and be paid by all parties in proportion to their then interests in the Unit Area.

29. FORCE MAJEURE

If any party is rendered unable, wholly or in part, by force majeure to carry out its obligations under this agreement, other than the obligation to make money payments, that party shall give to all other parties prompt written notice of the force majeure with reasonably full particulars concerning it; thereupon, the obligations of the party giving the notice, so far as they are affected by the force majeure, shall be suspended during, but no longer than, the continuance of the force majeure. The affected party shall use all possible diligence to remove the force majeure as quickly as possible.

The requirement that any force majeure shall be remedied with all reasonable dispatch shall not require the settlement of strikes, lockouts, or other labor difficulty by the party involved, contrary to its wishes; how all such difficulties shall be handled shall be entirely within the discretion of the party concerned.

The term "force majeure" as here employed shall mean an act of God, strike, lockout, or other industrial disturbance, act of the public enemy, war, blockade, public riot, lightning, fire, storm, flood, explosion, governmental restraint, unavailability of equipment, and any other cause, whether of the kind specifically enumerated above or otherwise, which is not reasonably within the control of the party claiming suspension.

30. NOTICES

All notices authorized or required between the parties, and required by any of the provisions of this agreement, shall, unless otherwise specifically provided, be given in writing by United States mail or Western Union Telegram, postage or charges prepaid, and addressed to the party to whom the notice is given at the addresses listed on Exhibit "A". The originating notice to be given under any provision hereof shall be deemed given only when received by the party to whom such notice is directed and the time for such party to give any notice in response thereto shall run from the date the originating notice is received. The second or any responsive notice shall be deemed given when deposited in the United States mail or with the Western Union Telegraph Company, with postage or charges prepaid. Each party shall have the right to change its address at any time, and from time to time, by giving written notice thereof to all other parties.

31. OTHER CONDITIONS, IF ANY, ARE!

This agreement may be signed in counterpart, and shall be binding upon the parties and upon their heirs, successors, representatives and assigns.

ATTEST:

_____   _____

OPERATOR

ATTEST:

_____   _____

ATTEST:

_____   _____

FOR GUIDANCE IN THE PREPARATION OF THIS CONTRACT, THE EXHIBITS TO BE ATTACHED TO IT, AND THEIR CONTENTS, ARE AS SHOWN BELOW:

Exhibit "A"— (1) (a) Lands subject to contract.

(b) Restrictions, if any, as to formations or depths.

(c) Drilling Unit for first well.

(Found in   (2) Percentage or fractional interests of parties under agreement.

Sections 1   (3) Leasehold and unleased interests of each party.

and 4)   (4) Addresses of parties to which notices should be sent.

Exhibit "B"— Form of lease.

(Found in
Section 3)

Exhibit "C"— Accounting Procedure
(Found in
Section 8)

Exhibit "D"— Insurance to be carried.
(Found in
Section 26)

BLANKS TO BE COMPLETED ARE FOUND
ON THE PAGES SHOWN BELOW:

Page   1— Date and name of Operator.

Page   1 "Joint Loss"—   Company to examine Operator's title and expiration date of lease submitted.

Page   2 "Joint Loss"—   Two blanks, number of days for submission of title reports and in which to satisfy requirements.

Page   3— Name of Operator, commencement date of test well and its location.

Page   3— Objective depth or formation of test well.

Page   4— Limitation of expenditure of Operator for single project.

Page   4— Amount of single project for which Operator may be called upon to furnish copies of its Authority for Expenditures.

Page   8— If reimbursement for less than 100% of delay rentals, strike 100% and insert proper percentage.

Page   12— Other conditions, if any.

OPERATING AGREEMENT

THIS AGREEMENT, entered into this_____ day of_____, 19____, between

_____

hereafter designated as "Operator", and the signatory parties other than Operator.

WITNESSETH, THAT:

WHEREAS, the parties to this agreement are owners of oil and gas leases covering and, if so indicated, unleased mineral interests in the tracts of land described in Exhibit "A", and all parties have reached an agreement to explore and develop these leases and interests for oil and gas to the extent and as hereinafter provided;

NOW, THEREFORE, it is agreed as follows:

1. DEFINITIONS

As used in this agreement, the following words and terms shall have the meanings here ascribed to them.

(1) The words "party" and "parties" shall always mean a party, or parties, to this agreement.

(2) The parties to this agreement shall always be referred to as "it" or "they", whether the parties be corporate bodies, partnerships, associations, or persons real.

(3) The term "oil and gas" shall include oil, gas, casinghead gas, gas condensate, and all other liquid or gaseous hydrocarbons, unless an intent to limit the inclusiveness of this term is specifically stated.

(4) The term "oil and gas interests" shall mean unleased fee and mineral interests in tracts of land lying within the Unit Area which are owned by parties to this agreement.

(5) The term "Unit Area" shall refer to and include all of the lands, oil and gas leasehold interests and oil and gas interests intended to be developed and operated for oil and gas purposes under this agreement. Such lands, oil and gas leasehold interests and oil and gas interests are described in Exhibit "A".

(6) The term "drilling unit" shall mean the area fixed for the drilling of one well by order or rule of any state or federal body having authority. If a drilling unit is not fixed by any such rule or order, a drilling unit shall be the drilling unit as established by the pattern of drilling in the Unit Area or as fixed by express agreement of the parties.

(7) All exhibits attached to this agreement are made a part of the contract as fully as though copied in full in the contract.

(8) The words "equipment" and "materials" as used here are synonymous and shall mean and include all oil field supplies and personal property acquired for use in the Unit Area.

2. TITLE EXAMINATION, LOSS OF LEASES AND OIL AND GAS INTERESTS

A. Title Examination:

There shall be no examination of title to leases, or to oil and gas interests, except that title to the lease covering the land upon which the exploratory well is to be drilled in accordance with Section 7, shall be examined on a complete abstract record by Operator's attorney, and the title to both the oil and gas lease and to the fee title of the lessors must be approved by the examining attorney, and accepted by all parties. A copy of the examining attorney's opinion shall be sent to each party immediately after the opinion is written, and, also, each party shall be given, as they are written, a copy of all subsequent supplemental attorney's reports. A good faith effort to satisfy the examining attorney's requirements shall be made by the party owning the lease covering the drillsite.

If title to the proposed drillsite is not approved by the examining attorney or the lease is not acceptable for a material reason, and all the parties do not accept the title, the parties shall select a new drillsite for the first exploratory well; provided, if the parties are unable to agree upon another drillsite, this agreement shall, in that case, come to an end and all parties shall then forfeit their rights and be relieved of obligations hereunder. If a new drillsite is selected, title to the oil and gas lease covering it and to the fee title of the lessor shall be examined, and title shall be approved or accepted or rejected in like manner as provided above concerning the drillsite first selected. If title to the oil and gas lease covering the second choice drillsite is not approved or accepted, other drillsites shall be successively selected and title examined, until a drillsite is chosen

— 1 —

to which title is approved or accepted, or until the parties fail to select another drillsite. As in the case of the drillsite first selected, so also with successive choices if the time comes that the parties have not approved title and are unable to agree upon an alternate drillsite, the contract shall, in that case and at that time, come to an end and all parties shall forfeit their rights and be relieved of obligations under this contract.

No well other than the first test shall be drilled in the Unit Area until after (1) the title to the lease covering the lands upon which such well is to be located has been examined by Operator's attorney, and (2) the title has been approved by the examining attorney and the title has been accepted by all of the parties who are to participate in the drilling of the well.

B. Failure of Title:

Should any oil and gas lease, or interest therein, be lost through failure of title, this agreement shall, nevertheless, continue in force as to all remaining leases and interests, and

(1) The party whose lease or interest is affected by the title failure shall bear alone the entire loss and it shall not be entitled to recover from Operator or the other parties any development or operating costs which it may have theretofore paid, but there shall be no monetary liability on its part to the other parties hereto by reason of such title failure; and

(2) There shall be no retroactive adjustment of expenses incurred or revenues received from the operation of the interest which has been lost, but the interests of the parties shall be revised on an acreage basis, as of the time it is determined finally that title failure has occurred, so that the interest of the party whose lease or interest is affected by the title failure will thereafter be reduced in the Unit Area by the amount of the interest lost; and

(3) If the proportionate interests of the other parties hereto in any producing well theretofore drilled on the Unit Area is increased by reason of the title failure, the party whose title has failed shall receive the proceeds attributable to the increase in such interests (less operating costs attributable thereto) until it has been reimbursed for unrecovered costs paid by it in connection with such well; and

(4) Should any person not a party to this agreement, who is determined to be the owner of any interest in the title which has failed, pay in any manner any part of the cost of operation, development, or equipment, or equipment previously paid under this agreement, such amount shall be proportionately paid to the party or parties hereto who in the first instance paid the costs which are so refunded; and

(5) Any liability to account to a third party for prior production of oil and gas which arises by reason of title failure shall be borne by the party or parties whose title failed in the same proportions in which they shared in such prior production.

C. Loss of Leases for Causes Other Than Title Failure:

If any lease or interest subject to this agreement be lost through failure to develop or because express or implied covenants have not been performed, or if any lease be permitted to expire at the end of its primary term and not be renewed or extended, or if any lease or interest therein is lost due to the fact that the production therefrom is shut in by reason of lack of market, the loss shall not be considered a failure of title and all such losses shall be joint losses and shall be borne by all parties in proportion to their interests and there shall be no readjustment of interests in the Unit Area.

3. UNLEASED OIL AND GAS INTERESTS

If any party owns an unleased oil and gas interest in the Unit Area, that interest shall be treated for the purpose of this agreement as if it were a leased interest under the form of oil and gas lease attached as "Exhibit "B" and for the primary term therein stated. As to such interests, the owner shall receive royalty on production as prescribed in the form of oil and gas lease attached hereto as Exhibit "B" Such party shall, however, be subject to all of the provisions of this agreement relating to lessees, to the extent that it owns the lessee interest.

4. INTERESTS OF PARTIES

Exhibit "A" lists all of the parties, and their respective percentage or fractional interests under this agreement. Unless changed by other provisions, all costs and liabilities incurred in operations under this contract shall be borne and paid, and all equipment and material acquired in operations on the Unit Area shall be owned, by the parties as their interests are given in Exhibit "A". All production of oil and gas from the Unit Area, subject to the payment of lessor's royalties, shall also be owned by the parties in the same manner.

"Individual Loss"
Revised 1967

17. DELAY RENTALS AND SHUT-IN WELL PAYMENTS

Delay rentals and shut-in well payments which may be required under the terms of any lease shall be paid by the party who has subjected such lease to this agreement, at its own expense. Proof of each payment shall be given to Operator at least ten (10) days prior to the rental or shut-in well payment date. Operator shall furnish similar proof to all other parties concerning payments it makes in connection with its leases. Any party may request, and shall be entitled to receive, proper evidence of all such payments. If, through mistake or oversight, any delay rental or shut-in well payment is not paid or is erroneously paid, and as a result a lease or interest therein terminates, there shall be no monetary liability against the party who failed to make such payment. Unless the party who failed to pay a rental or shut-in well payment secures a new lease covering the same interest within ninety (90) days from the discovery of the failure to make proper payment, the interests of the parties shall be revised on an acreage basis effective as of the date of termination of the lease involved, and the party who failed to make proper payment will no longer be credited with an interest in the Unit Area on account of the ownership of the lease which has terminated. In the event the party who failed to pay the rental or the shut-in well payment shall not have been fully reimbursed, at the time of the loss, from the proceeds of the sale of oil and gas attributable to the lost interest, calculated on an acreage basis, for the development and operating costs theretofore paid on account of such interest, it shall be reimbursed for unrecovered actual costs theretofore paid by it (but not for its share of the cost of any dry hole previously drilled or wells previously abandoned) from so much of the following as is necessary to effect reimbursement:

(1) Proceeds of oil and gas, less operating expenses, theretofore accrued to the credit of the lost interest, on an acreage basis, up to the amount of unrecovered costs;

(2) proceeds, less operating expenses thereafter incurred attributable to the lost interest on an acreage basis, of that portion of oil and gas thereafter produced and marketed (excluding production from any wells thereafter drilled) which would, in the absence of such lease termination, be attributable to the lost interest on an acreage basis, up to the amount of unrecovered costs, the proceeds of said portion of the oil and gas to be contributed by the other parties in proportion to their respective interests; and

(3) any moneys, up to the amount of unrecovered costs, that may be paid by any party who is, or becomes, the owner of the interest lost, for the privilege of participating in the Unit Area or becoming a party to this contract.

Operator shall attempt to notify all parties when a gas well is shut-in or returned to production, but assumes no liability whatsoever for failure to do so.

18. PREFERENTIAL RIGHT TO PURCHASE

Should any party desire to sell all or any part of its interests under this contract, or its rights and interests in the Unit Area, it shall promptly give written notice to the other parties, with full information concerning its proposed sale, which shall include the name and address of the prospective purchaser (who must be ready, willing and able to purchase), the purchase price, and all other terms of the offer. The other parties shall then have an optional prior right, for a period of ten (10) days after receipt of the notice, to purchase on the same terms and conditions the interest which the other party proposes to sell; and, if this optional right is exercised, the purchasing parties shall share the purchased interest in the proportions that the interest of each bears to the total interest of all purchasing parties. However, there shall be no preferential right to purchase in those cases where any party wishes to mortgage its interests, or to dispose of its interests by merger, reorganization, consolidation, or sale of all of its assets, or a sale or transfer of its interests to a subsidiary or parent company, or subsidiary of a parent company, or to any company in which any one party owns a majority of the stock.

19. SELECTION OF NEW OPERATOR

Should a sale be made by Operator of its rights and interests, the other parties shall have the right within sixty (60) days after the date of such sale, by majority vote in interest, to select a new Operator. If a new Operator is not so selected, the transferee of the present Operator shall assume the duties of and act as Operator. In either case, the retiring Operator shall continue to serve as Operator, and discharge its duties in that capacity under this agreement, until its successor Operator is selected and begins to function, but the present Operator shall not be obligated to continue the performance of its duties for more than 120 days after the sale of its rights and interests has been completed.

"Individual Loss"

A.A.P.L. FORM 610 - 1977

MODEL FORM OPERATING AGREEMENT

Use of this identifying mark is prohibited
except when authorized in writing by the
American Association of Petroleum Landmen

OPERATING AGREEMENT

DATED

_____ , 19____ ,

OPERATOR _____

CONTRACT AREA _____

_____

_____

COUNTY OR PARISH OF _____ STATE OF _____

GUIDANCE IN THE PREPARATION OF THIS AGREEMENT:

1. Title Page - Fill in blank as applicable.

2. Preamble, Page 1 - Name of Operator.

3. Article II - Exhibits:
   (a) Indicate Exhibits to be attached.
   (b) If it is desired that no reference be made to Non-discrimination, the reference to Exhibit "F" should be deleted.

4. Article IV.A - Title Examination - Select option as agreed to by the parties.

5. Article IV.B - Loss of Title - If "Joint Loss" of Title is desired, the following changes should be made:
   (a) Delete Articles IV.B.1 and IV.B.2.
   (b) Article IV.B.3 - Delete phrase "other than those set forth in Articles IV.B.1 and IV.B.2 above."
   (c) Article VII.F. - Change reference at end of the first grammatical paragraph from "Article IV.B.2" to "Article IV.B.3."

6. Article V - Operator - Enter name of Operator.

7. Article VI.A - Initial Well:
   (a) Date of commencement of drilling.
   (b) Location of well.
   (c) Obligation depth.

8. Article VI.B.2.(b) - Subsequent Operations - Enter penalty percentage as agreed to by parties.

9. Article VII.D.1. - Limitation of Expenditures - Select option as agreed to by parties.

10. Article VII.D.3. - Limitation of Expenditures - Enter limitation of expenditure of Operator for single project and amount above which Operator may furnish information AFE.

11. Article VII.E. - Royalties, Overriding Royalties and Other Payments - Enter royalty fraction as agreed to by parties.

12. Article X. - Claims and Lawsuits - Enter claim limit as agreed to by parties.

13. Article XIII. - Term of Agreement:
    (a) Select Option as agreed to by parties.
    (b) If Option No. 2 is selected, enter agreed number of days in two (2) blanks.

14. Signature Page - Enter effective date.

TABLE OF CONTENTS

OPERATING AGREEMENT

THIS AGREEMENT, entered into by and between _____
_____ , hereinafter designated and
referred to as "Operator", and the signatory party or parties other than Operator, sometimes hereinafter
referred to individually herein as "Non-Operator", and collectively as "Non-Operators",

WITNESSETH:

WHEREAS, the parties to this agreement are owners of oil and gas leases and/or oil and gas interests in the land identified in Exhibit "A", and the parties hereto have reached an agreement to explore
and develop these leases and/or oil and gas interests for the production of oil and gas to the extent and
as hereinafter provided:

NOW, THEREFORE, it is agreed as follows:

ARTICLE I.
DEFINITIONS

As used in this agreement, the following words and terms shall have the meanings here ascribed
to them:

A. The term "oil and gas" shall mean oil, gas, casinghead gas, gas condensate, and all other liquid
or gaseous hydrocarbons and other marketable substances produced therewith, unless an intent to
limit the inclusiveness of this term is specifically stated.

B. The terms "oil and gas lease", "lease" and "leasehold" shall mean the oil and gas leases covering tracts of land lying within the Contract Area which are owned by the parties to this agreement.

C. The term "oil and gas interests" shall mean unleased fee and mineral interests in tracts of
land lying within the Contract Area which are owned by parties to this agreement.

D. The term "Contract Area" shall mean all of the lands, oil and gas leasehold interests and oil
and gas interests intended to be developed and operated for oil and gas purposes under this agreement.
Such lands, oil and gas leasehold interests and oil and gas interests are described in Exhibit "A".

E. The term "drilling unit" shall mean the area fixed for the drilling of one well by order or rule
of any state or federal body having authority. If a drilling unit is not fixed by any such rule or order,
a drilling unit shall be the drilling unit as established by the pattern of drilling in the Contract Area
or as fixed by express agreement of the Drilling Parties.

F. The term "drillsite" shall mean the oil and gas lease or interest on which a proposed well is to
be located.

G. The terms "Drilling Party" and "Consenting Party" shall mean a party who agrees to join in
and pay its share of the cost of any operation conducted under the provisions of this agreement.

H. The terms "Non-Drilling Party" and "Non-Consenting Party" shall mean a party who elects
not to participate in a proposed operation.

Unless the context otherwise clearly indicates, words used in the singular include the plural, the
plural includes the singular, and the neuter gender includes the masculine and the feminine.

ARTICLE II.
EXHIBITS

The following exhibits, as indicated below and attached hereto, are incorporated in and made a
part hereof:

☐ A. Exhibit "A", shall include the following information:
    (1) Identification of lands subject to agreement,
    (2) Restrictions, if any, as to depths or formations,
    (3) Percentages or fractional interests of parties to this agreement,
    (4) Oil and gas leases and/or oil and gas interests subject to this agreement,
    (5) Addresses of parties for notice purposes.
☐ B. Exhibit "B", Form of Lease.
☐ C. Exhibit "C", Accounting Procedure.
☐ D. Exhibit "D", Insurance.
☐ E. Exhibit "E", Gas Balancing Agreement.
☐ F. Exhibit "F", Non-Discrimination and Certification of Non-Segregated Facilities.

If any provision of any exhibit, except Exhibit "E", is inconsistent with any provision contained
in the body of this agreement, the provisions in the body of this agreement shall prevail.

ARTICLE III.
INTERESTS OF PARTIES

A. Oil and Gas Interests:

If any party owns an unleased oil and gas interest in the Contract Area, that interest shall be treated for the purpose of this agreement and during the term hereof as if it were a leased interest under the form of oil and gas lease attached as Exhibit "B". As to such interest, the owner shall receive royalty on production as prescribed in the form of oil and gas lease attached hereto as Exhibit "B". Such party shall, however, be subject to all of the provisions of this agreement relating to lessees, to the extent that it owns the lessee interest.

B. Interest of Parties in Costs and Production:

Exhibit "A" lists all of the parties and their respective percentage or fractional interests under this agreement. Unless changed by other provisions, all costs and liabilities incurred in operations under this agreement shall be borne and paid, and all equipment and material acquired in operations on the Contract Area shall be owned by the parties as their interests are shown in Exhibit "A". All production of oil and gas from the Contract Area, subject to the payment of lessor's royalties which will be borne by the Joint Account, shall also be owned by the parties in the same manner during the term hereof; provided, however, this shall not be deemed an assignment or cross-assignment of interests covered hereby.

ARTICLE IV.
TITLES

A. Title Examination:

Title examination shall be made on the drillsite of any proposed well prior to commencement of drilling operations or, if the Drilling Parties so request, title examination shall be made on the leases and/or oil and gas interests included, or planned to be included, in the drilling unit around such well. The opinion will include the ownership of the working interest, minerals, royalty, overriding royalty and production payments under the applicable leases. At the time a well is proposed, each party contributing leases and/or oil and gas interests to the drillsite, or to be included in such drilling unit, shall furnish to Operator all abstracts (including Federal Lease Status Reports), title opinions, title papers and curative material in its possession free of charge. All such information not in the possession of or made available to Operator by the parties, but necessary for the examination of title, shall be obtained by Operator. Operator shall cause title to be examined by attorneys on its staff or by outside attorneys. Copies of all title opinions shall be furnished to each party hereto. The cost incurred by Operator in this title program shall be borne as follows:

☐ Option No. 1: Costs incurred by Operator in procuring abstracts and title examination (including preliminary, supplemental, shut-in gas royalty opinions and division order title opinions) shall be a part of the administrative overhead as provided in Exhibit "C," and shall not be a direct charge, whether performed by Operator's staff attorneys or by outside attorneys.

☐ Option No. 2: Costs incurred by Operator in procuring abstracts and fees paid outside attorneys for title examination (including preliminary, supplemental, shut-in gas royalty opinions and division order title opinions) shall be borne by the Drilling Parties in the proportion that the interest of each Drilling Party bears to the total interest of all Drilling Parties as such interests appear in Exhibit "A". Operator shall make no charge for services rendered by its staff attorneys or other personnel in the performance of the above functions.

Each party shall be responsible for securing curative matter and pooling amendments or agreements required in connection with leases or oil and gas interests contributed by such party. The Operator shall be responsible for the preparation and recording of Pooling Designations or Declarations as well as the conduct of hearings before Governmental Agencies for the securing of spacing or pooling orders. This shall not prevent any party from appearing on its own behalf at any such hearing.

No well shall be drilled on the Contract Area until after (1) the title to the drillsite or drilling unit has been examined as above provided, and (2) the title has been approved by the examining attorney or title has been accepted by all of the parties who are to participate in the drilling of the well.

B. Loss of Title:

1. _Failure of Title:_ Should any oil and gas interest or lease, or interest therein, be lost through failure of title, which loss results in a reduction of interest from that shown on Exhibit "A", this agreement, nevertheless, shall continue in force as to all remaining oil and gas leases and interests, and
   (a) The party whose oil and gas lease or interest is affected by the title failure shall bear alone the entire loss and it shall not be entitled to recover from Operator or the other parties any development

- 2 -

1   or operating costs which it may have theretofore paid, but there shall be no monetary liability on its
2   part to the other parties hereto for drilling, development, operating or other similar costs by reason of
3   such title failure; and

4   (b) There shall be no retroactive adjustment of expenses incurred or revenues received from the
5   operation of the interest which has been lost, but the interests of the parties shall be revised on an acre-
6   age basis, as of the time it is determined finally that title failure has occurred, so that the interest of
7   the party whose lease or interest is affected by the title failure will thereafter be reduced in the Contract
8   Area by the amount of the interest lost; and

9   (c) If the proportionate interest of the other parties hereto in any producing well theretofore drilled
10 on the Contract Area is increased by reason of the title failure, the party whose title has failed shall
11 receive the proceeds attributable to the increase in such interests (less costs and burdens attributable
12 thereto) until it has been reimbursed for unrecovered costs paid by it in connection with such well;
13 and

14   (d) Should any person not a party to this agreement, who is determined to be the owner of any in-
15 terest in the title which has failed, pay in any manner any part of the cost of operation, development,
16 or equipment, such amount shall be paid to the party or parties who bore the costs which are so refund-
17 ed; and

18   (e) Any liability to account to a third party for prior production of oil and gas which arises by
19 reason of title failure shall be borne by the party or parties in the same proportions in which they shared
20 in such prior production; and

21   (f) No charge shall be made to the joint account for legal expenses, fees or salaries, in connection
22 with the defense of the interest claimed by any party hereto, it being the intention of the parties
23 hereto that each shall defend title to its interest and bear all expenses in connection therewith.
24

25   2. <u>Loss by Non-Payment or Erroneous Payment of Amount Due:</u> If, through mistake or oversight,
26 any rental, shut-in well payment, minimum royalty or royalty payment, is not paid or is erroneously
27 paid, and as a result a lease or interest therein terminates, there shall be no monetary liability against
28 the party who failed to make such payment. Unless the party who failed to make the required payment
29 secures a new lease covering the same interest within ninety (90) days from the discovery of the fail-
30 ure to make proper payment, which acquisition will not be subject to Article VIII.B., the interests of
31 the parties shall be revised on an acreage basis, effective as of the date of termination of the lease in-
32 volved, and the party who failed to make proper payment will no longer be credited with an interest in
33 the Contract Area on account of ownership of the lease or interest which has terminated. In the event
34 the party who failed to make the required payment shall not have been fully reimbursed, at the time of
35 the loss, from the proceeds of the sale of oil and gas attributable to the lost interest, calculated on an
36 acreage basis, for the development and operating costs theretofore paid on account of such interest, it
37 shall be reimbursed for unrecovered actual costs theretofore paid by it (but not for its share of the
38 cost of any dry hole previously drilled or wells previously abandoned) from so much of the following
39 as is necessary to effect reimbursement:

40   (a) Proceeds of oil and gas, less operating expenses, theretofore accrued to the credit of the lost
41 interest, on an acreage basis, up to the amount of unrecovered costs;

42   (b) Proceeds, less operating expenses, thereafter accrued attributable to the lost interest on an
43 acreage basis, of that portion of oil and gas thereafter produced and marketed (excluding production
44 from any wells thereafter drilled) which, in the absence of such lease termination, would be attributable
45 to the lost interest on an acreage basis, up to the amount of unrecovered costs, the proceeds of said
46 portion of the oil and gas to be contributed by the other parties in proportion to their respective in-
47 terests; and

48   (c) Any monies, up to the amount of unrecovered costs, that may be paid by any party who is, or
49 becomes, the owner of the interest lost, for the privilege of participating in the Contract Area or be-
50 coming a party to this agreement.
51

52   3. <u>Other Losses:</u> All losses incurred, other than those set forth in Articles IV.B.1. and IV.B.2.
53 above, shall not be considered failure of title but shall be joint losses and shall be borne by all parties
54 in proportion to their interests. There shall be no readjustment of interests in the remaining portion of
55 the Contract Area.
56
57   **ARTICLE V.**
58   **OPERATOR**
59
60 **A. DESIGNATION AND RESPONSIBILITIES OF OPERATOR:**
61
62 _____ shall be the
63 Operator of the Contract Area, and shall conduct and direct and have full control of all operations on
64 the Contract Area as permitted and required by, and within the limits of, this agreement. It shall con-
65 duct all such operations in a good and workmanlike manner, but it shall have no liability as Operator
66 to the other parties for losses sustained or liabilities incurred, except such as may result from gross
67 negligence or willful misconduct.
68
69
70

B. Resignation or Removal of Operator and Selection of Successor:

1. Resignation or Removal of Operator: Operator may resign at any time by giving written notice thereof to Non-Operators. If Operator terminates its legal existence, no longer owns an interest in the Contract Area, or is no longer capable of serving as Operator, it shall cease to be Operator without any action by Non-Operator, except the selection of a successor. Operator may be removed if it fails or refuses to carry out its duties hereunder, or becomes insolvent, bankrupt or is placed in receivership, by the affirmative vote of two (2) or more Non-Operators owning a majority interest based on ownership as shown on Exhibit "A", and not on the number of parties remaining after excluding the voting interest of Operator. Such resignation or removal shall not become effective until 7:00 o'clock A.M. on the first day of the calendar month following the expiration of ninety (90) days after the giving of notice of resignation by Operator or action by the Non-Operators to remove Operator, unless a successor Operator has been selected and assumes the duties of Operator at an earlier date. Operator, after effective date of resignation or removal, shall be bound by the terms hereof as a Non-Operator. A change of a corporate name or structure of Operator or transfer of Operator's interest to any single subsidiary, parent or successor corporation shall not be the basis for removal of Operator.

2. Selection of Successor Operator: Upon the resignation or removal of Operator, a successor Operator shall be selected by the Parties. The successor Operator shall be selected from the parties owning an interest in the Contract Area at the time such successor Operator is selected. If the Operator that is removed fails to vote or votes only to succeed itself, the successor Operator shall be selected by the affirmative vote of two (2) or more parties owning a majority interest based on ownership as shown on Exhibit "A", and not on the number of parties remaining after excluding the voting interest of the Operator that was removed.

C. Employees:

The number of employees used by Operator in conducting operations hereunder, their selection, and the hours of labor and the compensation for services performed, shall be determined by Operator, and all such employees shall be the employees of Operator.

D. Drilling Contracts:

All wells drilled on the Contract Area shall be drilled on a competitive contract basis at the usual rates prevailing in the area. If it so desires, Operator may employ its own tools and equipment in the drilling of wells, but its charges therefor shall not exceed the prevailing rates in the area and the rate of such charges shall be agreed upon by the parties in writing before drilling operations are commenced, and such work shall be performed by Operator under the same terms and conditions as are customary and usual in the area in contracts of independent contractors who are doing work of a similar nature.

ARTICLE VI.
DRILLING AND DEVELOPMENT

A. Initial Well:

On or before the _____ day of _____ , 19 __ , Operator shall commence the drilling of a well for oil and gas at the following location:

and shall thereafter continue the drilling of the well with due diligence to

unless granite or other practically impenetrable substance or condition in the hole, which renders further drilling impractical, is encountered at a lesser depth, or unless all parties agree to complete or abandon the well at a lesser depth.

Operator shall make reasonable tests of all formations encountered during drilling which give indication of containing oil and gas in quantities sufficient to test, unless this agreement shall be limited in its application to a specific formation or formations, in which event Operator shall be required to test only the formation or formations to which this agreement may apply.

If, in Operator's judgment, the well will not produce oil or gas in paying quantities, and it wishes to plug and abandon the well as a dry hole, it shall first secure the consent of all parties and shall plug and abandon same as provided in Article VI.E.1. hereof.

1 **B. Subsequent Operations:**
2
3   1. <u>Proposed Operations</u>: Should any party hereto desire to drill any well on the Contract Area
4 other than the well provided for in Article VI.A., or to rework, deepen or plug back a dry hole drilled
5 at the joint expense of all parties or a well jointly owned by all the parties and not then producing
6 in paying quantities, the party desiring to drill, rework, deepen or plug back such a well shall give the
7 other parties written notice of the proposed operation, specifying the work to be performed, the loca-
8 tion, proposed depth, objective formation and the estimated cost of the operation. The parties receiv-
9 ing such a notice shall have thirty (30) days after receipt of the notice within which to notify the
10 parties wishing to do the work whether they elect to participate in the cost of the proposed operation.
11 If a drilling rig is on location, notice of proposal to rework, plug back or drill deeper may be given
12 by telephone and the response period shall be limited to forty-eight (48) hours, exclusive of Saturday,
13 Sunday or legal holidays. Failure of a party receiving such notice to reply within the period above fixed
14 shall constitute an election by that party not to participate in the cost of the proposed operation. Any
15 notice or response given by telephone shall be promptly confirmed in writing.
16
17   2. <u>Operations by Less than All Parties</u>: If any party receiving such notice as provided in Article
18 VI.B.1. or VI.E.1. elects not to participate in the proposed operation, then, in order to be entitled to
19 the benefits of this article, the party or parties giving the notice and such other parties as shall elect
20 to participate in the operation shall, within sixty (60) days after the expiration of the notice period of
21 thirty (30) days (or as promptly as possible after the expiration of the forty-eight (48) hour period
22 where the drilling rig is on location, as the case may be) actually commence work on the proposed
23 operation and complete it with due diligence. Operator shall perform all work for the account of the
24 Consenting Parties; provided, however, if no drilling rig or other equipment is on location, and if Op-
25 erator is a Non-Consenting Party, the Consenting Parties shall either: (a) request Operator to perform
26 the work required by such proposed operation for the account of the Consenting Parties, or (b) desig-
27 nate one (1) of the Consenting Parties as Operator to perform such work. Consenting Parties, when
28 conducting operations on the Contract Area pursuant to this Article VI.B.2., shall comply with all terms
29 and conditions of this agreement.
30
31   If less than all parties approve any proposed operation, the proposing party, immediately after the
32 expiration of the applicable notice period, shall advise the Consenting Parties of (a) the total interest
33 of the parties approving such operation, and (b) its recommendation as to whether the Consenting Par-
34 ties should proceed with the operation as proposed. Each Consenting Party, within forty-eight (48)
35 hours (exclusive of Saturday, Sunday or legal holidays) after receipt of such notice, shall advise the
36 proposing party of its desire to (a) limit participation to such party's interest as shown on Exhibit "A",
37 or (b) carry its proportionate part of Non-Consenting Parties' interest. The proposing party, at its
38 election, may withdraw such proposal if there is insufficient participation, and shall promptly notify
39 all parties of such decision.
40
41   The entire cost and risk of conducting such operations shall be borne by the Consenting Parties in
42 the proportions they have elected to bear same under the terms of the preceding paragraph. Consenting
43 Parties shall keep the leasehold estates involved in such operations free and clear of all liens and
44 encumbrances of every kind created by or arising from the operations of the Consenting Parties. If such
45 an operation results in a dry hole, the Consenting Parties shall plug and abandon the well at their sole
46 cost, risk and expense. If any well drilled, reworked, deepened or plugged back under the provisions
47 of this Article results in a producer of oil and/or gas in paying quantities, the Consenting Parties shall
48 complete and equip the well to produce at their sole cost and risk, and the well shall then be turned
49 over to Operator and shall be operated by it at the expense and for the account of the Consenting Parties.
50 Upon commencement of operations for the drilling, reworking, deepening or plugging back of any such
51 well by Consenting Parties in accordance with the provisions of this Article, each Non-Consenting Party
52 shall be deemed to have relinquished to Consenting Parties, and the Consenting Parties shall own and
53 be entitled to receive, in proportion to their respective interests, all of such Non-Consenting Party's
54 interest in the well and share of production therefrom until the proceeds of the sale of such share,
55 calculated at the well, or market value thereof if such share is not sold (after deducting production
56 taxes, royalty, overriding royalty and other interests existing on the effective date hereof, payable out of
57 or measured by the production from such well accruing with respect to such interest until it reverts)
58 shall equal the total of the following:
59
60   (a) 100% of each such Non-Consenting Party's share of the cost of any newly acquired **surface**
61 equipment beyond the wellhead connections (including, but not limited to, stock tanks, separators,
62 treaters, pumping equipment and piping), plus 100% of each such Non-Consenting Party's share of the
63 cost of operation of the well commencing with first production and continuing until each such Non-
64 Consenting Party's relinquished interest shall revert to it under other provisions of this Article, it being
65 agreed that each Non-Consenting Party's share of such costs and equipment will be that interest which
66 would have been chargeable to each Non-Consenting Party had it participated in the well from the be-
67 ginning of the operation; and
68
69   (b) _____% of that portion of the costs and expenses of drilling reworking, deepening, or plugging
70 back, testing and completing, after deducting any cash contributions received under Article VIII.C., and

_____% of that portion of the cost of newly acquired equipment in the well (to and including the well-head connections), which would have been chargeable to such Non-Consenting Party if it had participated therein.

Gas production attributable to any Non - Consenting Party's relinquished interest upon such Party's election, shall be sold to its purchaser, if available, under the terms of its existing gas sales contract. Such Non - Consenting Party shall direct its purchaser to remit the proceeds receivable from such sale direct to the Consenting Parties until the amounts provided for in this Article are recovered from the Non - Consenting Party's relinquished interest. If such Non - Consenting Party has not contracted for sale of its gas at the time such gas is available for delivery, or has not made the election as provided above, the Consenting Parties shall own and be entitled to receive and sell such Non-Consenting Party's share of gas as hereinabove provided during the recoupment period.

During the period of time Consenting Parties are entitled to receive Non-Consenting Party's share of production, or the proceeds therefrom, Consenting Parties shall be responsible for the payment of all production, severance, gathering and other taxes, and all royalty, overriding royalty and other burdens applicable to Non-Consenting Party's share of production.

In the case of any reworking, plugging back or deeper drilling operation, the Consenting Parties shall be permitted to use, free of cost, all casing, tubing and other equipment in the well, but the ownership of all such equipment shall remain unchanged; and upon abandonment of a well after such reworking, plugging back or deeper drilling, the Consenting Parties shall account for all such equipment to the owners thereof, with each party receiving its proportionate part in kind or in value, less cost of salvage.

Within sixty (60) days after the completion of any operation under this Article, the party conducting the operations for the Consenting Parties shall furnish each Non-Consenting Party with an inventory of the equipment in and connected to the well, and an itemized statement of the cost of drilling, deepening, plugging back, testing, completing, and equipping the well for production; or, at its option, the operating party, in lieu of an itemized statement of such costs of operation, may submit a detailed statement of monthly billings. Each month thereafter, during the time the Consenting Parties are being reimbursed as provided above, the Party conducting the operations for the Consenting Parties shall furnish the Non-Consenting Parties with an itemized statement of all costs and liabilities incurred in the operation of the well, together with a statement of the quantity of oil and gas produced from it and the amount of proceeds realized from the sale of the well's working interest production during the preceding month. In determining the quantity of oil and gas produced during any month, Consenting Parties shall use industry accepted methods such as, but not limited to, metering or periodic well tests. Any amount realized from the sale or other disposition of equipment newly acquired in connection with any such operation which would have been owned by a Non-Consenting Party had it participated therein shall be credited against the total unreturned costs of the work done and of the equipment purchased, in determining when the interest of such Non-Consenting Party shall revert to it as above provided; and if there is a credit balance, it shall be paid to such Non-Consenting party.

If and when the Consenting Parties recover from a Non-Consenting Party's relinquished interest the amounts provided for above, the relinquished interests of such Non-Consenting Party shall automatically revert to it, and, from and after such reversion, such Non-Consenting Party shall own the same interest in such well, the material and equipment in or pertaining thereto, and the production therefrom as such Non-Consenting Party would have been entitled to had it participated in the drilling, reworking, deepening or plugging back of said well. Thereafter, such Non-Consenting Party shall be charged with and shall pay its proportionate part of the further costs of the operation of said well in accordance with the terms of this agreement and the Accounting Procedure, attached hereto.

Notwithstanding the provisions of this Article VI.B.2., it is agreed that without the mutual consent of all parties, no wells shall be completed in or produced from a source of supply from which a well located elsewhere on the Contract Area is producing, unless such well conforms to the then-existing well spacing pattern for such source of supply.

The provisions of this Article shall have no application whatsoever to the drilling of the initial well described in Article VI.A. except (a) when Option 2, Article VII.D.1., has been selected, or (b) to the reworking, deepening and plugging back of such initial well, if such well is or thereafter shall prove to be a dry hole or non-commercial well, after having been drilled to the depth specified in Article VI.A.

C. **Right to Take Production in Kind:**

Each party shall have the right to take in kind or separately dispose of its proportionate share of all oil and gas produced from the Contract Area, exclusive of production which may be used in development and producing operations and in preparing and treating oil for marketing purposes and production unavoidably lost. Any extra expenditure incurred in the taking in kind or separate disposition by any party of its proportionate share of the production shall be borne by such party. Any party taking its share of production in kind shall be required to pay for only its proportionate share of such part of Operator's surface facilities which it uses.

Each party shall execute such division orders and contracts as may be necessary for the sale of its interest in production from the Contract Area, and, except as provided in Article VII.B., shall be entitled to receive payment direct from the purchaser thereof for its share of all production.

In the event any party shall fail to make the arrangements necessary to take in kind or separately dispose of its proportionate share of the oil and gas produced from the Contract Area, Operator shall have the right, subject to the revocation at will by the party owning it, but not the obligation, to purchase such oil and gas or sell it to others at any time and from time to time, for the account of the non-taking party at the best price obtainable in the area for such production. Any such purchase or sale by Operator shall be subject always to the right of the owner of the production to exercise at any time its right to take in kind, or separately dispose of, its share of all oil and gas not previously delivered to a purchaser. Any purchase or sale by Operator of any other party's share of oil and gas shall be only for such reasonable periods of time as are consistent with the minimum needs of the industry under the particular circumstances, but in no event for a period in excess of one (1) year. Notwithstanding the foregoing, Operator shall not make a sale, including one into interstate commerce, of any other party's share of gas production without first giving such other party thirty (30) days notice of such intended sale.

In the event one or more parties' separate disposition of its share of the gas causes split-stream deliveries to separate pipelines and/or deliveries which on a day-to-day basis for any reason are not exactly equal to a party's respective proportionate share of total gas sales to be allocated to it, the balancing or accounting between the respective accounts of the parties shall be in accordance with any Gas Balancing Agreement between the parties hereto, whether such Agreement is attached as Exhibit "E", or is a separate Agreement.

D. Access to Contract Area and Information:

Each party shall have access to the Contract Area at all reasonable times, at its sole risk to inspect or observe operations, and shall have access at reasonable times to information pertaining to the development or operation thereof, including Operator's books and records relating thereto. Operator, upon request, shall furnish each of the other parties with copies of all forms or reports filed with governmental agencies, daily drilling reports, well logs, tank tables, daily gauge and run tickets and reports of stock on hand at the first of each month, and shall make available samples of any cores or cuttings taken from any well drilled on the Contract Area. The cost of gathering and furnishing information to Non-Operator, other than that specified above, shall be charged to the Non-Operator that requests the information.

E. Abandonment of Wells:

1. Abandonment of Dry Holes: Except for any well drilled pursuant to Article VI.B.2., any well which has been drilled under the terms of this agreement and is proposed to be completed as a dry hole shall not be plugged and abandoned without the consent of all parties. Should Operator, after diligent effort, be unable to contact any party, or should any party fail to reply within forty-eight (48) hours (exclusive of Saturday, Sunday or legal holidays) after receipt of notice of the proposal to plug and abandon such well, such party shall be deemed to have consented to the proposed abandonment. All such wells shall be plugged and abandoned in accordance with applicable regulations and at the cost, risk and expense of the parties who participated in the cost of drilling of such well. Any party who objects to the plugging and abandoning such well shall have the right to take over the well and conduct further operations in search of oil and/or gas subject to the provisions of Article VI.B.

2. Abandonment of Wells that have Produced: Except for any well which has been drilled or reworked pursuant to Article VI.B.2. hereof for which the Consenting Parties have not been fully reimbursed as therein provided, any well which has been completed as a producer shall not be plugged and abandoned without the consent of all parties. If all parties consent to such abandonment, the well shall be plugged and abandoned in accordance with applicable regulations and at the cost, risk and expense of all the parties hereto. If, within thirty (30) days after receipt of notice of the proposed abandonment of such well, all parties do not agree to the abandonment of any well, those wishing to continue its operation shall tender to each of the other parties its proportionate share of the value of the well's salvable material and equipment, determined in accordance with the provisions of Exhibit "C", less the estimated cost of salvaging and the estimated cost of plugging and abandoning. Each abandoning party shall assign to the non-abandoning parties, without warranty, express or implied, as to title or as to quantity, quality, or fitness for use of the equipment and material, all of its interest in the well and related equipment, together with its interest in the leasehold estate as to, but only as to, the interval or intervals of the formation or formations then open to production. If the interest of the abandoning party is or includes an oil and gas interest, such party shall execute and deliver to the non-abandoning party or parties an oil and gas lease, limited to the interval or intervals of the formation or formations then open to production, for a term of one year and so long thereafter as oil and/or gas is produced from the interval or inter-

vals of the formation or formations covered thereby, such lease to be on the form attached as Exhibit "B". The assignments or leases so limited shall encompass the "drilling unit" upon which the well is located. The payments by, and the assignments or leases to, the assignees shall be in a ratio based upon the relationship of their respective percentages of participation in the Contract Area to the aggregate of the percentages of participation in the Contract Area of all assignees. There shall be no readjustment of interest in the remaining portion of the Contract Area.

Thereafter, abandoning parties shall have no further responsibility, liability, or interest in the operation of or production from the well in the interval or intervals then open other than the royalties retained in any lease made under the terms of this Article. Upon request, Operator shall continue to operate the assigned well for the account of the non-abandoning parties at the rates and charges contemplated by this agreement, plus any additional cost and charges which may arise as the result of the separate ownership of the assigned well.

ARTICLE VII.
EXPENDITURES AND LIABILITY OF PARTIES

A. Liability of Parties:

The liability of the parties shall be several, not joint or collective. Each party shall be responsible only for its obligations, and shall be liable only for its proportionate share of the costs of developing and operating the Contract Area. Accordingly, the liens granted among the parties in Article VII.B. are given to secure only the debts of each severally. It is not the intention of the parties to create, nor shall this agreement be construed as creating, a mining or other partnership or association, or to render the parties liable as partners.

B. Liens and Payment Defaults:

Each Non-Operator grants to Operator a lien upon its oil and gas rights in the Contract Area, and a security interest in its share of oil and or gas when extracted and its interest in all equipment, to secure payment of its share of expense, together with interest thereon at the rate provided in the Accounting Procedure attached hereto as Exhibit "C". To the extent that Operator has a security interest under the Uniform Commercial Code of the State, Operator shall be entitled to exercise the rights and remedies of a secured party under the Code. The bringing of a suit and the obtaining of judgment by Operator for the secured indebtedness shall not be deemed an election of remedies or otherwise affect the lien rights or security interest as security for the payment thereof. In addition, upon default by any Non-Operator in the payment of its share of expense, Operator shall have the right, without prejudice to other rights or remedies, to collect from the purchaser the proceeds from the sale of such Non-Operator's share of oil and or gas until the amount owed by such Non-Operator, plus interest has been paid. Each purchaser shall be entitled to rely upon Operator's written statement concerning the amount of any default. Operator grants a like lien and security interest to the Non-Operators to secure payment of Operator's proportionate share of expense.

If any party fails or is unable to pay its share of expense within sixty (60) days after rendition of a statement therefor by Operator, the non-defaulting parties, including Operator, shall, upon request by Operator, pay the unpaid amount in the proportion that the interest of each such party bears to the interest of all such parties. Each party so paying its share of the unpaid amount shall, to obtain reimbursement thereof, be subrogated to the security rights described in the foregoing paragraph.

C. Payments and Accounting:

Except as herein otherwise specifically provided, Operator shall promptly pay and discharge expenses incurred in the development and operation of the Contract Area pursuant to this agreement and shall charge each of the parties hereto with their respective proportionate shares upon the expense basis provided in the Accounting Procedure attached hereto as Exhibit "C". Operator shall keep an accurate record of the joint account hereunder, showing expenses incurred and charges and credits made and received.

Operator, at its election, shall have the right from time to time to demand and receive from the other parties payment in advance of their respective shares of the estimated amount of the expense to be incurred in operations hereunder during the next succeeding month, which right may be exercised only by submission to each such party of an itemized statement of such estimated expense, together with an invoice for its share thereof. Each such statement and invoice for the payment in advance of estimated expense shall be submitted on or before the 20th day of the next preceding month. Each party shall pay to Operator its proportionate share of such estimate within fifteen (15) days after such estimate and invoice is received. If any party fails to pay its share of said estimate within said time, the amount due shall bear interest as provided in Exhibit "C" until paid. Proper adjustment shall be made monthly between advances and actual expense to the end that each party shall bear and pay its proportionate share of actual expenses incurred, and no more.

D. Limitation of Expenditures:

1. <u>Drill or Deepen:</u> Without the consent of all parties, no well shall be drilled or deepened, except any well drilled or deepened pursuant to the provisions of Article VI.B.2. of this Agreement, it being understood that the consent to the drilling or deepening shall include:

☐ <u>Option No. 1:</u> All necessary expenditures for the drilling or deepening, testing, completing and equipping of the well, including necessary tankage and/or surface facilities.

☐ <u>Option No. 2:</u> All necessary expenditures for the drilling or deepening and testing of the well. When such well has reached its authorized depth, and all tests have been completed, Operator shall give immediate notice to the Non-Operators who have the right to participate in the completion costs. The parties receiving such notice shall have forty-eight (48) hours (exclusive of Saturday, Sunday and legal holidays) in which to elect to participate in the setting of casing and the completion attempt. Such election, when made, shall include consent to all necessary expenditures for the completing and equipping of such well, including necessary tankage and/or surface facilities. Failure of any party receiving such notice to reply within the period above fixed shall constitute an election by that party not to participate in the cost of the completion attempt. If one or more, but less than all of the parties, elect to set pipe and to attempt a completion, the provisions of Article VI.B.2. hereof (the phrase "reworking, deepening or plugging back" as contained in Article VI.B.2. shall be deemed to include "completing") shall apply to the operations thereafter conducted by less than all parties.

2. <u>Rework or Plug Back:</u> Without the consent of all parties, no well shall be reworked or plugged back except a well reworked or plugged back pursuant to the provisions of Article VI.B.2. of this agreement, it being understood that the consent to the reworking or plugging back of a well shall include consent to all necessary expenditures in conducting such operations and completing and equipping of said well, including necessary tankage and/or surface facilities.

3. <u>Other Operations:</u> Operator shall not undertake any single project reasonably estimated to require an expenditure in excess of_____ Dollars ($_____) except in connection with a well, the drilling, reworking, deepening, completing, recompleting, or plugging back of which has been previously authorized by or pursuant to this agreement; provided, however, that, in case of explosion, fire, flood or other sudden emergency, whether of the same or different nature, Operator may take such steps and incur such expenses as in its opinion are required to deal with the emergency to safeguard life and property but Operator, as promptly as possible, shall report the emergency to the other parties. If Operator prepares "Authority for Expenditures" for its own use, Operator, upon request, shall furnish copies of its "Authority for Expenditures" for any single project costing in excess of _____ Dollars ($_____).

E. Royalties, Overriding Royalties and Other Payments:

Each party shall pay or deliver, or cause to be paid or delivered, all royalties to the extent of _____due on its share of production and shall hold the other parties free from any liability therefor. If the interest of any party in any oil and gas lease covered by this agreement is subject to any royalty, overriding royalty, production payment, or other charge over and above the aforesaid royalty, such party shall assume and alone bear all such obligations and shall account for or cause to be accounted for, such interest to the owners thereof.

No party shall ever be responsible, on any price basis higher than the price received by such party, to any other party's lessor or royalty owner; and if any such other party's lessor or royalty owner should demand and receive settlements on a higher price basis, the party contributing such lease shall bear the royalty burden insofar as such higher price is concerned.

F. Rentals, Shut-in Well Payments and Minimum Royalties:

Rentals, shut-in well payments and minimum royalties which may be required under the terms of any lease shall be paid by the party or parties who subjected such lease to this agreement at its own expense. In the event two or more parties own and have contributed interests in the same lease to this agreement, such parties may designate one of such parties to make said payments for and on behalf of all such parties. Any party may request, and shall be entitled to receive, proper evidence of all such payments. In the event of failure to make proper payment of any rental, shut-in well payment or minimum royalty through mistake or oversight where such payment is required to continue the lease in force, any loss which results from such non-payment shall be borne in accordance with the provisions of Article IV.B.2.

Operator shall notify Non-Operator of the anticipated completion of a shut-in gas well, or the shutting in or return to production of a producing gas well, at least five (5) days (excluding Saturday, Sunday and holidays), or at the earliest opportunity permitted by circumstances, prior to taking such action, but assumes no liability for failure to do so. In the event of failure by Operator to so notify Non-Operator, the loss of any lease contributed hereto by Non-Operator for failure to make timely payments of any shut-in well payment shall be borne jointly by the parties hereto under the provisions of Article IV.B.3.

G. Taxes:

Beginning with the first calendar year after the effective date hereof, Operator shall render for ad valorem taxation all property subject to this agreement which by law should be rendered for such taxes, and it shall pay all such taxes assessed thereon before they become delinquent. Prior to the rendition date, each Non-Operator shall furnish Operator information as to burdens (to include, but not be limited to, royalties, overriding royalties and production payments) on leases and oil and gas interests contributed by such Non-Operator. If the assessed valuation of any leasehold estate is reduced by reason of its being subject to outstanding excess royalties, overriding royalties or production payments, the reduction in ad valorem taxes resulting therefrom shall inure to the benefit of the owner or owners of such leasehold estate, and Operator shall adjust the charge to such owner or owners so as to reflect the benefit of such reduction. Operator shall bill other parties for their proportionate share of all tax payments in the manner provided in Exhibit "C".

If Operator considers any tax assessment improper, Operator may, at its discretion, protest within the time and manner prescribed by law, and prosecute the protest to a final determination, unless all parties agree to abandon the protest prior to final determination. During the pendency of administrative or judicial proceedings, Operator may elect to pay, under protest, all such taxes and any interest and penalty. When any such protested assessment shall have been finally determined, Operator shall pay the tax for the joint account, together with any interest and penalty accrued, and the total cost shall then be assessed against the parties, and be paid by them, as provided in Exhibit "C".

Each party shall pay or cause to be paid all production, severance, gathering and other taxes imposed upon or with respect to the production or handling of such party's share of oil and/or gas produced under the terms of this agreement.

H. Insurance:

At all times while operations are conducted hereunder, Operator shall comply with the Workmen's Compensation Law of the State where the operations are being conducted; provided, however, that Operator may be a self-insurer for liability under said compensation laws in which event the only charge that shall be made to the joint account shall be an amount equivalent to the premium which would have been paid had such insurance been obtained. Operator shall also carry or provide insurance for the benefit of the joint account of the parties as outlined in Exhibit "D", attached to and made a part hereof. Operator shall require all contractors engaged in work on or for the Contract Area to comply with the Workmen's Compensation Law of the State where the operations are being conducted and to maintain such other insurance as Operator may require.

In the event Automobile Public Liability Insurance is specified in said Exhibit "D", or subsequently receives the approval of the parties, no direct charge shall be made by Operator for premiums paid for such insurance for Operator's fully owned automotive equipment.

ARTICLE VIII.

ACQUISITION, MAINTENANCE OR TRANSFER OF INTEREST

A. Surrender of Leases:

The leases covered by this agreement, insofar as they embrace acreage in the Contract Area, shall not be surrendered in whole or in part unless all parties consent thereto.

However, should any party desire to surrender its interest in any lease or in any portion thereof, and other parties do not agree or consent thereto, the party desiring to surrender shall assign, without express or implied warranty of title, all of its interest in such lease, or portion thereof, and any well, material and equipment which may be located thereon and any rights in production thereafter secured, to the parties not desiring to surrender it. If the interest of the assigning party includes an oil and gas interest, the assigning party shall execute and deliver to the party or parties not desiring to surrender an oil and gas lease covering such oil and gas interest for a term of one year and so long thereafter as oil and/or gas is produced from the land covered thereby, such lease to be on the form attached hereto as Exhibit "B". Upon such assignment, the assigning party shall be relieved from all obligations thereafter accruing, but not theretofore accrued, with respect to the acreage assigned and the operation of any well thereon, and the assigning party shall have no further interest in the lease assigned and its equipment and production other than the royalties retained in any lease made under the terms of this Article. The parties assignee shall pay to the party assignor the reasonable salvage value of the latter's interest in any wells and equipment on the assigned acreage. The value of all material shall be determined in accordance with the provisions of Exhibit "C", less the estimated cost of salvaging and the estimated cost of plugging and abandoning. If the assignment is in favor of more than one party, the assigned interest shall be shared by the parties assignee in the proportions that the interest of each bears to the interest of all parties assignee.

Any assignment or surrender made under this provision shall not reduce or change the assignor's or surrendering parties' interest, as it was immediately before the assignment, in the balance of the Contract Area; and the acreage assigned or surrendered, and subsequent operations thereon, shall not thereafter be subject to the terms and provisions of this agreement.

B. Renewal or Extension of Leases:

If any party secures a renewal of any oil and gas lease subject to this Agreement, all other parties shall be notified promptly, and shall have the right for a period of thirty (30) days following receipt of such notice in which to elect to participate in the ownership of the renewal lease, insofar as such lease affects lands within the Contract Area, by paying to the party who acquired it their several proper proportionate shares of the acquisition cost allocated to that part of such lease within the Contract Area, which shall be in proportion to the interests held at that time by the parties in the Contract Area.

If some, but less than all, of the parties elect to participate in the purchase of a renewal lease, it shall be owned by the parties who elect to participate therein, in a ratio based upon the relationship of their respective percentage of participation in the Contract Area to the aggregate of the percentages of participation in the Contract Area of all parties participating in the purchase of such renewal lease. Any renewal lease in which less than all parties elect to participate shall not be subject to this agreement.

Each party who participates in the purchase of a renewal lease shall be given an assignment of its proportionate interest therein by the acquiring party.

The provisions of this Article shall apply to renewal leases whether they are for the entire interest covered by the expiring lease or cover only a portion of its area or an interest therein. Any renewal lease taken before the expiration of its predecessor lease, or taken or contracted for within six (6) months after the expiration of the existing lease shall be subject to this provision; but any lease taken or contracted for more than six (6) months after the expiration of an existing lease shall not be deemed a renewal lease and shall not be subject to the provisions of this agreement.

The provisions in this Article shall apply also and in like manner to extensions of oil and gas leases.

C. Acreage or Cash Contributions:

While this agreement is in force, if any party contracts for a contribution of cash toward the drilling of a well or any other operation on the Contract Area, such contribution shall be paid to the party who conducted the drilling or other operation and shall be applied by it against the cost of such drilling or other operation. If the contribution be in the form of acreage, the party to whom the contribution is made shall promptly tender an assignment of the acreage, without warranty of title, to the Drilling Parties in the proportions said Drilling Parties shared the cost of drilling the well. If all parties hereto are Drilling Parties and accept such tender, such acreage shall become a part of the Contract Area and be governed by the provisions of this agreement. If less than all parties hereto are Drilling Parties and accept such tender, such acreage shall not become a part of the Contract Area. Each party shall promptly notify all other parties of all acreage or money contributions it may obtain in support of any well or any other operation on the Contract Area.

If any party contracts for any consideration relating to disposition of such party's share of substances produced hereunder, such consideration shall not be deemed a contribution as contemplated in this Article VIII.C.

D. Subsequently Created Interest:

Notwithstanding the provisions of Article VIII.E. and VIII.G., if any party hereto shall, subsequent to execution of this agreement, create an overriding royalty, production payment, or net proceeds interest, which such interests are hereinafter referred to as "subsequently created interest", such subsequently created interest shall be specifically made subject to all of the terms and provisions of this agreement, as follows:

1. If non-consent operations are conducted pursuant to any provision of this agreement, and the party conducting such operations becomes entitled to receive the production attributable to the interest out of which the subsequently created interest is derived, such party shall receive same free and clear of such subsequently created interest. The party creating same shall bear and pay all such subsequently created interests and shall indemnify and hold the other parties hereto free and harmless from any and all liability resulting therefrom.

2. If the owner of the interest from which the subsequently created interest is derived (1) fails to pay, when due, its share of expenses chargeable hereunder, or (2) elects to abandon a well under provisions of Article VI.E. hereof, or (3) elects to surrender a lease under provisions of Article VIII.A. hereof, the subsequently created interest shall be chargeable with the pro rata portion of all expenses hereunder in the same manner as if such interest were a working interest. For purposes of collecting such chargeable expenses, the party or parties who receive assignments as a result of (2) or (3) above shall have the right to enforce all provisions of Article VII.B. hereof against such subsequently created interest.

E. Maintenance of Uniform Interest:

For the purpose of maintaining uniformity of ownership in the oil and gas leasehold interests covered by this agreement, and notwithstanding any other provisions to the contrary, no party shall sell, encumber, transfer or make other disposition of its interest in the leases embraced within the Contract Area and in wells, equipment and production unless such disposition covers either:

1. the entire interest of the party in all leases and equipment and production; or

2. an equal undivided interest in all leases and equipment and production in the Contract Area.

Every such sale, encumbrance, transfer or other disposition made by any party shall be made expressly subject to this agreement, and shall be made without prejudice to the right of the other parties.

If, at any time the interest of any party is divided among and owned by four or more co-owners, Operator, at its discretion, may require such co-owners to appoint a single trustee or agent with full authority to receive notices, approve expenditures, receive billings for and approve and pay such party's share of the joint expenses, and to deal generally with, and with power to bind, the co-owners of such party's interests within the scope of the operations embraced in this agreement; however, all such co-owners shall have the right to enter into and execute all contracts or agreements for the disposition of their respective shares of the oil and gas produced from the Contract Area and they shall have the right to receive, separately, payment of the sale proceeds hereof.

F. Waiver of Right to Partition:

If permitted by the laws of the state or states in which the property covered hereby is located, each party hereto owning an undivided interest in the Contract Area waives any and all rights it may have to partition and have set aside to it in severalty its undivided interest therein.

G. Preferential Right to Purchase:

Should any party desire to sell all or any part of its interests under this agreement, or its rights and interests in the Contract Area, it shall promptly give written notice to the other parties, with full information concerning its proposed sale, which shall include the name and address of the prospective purchaser (who must be ready, willing and able to purchase), the purchase price, and all other terms of the offer. The other parties shall then have an optional prior right, for a period of ten (10) days after receipt of the notice, to purchase on the same terms and conditions the interest which the other party proposes to sell; and, if this optional right is exercised, the purchasing parties shall share the purchased interest in the proportions that the interest of each bears to the total interest of all purchasing parties. However, there shall be no preferential right to purchase in those cases where any party wishes to mortgage its interests, or to dispose of its interests by merger, reorganization, consolidation, or sale of all or substantially all of its assets to a subsidiary or parent company or to a subsidiary of a parent company, or to any company in which any one party owns a majority of the stock.

ARTICLE IX.
INTERNAL REVENUE CODE ELECTION

This agreement is not intended to create, and shall not be construed to create, a relationship of partnership or an association for profit between or among the parties hereto. Notwithstanding any provisions herein that the rights and liabilities hereunder are several and not joint or collective, or that this agreement and operations hereunder shall not constitute a partnership, if, for Federal income tax purposes, this agreement and the operations hereunder are regarded as a partnership, each party hereby affected elects to be excluded from the application of all of the provisions of Subchapter "K", Chapter 1, Subtitle "A", of the Internal Revenue Code of 1954, as permitted and authorized by Section 761 of the Code and the regulations promulgated thereunder. Operator is authorized and directed to execute on behalf of each party hereby affected such evidence of this election as may be required by the Secretary of the Treasury of the United States or the Federal Internal Revenue Service, including specifically, but not by way of limitation, all of the returns, statements, and the data required by Federal Regulations 1.761. Should there be any requirement that each party hereby affected give further evidence of this election, each such party shall execute such documents and furnish such other evidence as may be required by the Federal Internal Revenue Service or as may be necessary to evidence this election. No such party shall give any notices or take any other action inconsistent with the election made hereby. If any present or future income tax laws of the state or states in which the Contract Area is located or any future income tax laws of the United States contain provisions similar to those in Subchapter "K", Chapter 1, Subtitle "A", of the Internal Revenue Code of 1954, under which an election similar to that provided by Section 761 of the Code is permitted, each party hereby affected shall make such election as may be permitted or required by such laws. In making the foregoing election, each such party states that the income derived by such party from Operations hereunder can be adequately determined without the computation of partnership taxable income.

ARTICLE X.
CLAIMS AND LAWSUITS

Operator may settle any single damage claim or suit arising from operations hereunder if the expenditure does not exceed _____ Dollars ($ _____) and if the payment is in complete settlement of such claim or suit. If the amount required for settlement exceeds the above amount, the parties hereto shall assume and take over the further handling of the claim or suit, unless such authority is delegated to Operator. All costs and expense of handling, settling, or otherwise discharging such claim or suit shall be at the joint expense of the parties. If a claim is made against any party or if any party is sued on account of any matter arising from operations hereunder over which such individual has no control because of the rights given Operator by this agreement, the party shall immediately notify Operator, and the claim or suit shall be treated as any other claim or suit involving operations hereunder.

ARTICLE XI.
FORCE MAJEURE

If any party is rendered unable, wholly or in part, by force majeure to carry out its obligations under this agreement, other than the obligation to make money payments, that party shall give to all other parties prompt written notice of the force majeure with reasonably full particulars concerning it; thereupon, the obligations of the party giving the notice, so far as they are affected by the force majeure, shall be suspended during, but no longer than, the continuance of the force majeure. The affected party shall use all reasonable diligence to remove the force majeure situation as quickly as practicable.

The requirement that any force majeure shall be remedied with all reasonable dispatch shall not require the settlement of strikes, lockouts, or other labor difficulty by the party involved, contrary to its wishes; how all such difficulties shall be handled shall be entirely within the discretion of the party concerned.

The term "force majeure", as here employed, shall mean an act of God, strike, lockout, or other industrial disturbance, act of the public enemy, war, blockade, public riot, lightning, fire, storm, flood, explosion, governmental action, governmental delay, restraint or inaction, unavailability of equipment, and any other cause, whether of the kind specifically enumerated above or otherwise, which is not reasonably within the control of the party claiming suspension.

ARTICLE XII.
NOTICES

All notices authorized or required between the parties, and required by any of the provisions of this agreement, unless otherwise specifically provided, shall be given in writing by United States mail or Western Union telegram, postage or charges prepaid, or by teletype, and addressed to the party to whom the notice is given at the addresses listed on Exhibit "A". The originating notice given under any provision hereof shall be deemed given only when received by the party to whom such notice is directed, and the time for such party to give any notice in response thereto shall run from the date the originating notice is received. The second or any responsive notice shall be deemed given when deposited in the United States mail or with the Western Union Telegraph Company, with postage or charges prepaid, or when sent by teletype. Each party shall have the right to change its address at any time, and from time to time, by giving written notice hereof to all other parties.

ARTICLE XIII.
TERM OF AGREEMENT

This agreement shall remain in full force and effect as to the oil and gas leases and/or oil and gas interests subjected hereto for the period of time selected below; provided, however, no party hereto shall ever be construed as having any right, title or interest in or to any lease, or oil and gas interest contributed by any other party beyond the term of this agreement.

☐ Option No. 1: So long as any of the oil and gas leases subject to this agreement remain or are continued in force as to any part of the Contract Area, whether by production, extension, renewal or otherwise, and/or so long as oil and/or gas production continues from any lease or oil and gas interest.

☐ **Option No. 2:** In the event the well described in Article VI.A., or any subsequent well drilled under any provision of this agreement, results in production of oil and/or gas in paying quantities, this agreement shall continue in force so long as any such well or wells produce, or are capable of production, and for an additional period of _____ days from cessation of all production; provided, however, if, prior to the expiration of such additional period, one or more of the parties hereto are engaged in drilling or reworking a well or wells hereunder, this agreement shall continue in force until such operations have been completed and if production results therefrom, this agreement shall continue in force as provided herein. In the event the well described in Article VI.A., or any subsequent well drilled hereunder, results in a dry hole, and no other well is producing, or capable of producing oil and/or gas from the Contract Area, this agreement shall terminate unless drilling or reworking operations are commenced within _____ days from the date of abandonment of said well.

It is agreed, however, that the termination of this agreement shall not relieve any party hereto from any liability which has accrued or attached prior to the date of such termination.

ARTICLE XIV.
COMPLIANCE WITH LAWS AND REGULATIONS

A. **Laws, Regulations and Orders:**

This agreement shall be subject to the conservation laws of the state in which the committed acreage is located, to the valid rules, regulations, and orders of any duly constituted regulatory body of said state; and to all other applicable federal, state, and local laws, ordinances, rules, regulations, and orders.

B. **Governing Law:**

The essential validity of this agreement and all matters pertaining thereto, including, but not limited to, matters of performance, non-performance, breach, remedies, procedures, rights, duties and interpretation or construction, shall be governed and determined by the law of the state in which the Contract Area is located. If the Contract Area is in two or more states, the law of the state where most of the land in the Contract Area is located shall govern.

ARTICLE XV.
OTHER PROVISIONS

ARTICLE XVI.
MISCELLANEOUS

This agreement shall be binding upon and shall inure to the benefit of the parties hereto and to their respective heirs, devisees, legal representatives, successors and assigns.

This instrument may be executed in any number of counterparts, each of which shall be considered an original for all purposes.

IN WITNESS WHEREOF, this agreement shall be effective as of _____ day of _____, 19_____ .

OPERATOR

_____   _____

NON-OPERATORS

_____   _____

_____   _____

Appendix D

SAMPLE DIVISION AND TRANSFER ORDERS

Oil and Gas Division Order

To: Amoco Production Company
P. O. Box 591, Tulsa, Oklahoma 74102

Property No. _____

_____, 19___

Each of the undersigned OWNERS guarantees and warrants he is the owner of the interest set out opposite his name on the reverse side hereof in oil and gas or the proceeds from the sale of oil and gas from the property described on the reverse side hereof, and until further written notice either from you or from us, the undersigned owner and all other parties executing this instrument hereby authorize you, your successors or assigns to receive and measure such sales in accordance with applicable governmental rules and regulations and to give credit as set forth on the reverse side hereof.

The following covenants are parts of this instrument and shall be binding on the undersigned, their successors, legal representatives, and assigns:

Oil: Oil sold hereunder shall be delivered f.o.b. to the carrier designated to gather and receive such oil, and shall become your property upon receipt thereof by the carrier designated by you or by any other purchaser to whom you may resell such oil. The term "oil" as used in this division order shall include all marketable liquid hydrocarbons.

Should the oil produced from the herein described land be commingled with oil produced from one or more other separately owned tracts of land prior to delivery to the designated carrier, the commingled oil sold hereunder shall be deemed to be the interest of the undersigned in that portion of the total commingled oil delivered which is allocated to the herein described land on the basis of lease meter readings or any other method generally accepted in the industry as an equitable basis for determining the quantity and quality of oil sold from each separately owned tract. Such formula shall be uniformly applied to all owners of an interest in the tracts of land involved.

Should the interest of the undersigned in the oil produced from the herein described land be unitized with oil produced from one or more other tracts of land, this instrument shall thereafter be deemed to be modified to the extent necessary to conform with the applicable unitization agreement or plan of unitization, and all revisions or amendments thereto, but otherwise to remain in force and effect as to all other provisions. In such event, the portion of the unitized oil sold hereunder shall be the interest of the undersigned in that portion of the total unitized oil delivered which is allocated to the herein described land and shall be deemed for all purposes to have been actually produced from said land.

You agree to pay for the oil sold hereunder at the price posted by you for oil of the same grade and gravity in the same producing field or area on the date said oil is received by you or the designated carrier. If you do not currently post such a price, then until such time as you do so, you agree to pay the price established by you. You are authorized to reduce the price by those truck, barge, tankcar, or pipe line transportation charges as determined by you.

Should the oil sold hereunder be resold by you to another purchaser accepting delivery thereof at the same point at which you take title, you agree to pay for such oil based upon the volume computation made by such purchaser and at the price received by you

298

for such oil, reduced by any transportation charges deducted by such purchaser. Quality and quantity shall be determined in accordance with the conditions specified in the price posting. You may refuse to receive any oil not considered merchantable by you.

Gas: Settlements for gas shall be based on the net proceeds realized from the sale thereof, after deducting a fair and reasonable charge for compressing and making it merchantable and for transporting if the gas is sold off the property. Where gas is sold subject to regulation by the Federal Power Commission or other governmental authority, the price applicable to such sale approved by order of such authority shall be used to determine the net proceeds realized from the sale.

Settlements: Settlements shall be made monthly by check mailed to the respective parties according to the division of interest herein specified at the latest address known by you, less any taxes required by law to be deducted and paid by you applicable to owner's interest.

Evidence of Title: The oil and gas lease or leases, and any amendments, ratifications, or corrections thereof, under which said gas and/or oil is produced are hereby adopted, ratified, and confirmed as herein and heretofore amended. In the event any dispute or question arises concerning the title to the interest of the undersigned in said land and/or the oil or gas produced therefrom or the proceeds thereof, you will be furnished satisfactory abstracts or other evidence of title upon demand. Until such evidence of title has been furnished and/or such dispute, defect, or question of title is corrected or removed to your satisfaction, or until indemnity satisfactory to you has been furnished, you are authorized to withhold the proceeds of such oil or gas received and run, without interest. In the event any action or suit is filed in any court affecting the title to the interest of the undersigned in the herein described land or the oil or gas produced therefrom or the proceeds thereof to which the undersigned is a party, written notice of the filing of such suit or action shall be immediately furnished you by the undersigned, stating the court in which the same is filed and the title of such suit or action. You will not be responsible for any change of ownership in the absence of actual notice and satisfactory proof thereof.

Contingent Interests: Whether or not any contingency is expressly stated in this instrument, you are hereby relieved of any responsibility for determining when any of the interests herein shall increase, diminish, terminate, be extinguished or revert to other parties as a result of the completion or discharge of money or other payments from said interest, or as a result of the expiration of any time or term limitation (either definite or indefinite), and, unless you are also the operator of the property, as a result of an increase or decrease in production, or as a result of a change in the depth, the methods or the means of production, or as a result of a change in the allocation of production affecting the herein described land or any portion thereof under any agreement or by order of Governmental authority, and until you receive notice in writing to the contrary, you are hereby authorized to continue to remit without liability pursuant to the division of interest shown herein.

Warranties: Working Interest Owners and/or Operators, and each of them, by signature to this instrument, certify, guarantee and warrant, for your benefit and that of any pipe line or other carrier designated to run or transport said oil or gas, that all oil or gas tendered here-under has been and shall be produced from or lawfully allocated to the herein described land in accordance with all applicable Federal, state and local laws, orders, rules and regulations. This instrument may be executed by one or more, but all covenants herein shall be binding upon any party executing same and upon his heirs, devisees, successors, and assigns irrespective of whether other parties have executed this instrument.

299

Witness of Signature

Witness of Signature

Witness of Signature

Name

Social Security (or Tax ID) Number

Name

Social Security (or Tax ID) Number

Name

Social Security (or Tax ID) Number

Street or Box No

City. State. Zip

Street or Box No

City. State. Zip

Street or Box No

City. State. Zip

Form 697 2-77

300

Oil and Gas Division Order

Property described as:

in_____ Parish

_____ County. State of _____, and

commencing with First Runs, or at 7 a.m. _____

_____   _____
Credit To (Name of Owner)   Division of Interest

(See Division of Interest Digest of even date attached and made a part hereof)

NATURAL GAS DIVISION ORDER

Property No. _____

To: **Amoco Production Company**
P. O. Box 591, Tulsa, Oklahoma

_____, 19___

The undersigned, and each of them, guarantee and warrant that they are the legal owners in the proportions set out below of the proceeds of all gas produced from the _____ property described as:

in _____ Parish,

_____ County, State of _____, and commencing at 7 a.m. _____,

and until further written notice either from you or us you are authorized to receive natural gas therefrom, purchase it and pay therefor as follows:

Credit To	Division of Interest	Address
	(See Division of Interest Digest of even date attached and made a part hereof)	

302

The following covenants are also parts of this division order, and shall be binding upon the undersigned, their successors, legal representatives and assigns:

FIRST: The natural gas produced from said wells and received in pursuance of this division order shall become your property as soon as the same is received into your custody or that of any pipe line company designated by you.

SECOND: The natural gas purchased and received in pursuance of this division order shall be delivered f.o.b. to any pipe line designated by you which gathers and receives said natural gas and shall be paid for to the well owners or their assigns in proportion to their respective interests as above shown, at the price and under the provisions specified in the written contract between _____

_____ , and all renewals or modifications thereof

THIRD: Settlement for the natural gas taken and received in pursuance of this division order shall be made monthly by check mailed to the respective parties according to the division of interest herein specified at the addresses herein given.

FOURTH: At any time on demand the undersigned agree to furnish Amoco Production Company satisfactory abstracts and/or other evidence of title, and in case of failure to do so, or in case of any adverse claim or dispute of title to the natural gas sold hereunder or to the land from which produced said company is authorized to hold the proceeds of all natural gas taken hereunder until such defect of title is corrected or such adverse claim or dispute is fully settled and determined to the satisfaction of said company, unless indemnity acceptable to said company shall be furnished.

FIFTH: This division order shall become effective as to each and every seller above named as soon as signed by him, irrespective of whether the other above named sellers have so signed.

SIXTH: Working Interest Owners and/or Operators, and each of them, by signature to this division order, certify, guarantee and warrant, for your benefit and that of any pipe line or other carrier designated to gather and receive said natural gas that all natural gas tendered hereunder has been and shall be produced from or lawfully allocated to the above-described land in accordance with all applicable Federal, state and local laws, orders, rules and regulations.

Witness of Signatures

Signatures

Taxpayer Identification or
Social Security Number

Address _____

Address _____

Address _____

BE SURE TO INSERT YOUR TAXPAYER IDENTIFICATION OR SOCIAL SECURITY NUMBER IN THE SPACE PROVIDED ABOVE

Form 420 6-71

303

CASINGHEAD GAS DIVISION ORDER

Property No. _____

_____, 19____

To **Amoco Production Company**
P. O. Box 591, Tulsa, Oklahoma 74102

The undersigned, and each of them, guarantee and warrant that they are entitled to receive, in the proportions set out below, the proceeds of all casinghead gas produced from the following described property:

in _____ Parish,
_____ County, State of _____, and commencing at 7 a.m., _____,
and until further written notice, either from you or us, you are authorized to pay for casinghead gas received therefrom to the undersigned in the proportions shown.

Credit To	Division of Interest	Address

(See Division of Interest Digest of even date attached and made a part hereof)

304

assigns.

FIRST: The casinghead gas produced from said lease and received in pursuance of this division order shall become your property as soon as the same is received into your custody or that of any pipe line company designated by you:

SECOND: The casinghead gas purchased and received in pursuance of this division order shall be delivered f.o.b. to any pipe line designated by you which gathers and receives said casinghead gas and shall be paid for to the well owners or their assigns in proportion to their respective interests as above shown, at the price and under the provisions specified in the written contract between _____, and all renewals or modifications thereof.

THIRD: Settlement for the casinghead gas taken and received in pursuance of this Division Order shall be made monthly by check mailed to the respective parties according to the division of interest herein specified at the addresses herein given.

FOURTH: At any time on demand the undersigned agree to furnish Amoco Production Company satisfactory abstracts and/or other evidence of title, and in case of failure to do so, or in case of any adverse claim or dispute respecting title to the casinghead gas sold hereunder or the proceeds thereof or to the land from which produced, said company is authorized to hold the proceeds of all casinghead gas taken hereunder until such defect in title is corrected or such adverse claim or dispute is fully settled and determined to the satisfaction of said company, unless indemnity acceptable to said company shall be furnished.

FIFTH: This division order shall become effective as to each and every seller above named as soon as signed by him, irrespective of whether the other above named sellers have so signed.

Taxpayer Identification or
Social Security Number

Signatures

Address

Address

Address

Witness of Signatures

BE SURE TO INSERT YOUR TAXPAYER IDENTIFICATION OR SOCIAL SECURITY NUMBER IN THE SPACE PROVIDED ABOVE

Form 200 5-71

305

Oil and Gas Transfer Order

To: Amoco Production Company
P. O. Box 591, Tulsa, Oklahoma 74102

Property No. _____

_____, 19____

Beginning at 7 a.m., on the first day of _____, 19____ the undersigned who are credited with the interest or interests set out in Schedule "A" hereof authorize and direct you to give credit for said interest or interests to those named and in the proportions set out under Schedule "B" hereof.

Each of the undersigned OWNERS who are credited with the interest or interests in Schedule "B" guarantees and warrants he is the owner of the interest set out opposite his name on the reverse side hereof in oil and gas or the proceeds from the sale of oil and gas from the property described on the reverse side hereof, and until further written notice either from you or from us, the undersigned owners and all other parties executing this instrument hereby authorize you, your successors or assigns to receive and measure such sales in accordance with applicable governmental rules and regulations and to give credit as set forth on the reverse side hereof.

The following covenants are parts of this instrument and shall be binding on the undersigned, their successors, legal representatives, and assigns:

Oil: Oil sold hereunder shall be delivered f.o.b. to the carrier designated to gather and receive such oil, and shall become your property upon receipt thereof by the carrier designated by you or by any other purchaser to whom you may resell such oil. The term "oil" as used in this division order shall include all marketable liquid hydrocarbons.

Should the oil produced from the herein described land be commingled with oil produced from one or more other separately owned tracts of land prior to delivery to the designated carrier, the commingled oil sold hereunder shall be deemed to be the interest of the undersigned in that portion of the total commingled oil delivered which is allocated to the herein described land on the basis of lease meter readings or any other method generally accepted in the industry as an equitable basis for determining the quantity and quality of oil sold from each separately owned tract. Such formula shall be uniformly applied to all owners of an interest in the tracts of land involved.

Should the interest of the undersigned in the oil produced from the herein described land be unitized with oil produced from one or more other tracts of land, this instrument shall thereafter be deemed to be modified to the extent necessary to conform with the applicable unitization agreement or plan of unitization, and all revisions or amendments thereto, but otherwise to remain in force and effect as to all other provisions. In such event, the portion of the unitized oil sold hereunder shall be the interest of the undersigned in that portion of the total unitized oil delivered which is allocated to the herein described land and shall be deemed for all purposes to have been actually produced from said land.

You agree to pay for the oil sold hereunder at the price posted by you for oil of the same grade and gravity in the same producing such time as you do so, you agree to pay the price established by you. You are authorized to reduce the price by those truck, barge, tankcar, or pipe line transportation charges as determined by you.

Should the oil sold hereunder be resold by you to another purchaser accepting delivery thereof at the same point at which you take title, you agree to pay for such oil based upon the volume computation made by such purchaser and at the price received by you for such oil, reduced by any transportation charges deducted by such purchaser. Quality and quantity shall be determined in accordance with the conditions specified in the price posting. You may refuse to receive any oil not considered merchantable by you.

Gas: Settlements for gas shall be based on the net proceeds at the wells, after deducting a fair and reasonable charge for compressing and making it merchantable and for transporting if the gas is sold off the property. Where gas is sold subject to regulation by the Federal Power Commission or other governmental authority, the price applicable to such sale approved by order of such authority shall be used to determine the net proceeds at the wells.

Settlements: Settlements shall be made monthly by check mailed to the respective parties according to the division of interest herein specified at the latest address known by you, less any taxes required by law to be deducted and paid by you applicable to owner's interest.

Evidence of Title: In the event any dispute or question arises concerning the title to the interest of the undersigned in said land and/or the oil or gas produced therefrom or the proceeds thereof, you will be furnished satisfactory abstracts or other evidence of title upon demand. Until such evidence of title has been furnished and/or such dispute, defect, or question of title is corrected or removed to your satisfaction, or until indemnity satisfactory to you has been furnished, you are authorized to withold the proceeds of such oil or gas received and run, without interest. In the event any action or suit is filed in any court affecting the title to the interest of the under- signed in the herein described land or the oil or gas produced therefrom or the proceeds thereof to which the undersigned is a party, written notice of the filing of such suit or action shall be immediately furnished you by the undersigned, stating the court in which the same is filed and the title of such suit or action. You will not be responsible for any change of ownership in the absence of actual notice and satisfactory proof thereof.

Contingent Interests: Whether or not any contingency is expressly stated in this instrument, you are hereby relieved of any responsibility for determining when any of the interests herein shall increase, diminish, terminate, be extinguished or revert to other parties as a result of the completion or discharge of money or other payments from said interest, or as a result of the expiration of any time or term limitation (either definite or indefinite), and, unless you are also the operator of the property, as a result of an increase or decrease in production, or as a result of a change in the depth, the methods or the means of production, or as a result of a change in the allocation of production affecting the herein described land or any portion thereof under any agreement or by order of Governmental authority, and until you receive notice in writing to the contrary, you are hereby authorized to continue to remit without liability pursuant to the division of interest shown herein.

Warranties: Working Interest Owners and/or Operators, and each of them, by signature to this instrument, certify, guarantee and warrant, for your benefit and that of any pipe line or other carrier designated to run or transport said oil or gas, that all oil or gas tendered here- under has been and shall be produced from or lawfully allocated to the herein described land in accordance with all applicable Federal, state and local laws, orders, rules and regulations.

This instrument may be executed by one or more, but all covenants herein shall be binding upon any party executing same and upon his heirs, devisees, successors, and assigns irrespective of whether other parties have executed this instrument

307

Witness of Signature _____

Name _____

Street or Box No. _____

Social Security (or Tax ID) Number _____

City, State, Zip _____

Witness of Signature _____

Name _____

Street or Box No. _____

Social Security (or Tax ID) Number _____

City, State, Zip _____

Witness of Signature _____

Name _____

Street or Box No. _____

Social Security (or Tax ID) Number _____

City, State, Zip _____

Form 698 2-74

308

Chapter Five

EDITOR'S COMMENTS

Whether or not you are presently involved with Canadian oil and gas operations, you need to understand how the petroleum search in Canada is conducted.

The Canadian system of government, Canadian title law and Canadian leasing vary significantly from comparable concepts and practices in the United States. This is particularly an omalous, since the legal systems of both Canada and the United States are Anglo-Saxon in origin. The closest parallel with the United States is the leasing of Federal lands.

In "Canadian Petroleum and Natural Gas Land Operations", Robert A. Seaton traces the history of oil and gas development in Canada; the significance of the governmental structure in Canada on oil and gas development; the distinction between mineral ownership in the Canadian provinces, Canadian Federal territories and offshore areas, and "freehold" minerals; the types of petroleum and natural gas "agreements" available in Canada; and the many other features of Canadian law affecting land operations in Canada.

The following will prove of particular importance in understanding land operations in Canada:

- Why regulations play such a significant role in Canadian petroleum operations vis-a-vis statutes (see § **5.02(d)(2)**).

- The distinction in Canada between petroleum and natural gas leases and petroleum and natural gas exploratory agreements (see § **5.05(a)**).

• The easy-to-follow schedule of the various types of petroleum and natural gas "agreements" granted by the governments of key Canadian provinces (see § **5.05(b)**, Figure 8; § **5.05(c)**, Figures 9 and 10).

• The manner in which authority for surface access may, and must be acquired in Canada before a well can be drilled (see § **5.06**).

• The system of "incentive regulations" in the province of Alberta which provide economic encouragements to drill exploratory wells or conduct seismic operations through a system of "credits" (see § **5.09**).

• The Canadian tax system, including the special taxes applicable to the oil and gas industry and to Canadian branches of foreign corporations (§ **5.11(a)**).

— *L.G.M.*

Chapter Five

CANADIAN PETROLEUM AND NATURAL GAS LAND OPERATIONS

By Robert A. Seaton *

§ 5.01 Introduction

The major factor distinguishing Canadian petroleum land operations from those of the United States is the manner in which oil and gas rights are owned in Canada. The rule in Canada, subject to certain exceptions, is that oil and gas rights both on and offshore are owned by either the Federal government or one of the provincial governments.

The ownership of these oil and gas rights and the development of petroleum and natural gas land tenure systems by these governments relates directly to Canada's Constitution — the British North America Act. With the passing of this act by the British Parliament in 1867 Canada

* *Editor's note:* Seaton, president of the mineral management consultant firm of Seaton-Jordan & Associates Ltd., with offices in Calgary and Edmonton, Alberta, Canada, capped an illustrious 20-year career with the Department of Mines and Minerals of the Government of Alberta by serving as Alberta's director of minerals. Since leaving the government in 1970, he has served as a consultant to both industry and government in both Canadian exploration and worldwide offshore drilling. Seaton is co-editor of and the principal author of *Canadian Petroleum Land Operations* (The Institute for Energy Development, 1977).

became a self-governing nation. This Constitution provides that the elected governments would function under a parliamentary system and the several provinces would be joined together in a federal system.

A federal system implies a measure of self government by local units — provinces — with the central or federal government controlling matters of interest to all. One of the rights given to the provinces is the ownership of the minerals within provincial boundaries.

At present there are 10 provinces, each owning its own minerals and each having oil and gas laws providing the terms and conditions under which oil and gas rights may be required, held and continued during exploration and development. There are large areas of northern Canada that do not have provincial status because of the limited population. The Federal territories and the offshore areas are administered by the Federal government. In addition to this ownership of the minerals by the governments the structure of government in Canada also influences the petroleum and natural gas land tenure systems.

§ 5.02 Canada's System of Government: Federal Government; Governor General

§ 5.02(a) Governor General.

A casual reading of the British North America Act could convey a completely wrong idea as to how Canada is governed. The act provides that executive authority in Canada is vested in the Queen who is represented by a governor general appointed by her. This governor general summons and removes cabinet ministers who advise him in his work. He appoints the lieutenant governors of the provinces, the speaker of the Senate and the judges. He appoints the members of the Senate, and the House of Commons is called together by him and he can dissolve it at any time. No act of Parliament can become law unless the governor general assents to that law. While it would appear that the governor general occupies the central seat of authority in Canada, the opposite is the case, as he has little real power. Everything that is done in the name of the Queen by the governor general can be done by permission of the elected representatives of the people who are the members of the House of Commons.

§ **5.02(b) Cabinet.** Under the Constitution the governor general does not act according to his own judgment at any time, but only on the advice of his council, which is the cabinet. The cabinet is chosen by the prime minister who is the leader of the political party having a majority of the seats in the House of Commons. All members of the cabinet must have seats in the House of Commons or in the Senate, and most cabinet ministers are executive heads of departments of government. The prime minister and his cabinet must always have the support of the House of Commons or they must resign.

§ **5.02(c) Senate and House of Commons.** The legislative branch of government is made up of the Senate and the House of Commons. Representation in the House of Commons is given to each province in proportion to its population. In the case of the Senate, representation is so arranged to give some measure of equality to the different regions of the country.

§ **5.02(c)(1) Senate.** The Senate has 101 seats; 24 are allotted to each of the four major sections of the country — Ontario, Quebec, the Maritime Provinces and the four western provinces — and Newfoundland has six.

Senators are appointed by the cabinet and hold their office until retirement age of 75 years. The Senate has always been intended to be a secondary body, one that would represent the regions of the country and ensure that all legislation of the House of Commons would get a second look.

§ **5.02(c)(2) House of Commons.** The House of Commons is the principle law-making body and the dominant voice in Canadian political life. The Commons is a body of 265 members (MPs) elected from 263 electoral districts, two of them being dual-constituencies and each electing two members. It is the body through which the public exercises its political power and the body to which the executive or cabinet must turn for approval of any of its proposals.

The basic procedure in the passage of a bill from which an act of the Parliament results is that a bill receives three readings in the House of Commons and three in the Senate and then goes to the governor general for signature.

The House of Commons in the last resort controls all branches of the government — legislative, executive and judicial. No government can continue in existence without the support of the majority of the Commons.

§ 5.02(d) Provinces
§ 5.02(d)(1) Legislative Assembly.
For the provinces, government organization is similar to that of Federal government; however, there is one important difference in that there is only one legislative body. This body is generally called the legislative assembly or legislature. The legislature is made up of members (MLAs) elected by the various constituencies. In Alberta there are 75 constituencies, thus the legislature has 75 members.

Each provincial government has a cabinet with the ministers of each department of government being members of the cabinet, who must be elected members of the legislature. The cabinet is chaired by the Premier, who is the leader of the party having the majority of seats in the legislative assembly. The legislative assembly of the province is the only body capable of making law in the province and, within the framework of its constitutional powers, it is the supreme legislative authority in the province.

§ 5.02(d)(2) Statutes and Regulations.
While only the House of Commons at the Federal level and the legislature for the provinces can pass statutes, provision is usually made for the delegation of power to make regulations in these statutes to the governor general in council or the lieutenant governor in council. These statutes usually contain sections specifically allowing the governor general in council or the lieutenant governor in council to establish regulations from time to time dealing with matters referred to in the section.

The lieutenant governor in council is, by definition, the cabinet of the provincial government, and this means that this authority to make regulations is exercised by the cabinet.

This delegation of power to the cabinet provides a flexibility that would not be had otherwise as the lieutenant governor in council is able to exercise these delegated authorities at any time. The legislature meets infrequently, usually not more than twice a year, and then only for relatively short periods of time, so that a statute can only be revised or amended during these sessions of the legislature, whereas a regulation can be established, amended or withdrawn at any time that the cabinet cares to meet.

§ 5.03 Freehold Minerals

Most minerals are still owned by one of the governments, either provincial or Federal; however, title to some minerals has passed to private ownership, and the more important of these are the Hudson's Bay Lands, Railway Lands and Homestead Lands acquired before 1887. Each of these is discussed below.

§ 5.03(a) Hudson's Bay Lands. In 1670 Charles II granted to the Hudson's Bay Company all the lands draining into Hudson's Bay. This grant amounted to some one and one-half million square miles westerly from Hudson's Bay in much of what is now Manitoba, Saskatchewan, Alberta and the Northwest Territories.

In 1869 the Hudson's Bay Company entered into an agreement with the newly-formed government of Canada whereby the company surrendered all of its lands to Canada in consideration of: (1) a payment of 300,000 pounds sterling; (2) return of lands at the company's different trading posts and settlements totaling about 45,000 acres; and (3) a grant of one-twentieth of the land surveyed and settled during the next 50 years in what was referred to as the "Fertile Belt", an area described in the Deed of Surrender as being bounded on the south by the United States, on the west by the Rocky Mountains, on the east by Lake Winnipeg, Lake of the Woods and the waters connecting them, and on the north by the North Saskatchewan River.

315

As a result of this agreement, the people of Canada in 1869 owned 95 percent of the surface and minerals in the area known as the Fertile Belt and, of course, the surface and minerals in the balance of the area. This ownership and control made possible the settlement of the west.

§ **5.03(b) Railway Lands.** Following confederation in 1867, it became Federal policy to effect settlement of the west as quickly as possible, partly at least to frustrate the doctrine of "Manifest Destiny" which was being heard from the south. To implement this policy, subsidies of land which included the minerals were granted to those building railroads.

When the Canadian Pacific Railway Company was organized, the Dominion of Canada granted to it by way of subsidy for building a transcontinental railway, a large area including surface and minerals along the right-of-way of the railway. Similar grants were made to smaller railways, some of which were subsequently taken over by the Canadian Pacific Railway Company.

In the four western provinces approximately 3,500 miles of railway were constructed and some 31.6 million acres of both surface and minerals were transferred to the different companies as subsidies.

§ 5.03(c) **Homestead Lands.** Settlers acquiring homesteads obtained the minerals with the surface until October 31, 1887, when the Federal government, by Order in Council 1070, excepted minerals from such grants.

§ **5.03(d) Mineral Sales.** For a period of time after 1887, when the government commenced reserving minerals from surface dispositions, it was the practice to grant title to minerals for development purposes. In some instances the titles comprised all minerals; in other instances just specific minerals were granted, i.e. coal. Many of these mineral sales reserved a royalty and, even though the rights are held under freehold title, there may be royalties payable to the respective provincial governments.

§ **5.03(e) National Parks and Indian Reserves.** The minerals within the national parks scattered about Canada are owned by the Federal government and no exploration for or development of minerals is allowed within these parks.

The minerals within Indian reserves are owned by the Indians living on the reserves and are administered for them by a Federal department of government known as the Department of Indian Affairs and Northern Development.

§ 5.04 The Oil and Gas Industry in Canada

Oil has been produced in Canada for more than 100 years with the first oil being produced from wells in southwestern Ontario near Sarnia. It was not until 1914, however, that Canada's first significant discovery was made when wet gas was discovered at Turner Valley in southern Alberta. Crude oil was later discovered along the flank of Turner Valley, however, and no further major discoveries were made for many years.

The discovery of light oil at Leduc in central Alberta in 1947 heralded a new era for oil and gas exploration in western Canada. This discovery at Leduc opened the way for an exploratory effort which eventually extended throughout the whole of the western Canadian sedimentary basin and led to the establishment of the industry as we know it today. This western Canadian sedimentary basin extends for more than 1,800 miles from the United States border to the Arctic Ocean and then northeasterly for another 1,400 miles through the Arctic Islands.

The industry has developed to the point where today there are in excess of 16,000 wells capable of producing oil and in excess of 9,000 wells capable of producing gas. Annual expenditures of approximately $1 billion have been incurred in exploration and development by the industry in Alberta alone during the past few years.

Crude oil production in Alberta averages approximately one and one-half million barrels a day. Approximately 5,000 wells in 1977 have been drilled in Alberta. The footage drilled has consistently been in excess of 10 million feet per year.

Some 50,000 barrels of oil a day are being produced from an oil sands area of northeastern Alberta and, at present, there is under construction a second plant for the extraction of oil from the same oil sands that will have a daily capacity of approximately 100,000 barrels of oil a day.

Direct revenues to the various governments of Canada have also increased. As most oil and gas activity is centered in Alberta, it is the government of the province of Alberta that reaps the major benefits.

This revenue comes from three major sources — the sale of petroleum and natural gas rights, rentals and royalties. The total of these three revenue sources has been well in excess of $2 billion a year over the past few years.

It has been estimated that as of January 1, 1976 there were approximately 7.1 billion barrels of conventional crude oil remaining to be recovered in the whole of Canada. At the same time it has been estimated that some 300 billion barrels of oil may be recovered from the oil sands areas of northern Alberta.

§ 5.05 Overview of Petroleum and Natural Gas Land Tenure Systems

§ 5.05(a) Petroleum and Natural Gas Agreements.
Each of the provinces has legislation pursuant to which they grant the right to the oil and gas they own within their provinces. These oil and gas rights are granted pursuant to agreements that are referred to as leases, licenses, reservations or permits.

A distinction is usually made between petroleum and natural gas leases and all of the other types of agreements. The petroleum and natural gas lease is the agreement pursuant to which rights are held for the purposes of having the oil and gas developed or exploited.

All of the other agreements — reservations, permits, licenses, drilling reservations — are what are referred to as exploratory agreements. These agreements comprise relatively large areas and have relatively short terms during which the holders are obligated to conduct specific exploration work. Upon having completed this specific work the oil company is entitled to acquire petroleum and natural gas leases comprising all or some part of the exploratory agreement.

§ 5.05(b) Petroleum and Natural Gas Leases. In Alberta there are in existence three types of petroleum and natural gas leases. The main difference between these leases is the length of the term. Leases issued prior to June 1, 1962 had a term of 21 years and many of these leases are still in existence. Leases issued between June 1, 1962 and July 1, 1976 had a term of 10 years. Petroleum and natural gas leases issued after July 1976 (excepting in certain instances) will have terms of five years. The terms and conditions of these three different types of petroleum and natural gas leases and petroleum and natural gas leases issued by the other provinces are summarized in Figure 8.

§ 5.05(c) Exploratory Agreements. In Alberta under the new petroleum and natural gas land tenure system introduced on July 1, 1976, the only exploratory agreement that can now be acquired is the Petroleum and Natural Gas License. This license can be held for a term of from two to five years depending upon the location of the license. Licenses located in the plains area of the province can be held for a maximum of two years, in the northern area for a maximum of four years and in the foothills area for a maximum of five years.

In the same way the maximum size of the licenses varies with its location. The maximum size of a plains area license is 29 sections, of a northern area license is 32 sections and of a foothills area license is 36 sections.

In order to earn petroleum and natural gas leases out of one of these exploratory licenses it is necessary that a well be drilled. The size of the area that can be acquired under petroleum and natural gas leases is determined by the number of wells that are drilled on the license and the depths of such wells. The deeper the wells the more acreage that can be earned. By drilling either enough wells or wells that are deep enough, the whole of the licensed

319

P & N G LEASES

PROVINCE	TERM	ANNUAL RENTAL	MAXIMUM SIZE
Alberta			
1. Issued before June 1, 1962	21 years	$1.00	9 sections
2. Issued between June 1, 1962 and July 1, 1976	10 years	$1.00	9 sections
3. Issued after July 1, 1976	5 years	$1.00	36 sections
Saskatchewan	5 years	$1.00	12¼ sections
British Columbia	10 years	$2.00	36 units

Figure 8

320

area can be acquired under a petroleum and natural gas lease.

Under the petroleum and natural gas land tenure system in existence prior to July 1, 1976, petroleum and natural gas reservations, petroleum and natural gas permits, Crown reserve drilling reservations, natural gas licenses and Crown reserve natural gas licenses could be issued for exploratory purposes. Any of these agreements that were in existence on July 1, 1976, when the new system was introduced, can be continued to the end of their term. At that time the holders will be entitled to acquired petroleum and natural gas leases under the regulations as they existed on July 1, 1976.

The other provinces also have provision for the granting of exploratory agreements and these are usually referred to as petroleum and natural gas permits. Some of the provinces also provide for natural gas licenses and Crown reserve drilling reservations. The terms and conditions of the various types of petroleum and natural gas exploration agreements for the various provinces are summarized in Figures 9 and 10.

§ 5.06 Acquisition of Surface Rights

§ 5.06(a) In General. In order to drill a well it is necessary to have access to enough of the surface area to enable a drilling rig to be moved onto the location. This right to the surface is required in addition to the petroleum and natural gas lease that the person drilling the well would have acquired either from the government or the mineral owner.

The three western provinces have provided special legislation which makes it possible for the mineral owner, or the person having the right to recover the oil and gas, to acquire that part of the surface necessary for his drilling operations. This legislation takes away the common law right of a mineral owner to work his mineral but, at the same time, provides a procedure enabling him to acquire surface rights without litigation.

Under these laws owners are compensated for loss of the surface, damage to the land and any inconvenience they might experience arising from the drilling operations. In Alberta the act is known as the Surface

PETROLEUM & NATURAL GAS EXPLORATORY AGREEMENTS

PROVINCE	TERM	MAXIMUM SIZE	WORK REQRMNT.	LEASE ENTLMNT.	GROUPING	RENT
Alberta P&NG License	2 years - Plains, 4 years - Northern, 5 years - Foothills	2 secs. - Plains, 32 secs. - Northern, 37 secs. Foothills	At least one well required	100% of license area if enough wells drilled or if wells drilled deeply enough	Maximum of 2 licences within distance of 2 miles	$1.00 an acre a year
P&NG Reservations	6½ years	100,000 acres (156 secs.)	Geophysical or drilling	50% of reservation area in each township in approved pattern	Any number of reservations up to a maximum of 200,000 acres	1st year - Nil, 2nd & 3rd years - 20¢/acre, 4th year - 20¢/acre, 5th year - 30¢/acre, 6th year - 70¢/acre, last 6 months - 50¢/acre
P&NG Permits	3 years	36 secs.	A well to test for oil or gas must be drilled	100% of permit area for the drilling of one well	Not allowed	$1.00 an acre a year with provision for rebate if early drilling undertaken
British Columbia P&NG Permit	8 years	One grid area 96,370 acres to 126,230 acres	Geophysical or drilling	50% of permit area in approved pattern	Permits comprising not more than 5 grid areas within a radius of 50 miles	(see rate table below)
Saskatchewan P&NG Permit	5 years	100,000 acres	Geophysical or drilling	60% if lease more than 2x2 mi. 50% if lease no more than 3x3 mi. 40% if lease no more than 3½x3½ mi.	Any number of permits not separated by more than 6 miles	10¢ an acre a year

British Columbia P&NG Permit rent:

	Class A	Class B	Class C	Class D
Year 1-2	10¢	10¢	10¢	5¢
Year 4-5	20¢	20¢	20¢	10¢
Year 6	40¢	30¢	20¢	10¢
Year 7	50¢	30¢	20¢	15¢
Year 8	60¢	50¢	20¢	15¢

Figure 9

322

DRILLING RESERVATIONS

PROVINCE	TERM	SIZE	WORK REQUIREMENT	LEASE ENTITLEMENT	RENT
Alberta	3 years	10 sections to 16 sections	Holder must drill to specified zone	Approximately 25% drilling reservation area	25¢ an acre for each 6 month period
British Columbia	3½ years	32 units to 144 units	Holder must drill to specified zone	25% of drilling reservation area	1st year - 20¢/acre 1st 6 mo. - 10¢/acre Successive 6 mo. - 20¢/acre renewal
Saskatchewan	3 years	20,000 acres	Holder must drill to specified zone	Up to maximum of 50% of drilling reservation area	50¢ an acre a year

Figure 10

Rights Act and, under the provisions of the act, there is a Surface Rights Board established. The act provides that no mineral operator may enter on the surface of the land without having either the consent of the surface owner and the occupant or an order of the Surface Rights Board granting the operator the right to enter upon the surface in order to conduct his operation.

Applications to the Board must be accompanied by a plan or a description of the land required and copies must be served on all persons having an interest in the land to be affected by the operations.

The amount of compensation granted to the surface owner varies from case to case and depends upon the nature of the soil, damages to growing crops, inconvenience caused by severence, noise and any other factors which the Board deems advisable to consider. The Board orders for right of entry may be acquired for well sites, access roads, flow lines and other pipelines.

§ 5.06(b) **Crown Lands in Alberta.** Where the surface is not held under title but is still owned by the government of the province of Alberta, then applications for the use of the area required for well sites or the area required for roadways would be submitted to the Department of Energy and Natural Resources, Land Use Branch.

§ 5.06(c) **British Columbia and Saskatchewan.** Provision for right of entry in British Columbia is made in Part III of the Petroleum and Natural Gas Act of 1965. It provides for the acquisition of the necessary surface for oil and gas operations whether or not the surface is owned by the Crown, or where it has been granted to a freehold owner. The Board administering this part of the act in British Columbia is known as the Mediation and Arbitration Board.

There is a similar Board in the province of Saskatchewan.

§ 5.07 Drilling and Production Regulations

For each of the provinces there is legislation that provides that some agency of the government shall assure that:

(1) The oil and gas resources are produced in such a way so as to provide for a minimum of waste.

(2) Safe and efficient drilling completion and abandonment practices are adhered to.

(3) Each owner of oil and gas rights has the opportunity of obtaining oil or gas from any pool.

(4) The recording and dissemination of information relating to oil and gas is provided for.

All three of the western provinces have such legislation, and in Alberta the agency of the government charged with this responsibility is the Energy Resources Conservation Board.

It is under these statutes that licenses for the drilling of wells are required. In Alberta an application for a license to drill a well anywhere in Alberta, on either Crown or freehold lands, must be submitted to the Conservation Board.

In Saskatchewan a drilling license is required under the provisions of that Province's Oil and Gas Conservation Act. In British Columbia there is a similar requirement and there the authority for the drilling of a well is not called a license but a "well authorization".

Each of the three western provinces has drilling and production regulations pursuant to which, after the issue of either drilling licenses or well authorizations, the drilling, completion and abandonment of wells are administered.

§ 5.08 Petroleum and Natural Gas Sales

Most petroleum and natural gas rights acquired from the Crown in western Canada today are acquired only after they have been advertised for sale.

In Alberta sales of petroleum and natural gas rights, which include both petroleum and natural gas leases and petroleum and natural gas licenses, are held weekly in Calgary. The other provinces hold sales of petroleum and natural gas rights less frequently.

The sales are conducted by the departments of the government administering the mineral resources for the provinces, and requests to have petroleum and natural gas rights advertised must be forwarded to the department having jurisdiction well in advance of the time that the rights are requested to be advertised. In Alberta at the present time a minimum of 12 weeks is required in order to have the government do the necessary work preliminary to having the rights advertised.

Each sale of petroleum and natural gas rights, regardless of the province, is conducted pursuant to the instructions set out in a sale notice. These sale notices should be carefully scrutinized to make sure that the instructions are followed in detail.

When a reference is made to the posting of petroleum and natural gas rights, this means the advertising of such rights and the sale notices are often referred to as posting notices. The words "posting" and "advertising" in this context are interchangeable.

§ 5.09 Overview of Incentive Regulations in Alberta

There are two sets of what are termed incentive regulations in the province of Alberta — the exploratory drilling incentive regulations of 1974 and the geophysical incentive program regulations.

Both these sets of regulations are intended to provide incentives to the oil industry to undertake the drilling of exploratory wells and seismic operations in the province of Alberta.

The regulations were established by the Alberta government as the result of circumstances that arose following the last war in the Middle East which resulted in the OPEC nations placing a partial embargo upon crude oil being shipped to certain countries. This in turn resulted in the rapid escalation of the international price for crude oil which rose from approximately $3 to something in excess of $11.

The immediate response of governments around the world was that the "windfall profits" accruing to the oil companies as a result of these rapidly increasing prices should more properly go to the governments. In Canada the provincial governments all, in very short order, attempted to appropriate what they thought was their fair share of these increased prices. In Alberta we saw that in a short period of time the royalties increased from a maximum of 16 ⅔ percent to rates exceeding 40 percent. The situation was similar in the other producing provinces.

Following the increase in the royalty rates by the provinces the Federal government suggested that their share of the crude oil barrel was now even lower than it had been prior to the increased prices, because the Federal government traditionally took its share through income taxes. As royalties were allowable deductions in the calculations of these taxes, the Federal government contended that following the rapid increase in royalty rates their tax base had been eroded to the point where their return was lower than it had been before the increase in prices.

Oil companies now found that in addition to paying the very high royalties amounting to — in the case of Alberta — 40 percent, they also had to pay taxes on these royalties paid to the provincial governments. In some instances companies indicated that they were in a position where they were losing money for each barrel of oil produced, following the imposition of the higher prices and with the subsequent increase in royalty rates and the denial of royalty deductions for tax purposes.

This situation led to a marked decrease in the amount of exploration drilling done in western Canada. In fact, it was suggested that many drilling companies were moving equipment to the United States and seismic operations were being curtailed. The Alberta government, in response to this situation, provided an incentive package for the oil industry, part of which were the two regulations referred to earlier.

These regulations provide, in essence, for the establishing of a credit by the government where oil companies drill certain kinds of exploratory wells or where they conduct reflection seismic work in Alberta. A credit is established for the expenditures incurred in doing this work and then these credits can be utilized by the companies in defraying certain cash obligations that they might have to the government.

The credit that is established for both the expenditures under the incentive drilling regulations and the geophysical regulations can be utilized to pay rentals on leases, reservations, permits or licenses, or they can be utilized in the payment of royalties for both oil and gas. They can also be utilized as bonuses when bidding on leases or licenses at sales conducted by the government.

There is also provision, where an incentive well is drilled and discovers either oil or gas, for a waiver of royalty on the production from the well for a period of five years in the case of an oil discovery, and for a period of two years in the case of a gas discovery.

§ 5.10 Survey Systems

§ 5.10(a) **Western Canada.** The settled areas of western Canada are surveyed under a system which divides the province into townships comprising 36 sections.

This type of survey uses meridians running north and south, range lines running north and south and township lines running east and west.

The principal meridian passes about 12 miles west of the city of Winnipeg with each successive meridian being four degrees west of the preceeding one. The fourth meridian, for example, is the border line between the provinces of Alberta and Saskatchewan.

The areas between the meridians are divided by ranges six miles apart. To complete the grid, township lines are also six miles apart. Thus, each township comprises 36 sections of approximately one square mile each.

The townships are numbered northerly from the international border — 49th parallel of latitude — and the ranges are numbered westerly from each meridian.

Between each section there is a road allowance running north and south, and between every second section there is an east-west roadway. Each section is divided into quarter sections and there is a further division called a legal subdivision comprising 40 acres. In the example attached the area cross-hatched would be described as Legal Subdivision 6, Section 9,

Township 45, Range 10, west of the 4th Meridian (see Figure 11).

§ 5.10(b) Federal Lands. For the purposes of the Canada Oil and Gas Land Regulations pursuant to which oil and gas rights in the Northwest and Yukon Territories, Arctic Islands and offshore areas are administered, lands are divided into grid areas.

Grid areas south of latitude 70 degrees are bounded on the east and west by successive lines of longitude in the series: 50° 00'00", 50° 15' 00", 51° 30' 00"; the north and south sides of these grid areas bounded by successive parallels of latitude in the series: 40° 00' 00", 40° 10' 00", 40° 20' 00".

The grid areas lying north of latitude 70 degrees are bounded on the east and west by successive lines of longitude in the series: 50° 00' 00", 50° 15' 00", 51° 30' 00"; the north and south sides of these grid areas bounded by successive parallels of latitude in the series: 70° 00' 00", 70° 10' 00", 70° 20' 00".

The grid area is identified by the latitude and longitude of the northeast corner of that grid area.

Each grid area is divided into sections. In the case of a grid area lying between latitudes 40 degrees and 60 degrees, or between latitudes 70 degrees and 75 degrees, the grid is divided into 100 sections.

In the case of a grid area lying between latitudes 60 degrees and 68 degrees, or between latitudes 75 degrees and 78 degrees, the grid is divided into 80 sections.

In the case of a grid area lying between latitudes 68 degrees and 70 degrees, or between latitudes 78 degrees and 85 degrees, the grid is divided into 60 sections.

Each section is divided into 16 units.

Figure 12 indicates the manner of identifying the sections and units.

§ 5.10(c) British Columbia. A large part of British Columbia, including much of the area of interest for petroleum and natural gas, is unsurveyed. Consequently, a survey system has been set up under the Petroleum and

SKETCH OF
TOWNSHIP NO. 45 RANGE 10 WEST OF 4th MERIDIAN

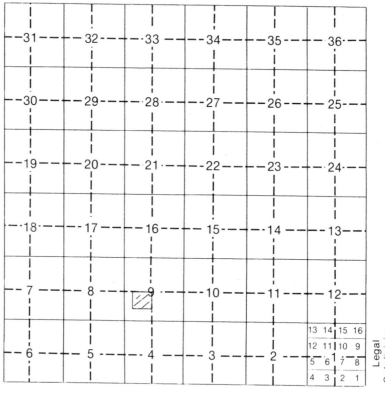

Township - 36 sections / Section - 640 acres / Quarter Section - 160 acres
Legal Subdivision - 40 acres

Figure 11

330

UNITS

M	N	O	P
L	K	J	I
E	F	G	H
D	C	B	A

Grid area lying between lat. 40° and 60°, or between lat. 70° and 75°.

100	90	80	70	60	50	40	30	20	10
					49				
					48				
					47				
					46				
95	85	75	65	55	45	35	25	15	5
					44				
					43				
					42				
91	81	71	61	51	41	31	21	11	1

Grid area lying between lat. 60° and 68°, or between lat. 75° and 78°.

80	70	60	50	40	30	20	10
				39			
				38			
				37			
				36			
75	65	55	45	35	25	15	5
				34			
				33			
				32			
71	61	51	41	31	21	11	1

Grid area lying between lat. 68° and 70°, or between 78° and 85°.

60	50	40	30	20	10
			29		
			28		
			27		
			26		
55	45	35	25	15	5
			24		
			23		
			22		
51	41	31	21	11	1

Figure 12

331

Natural Gas Act of British Columbia pursuant to which the unsurveyed areas of British Columbia are divided into grid areas:

(1) Grid area — A grid area is defined in Section 147 of the act as being a rectangular figure bounded on the north and south by 20 successive chords of parallels of latitude of a 15-minute series commencing at any degree of north latitude, and bounded on the east and west by successive meridians of longitude of a 15-minute series commencing at any degree of west longitude.

(2) Block — Each grid area is divided into six blocks. Each block is a rectangle bounded by five minutes of latitude and by seven minutes and 30 seconds of longitude.

(3) Units — Each block is subdivided into 100 units.

The numbering of the blocks and units for this system are indicated on Figure 13.

The size of the grid area varies with its location. From south to north the acreage of a grid area would vary from 127,230 acres at the south portion of the province to 96,368 acres at the north boundary of the province. A block would vary in size from 21,239 acres to 16,143 acres. The units would vary from 212 acres at the south to 160 acres at the north boundary of the province.

§ 5.11 Overview of Canadian Taxes

§ 5.11(a) Income Taxes.

§ 5.11(a)(1) In General. Both the Federal and provincial governments impose an income tax on individuals and corporations. The income tax system, except for the provinces of Ontario and Quebec, is administered by the Federal government. Total Federal and provincial corporation income taxes approximate 50 percent. Individuals are taxed on a graduated scale with the approximate maximum of 62 percent being reached at about $70,000 of taxable income. Non-capital losses of the five preceding years and the following year may be deducted in computing taxable income.

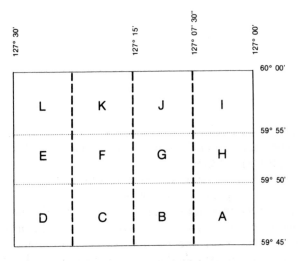

A map-sheet in the National Topographic Series is divided into 12 blocks as illustrated above for Map 94-M-14. A grid consists of the six blocks in the east or west half of the sheet.

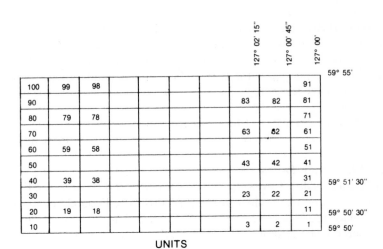

UNITS

Figure 13

333

Capital losses may be carried back one year and forward indefinitely, but may be deducted only against capital gains.

§ **5.11(a)(2) Capital Gains.** Effective January 1, 1972 one-half of capital gains are included in taxable income. One-half of capital losses are deductible, but only against capital gains.

§ **5.11(a)(3) Resource Industries.** The basic rate of Federal income tax on oil and gas profits is 46 percent, but in recognition of the provinces' right to impose separate taxes and royalties special abatements for provincial income taxes and royalties reduce the Federal levy on oil and gas profits to approximately 25 percent.

Provincial royalties on oil and gas are not deductible in computing Federal income taxes.

There is an "earned depletion allowance" allowing a deduction of $1 for each $3 spent on exploration and development and on certain new processing facilities. The maximum annual depletion allowance is 25 percent of resource income.

§ **5.11(a)(4) Branch Operation of a Foreign Corporation.** A Canadian branch of a nonresident corporation is subject to the same income taxes as a Canadian company. Taxable income is determined on the basis of separate accounting as if the branch were an independent enterprise.

Allocations for a reasonable portion of home office expense are deductible.

Canadian branches of foreign corporations are subject to a special branch tax of 25 percent on profits that are not reinvested in Canada.

§ **5.11(b) Other Taxes**

§ **5.11(b)(1) Sales Taxes.** Sales taxes are imposed by both the Federal government and all provinces except Alberta. The Federal tax applies to most goods produced or imported into Canada. The tax is generally levied at 12 percent on the selling price, but is reduced for some products such as building materials.

The provincial sales taxes are generally levied on purchases of tangible goods purchased for personal consumption and range from five to eight percent.

§ 5.11(b)(2) Excise Tax and Custom Duties. These taxes and duties are levied by the Federal government on specified goods at various rates as listed in lengthy tariff schedules to the Excise Tax Act and the Customs Act.

§ 5.11(b)(3) Property Taxes. Property taxes are imposed by municipal governments on the value of real property and by the provinces upon land not situated in a municipality. In general, a property tax is levied on the owner and a business tax is levied on the occupant if used for business purposes.

School taxes based on the value of real property are levied by local school boards.

Chapter Six

EDITOR'S COMMENTS

You cannot understand the economic effects of a transaction on your company unless you understand its tax consequences; it is the *after-tax*, not the *pre-tax*, profit which determines the "bottom line" of how well the company did (see § **6.01**).

An understanding of the tax consequences of the transaction *to the other party* also can help you make a deal when others have failed, or make a more advantageous arrangement for your company by structuring the lease, farmout or other trade so that advantage is taken of otherwise ignored tax benefits.

In "Basic Tax for the Landman", Joe Shannonhouse explains the ABC's of Federal income taxation as it applies to oil and gas transactions, including extensive illustrations of the tax consequences of the various types of transactions. You should find the following points particularly useful:

- How and under what circumstances "depletion" can still be claimed despite the Tax Reduction Act of 1975 (see § **6.02**).

- The distinction between a "sale" and a "sublease", including their differing tax consequences (see § **6.03**).

- Special problems which may be created if a geologist or engineer takes an interest in the property to which his services related in lieu of cash payment for his services (see § **6.04**).

• How percentage depletion, otherwise available to a small producer, may be inadvertently lost if a farm-out agreement is improperly drafted (see § **6.04**).

• How to avoid problems concerning "IDC" deductions under "turnkey" drilling contracts (see § **6.05**).

• When prepayments will stand up for tax purposes (see § **6.05**).

• How to avoid the potentially disastrous tax results of a joint oil and gas operation being classified as an "association taxable as a corporation" (see § **6.06**).

— *L.G.M.*

Chapter Six

BASIC TAX FOR THE LANDMAN

By Joseph G. Shannonhouse *

§ 6.01 Introduction

Typically, the oil and gas Landman will not be expected to be an expert in the area of oil and gas taxation. However, in many instances it could be of significant benefit to the Landman to have a basic knowledge of oil and gas taxation. For example, a basic understanding of the tax treatment of farmouts may allow the Landman to enter into a farmout transaction on a basis that will minimize current tax liability resulting from the transaction, and also insure the availability of future anticipated tax benefits. A basic knowledge of the concepts of oil and gas taxation is also very important in attempting to evaluate the economics on which a property may be acquired.

This chapter is designed to introduce the Landman to some of the basic concepts of oil and gas taxation so that he will be in a position to maximize the available tax benefits in connection with the acquisition of oil and gas properties.

---

* *Editor's note:* Shannonhouse is a member of the Oklahoma City-based law firm of Andrews, Davis, Elam, Legg & Bixler, Inc. A member of the Section of Taxation of the American Bar Association, Shannonhouse holds a Master of Laws Degree in taxation from New York University.

§ 6.02 Depletion

§ 6.02(a) In General. Depletion is one of the most fundamental areas of oil and gas taxation and, at the same time, is one of the most difficult to understand fully. Simply stated, depletion is the gradual using up or destruction of a capital asset. In connection with oil and gas, depletion may be defined as the exhaustion of oil and gas reserves as a result of the drilling of wells and production therefrom. For Federal income tax purposes it is a deduction from gross income provided by the Internal Revenue Code (the "Code") to compensate for the taxpayer's capital depletion which is the result of production. Depletion may be thought of as being somewhat analogous to depreciation.

Annual depletion deductions are allowed only to the owner of an economic interest in the mineral deposit. Under U.S. Treasury regulations, an economic interest is deemed to be possessed in every case in which the taxpayer has acquired by investment any interest in minerals in place and has secured, by any form of legal relationship, the income derived from the extraction of the minerals to which he must look for a return of his capital.[1]

The existence of an economic interest in the minerals is the *sine qua non* for claiming the deduction for depletion; otherwise, the taxpayer has no capital investment that has suffered depletion and is not entitled to the statutory allowance.[2] The existence of an economic interest is, in turn, based on capital invested in the oil and gas in place. Without capital invested the taxpayer has not suffered the depletion for which the deduction is intended to compensate.

How does an Oil and Gas Lease fit in with the concept of ownership of an economic interest in oil and gas in place? The usual form of Oil and Gas Lease provides that the landowner shall be entitled to receive one-eighth of

---

[1] See U.S. Treasury Regulations, § 1.611-1(b)(1).

[2] See *Kirby Petroleum Co. v. Commissioner*, 46-1 USTC § 9149 (U.S. Sup. Ct. 1946).

the oil and gas produced. In GCM 22730[3], the Internal Revenue Service (IRS) considered the reserved royalty equivalent to the entire interest in oil and gas in place, subject to the obligation to develop. On a lease, the Lessor landowner is treated as not having parted with a capital investment upon execution of the lease, even though the Lessee is deemed to have made a capital investment in the oil and gas in place. The Lessee acquires an economic interest represented by the bonus paid upon execution of the lease, if any, and by assuming the obligation to develop. The Lessee is treated as having made a purchase of an economic interest under the Oil and Gas Lease, but the Lessor from whom the Lessee made his purchase is not treated as having made a sale.

U.S. Treasury regulations provide that any form of legal relationship can be utilized in the acquisition of an economic interest.[4] The determinative factor is that income from the extraction of the oil and gas must be looked to for a return of capital. It is not necessary to hold a working or operating interest. Nor is it necessary to be present in the chain of title. In *Comm. v. Southwest Exploration Co.*[5], the court stated as follows:

> "It is to be noted that in each of the prior cases where the taxpayer has had a sufficient economic interest to entitle him to depletion, he has once had at least a fee or leasehold in the oil producing properties themselves. No prior depletion case decided by this court has presented a situation... where a fee owner of adjoining lands necessary to the extraction of oil is claiming a depletion allowance.
>
> "Southwest contends that there can be no economic interest separate from the right to enter and drill for oil on the land itself. Since the upland owners did not themselves have the right to drill for offshore oil, it is argued that respondent — who has the sole right to drill — has the sole economic interest. It is true that the exclusive right to drill was granted to Southwest, and it is also true that the

---

[3] 1941-1 CB 214.

[4] See U.S. Treasury Regulations § 1.611-1(b)(1); *Palmer v. Bender*, 3 USTC § 1026 (U.S. Sup. Ct. 1932).

[5] 56-1 USTC § 9304 (U.S. Sup. Ct. 1956).

agreements expressly create no interest in the oil in the upland owners. But the tax law deals in economic realities, not legal abstractions, and upon closer analysis it becomes clear that these factors do not preclude an economic interest in the upland owners.

"Southwest's right to drill was clearly a conditional rather than an absolute grant. Without the prior agreements with the upland owners, Southwest could not even have qualified as a bidder for a state lease. Permission to use the upland sites was the express condition precedent to the State's consideration of Southwest's bid.... For a default in that condition the State retained the right to re-enter or to cancel the lease. Thus, it is seen that the upland owners have played a vital role at each successive stage of the proceedings. Without their participation there could have been no bid, no lease, no wells and no production."

A royalty interest also constitutes an economic interest. This is true whether the royalty owner is paid in cash or by a share of production in kind.[6] Similarly, an overriding royalty constitutes an economic interest in oil and gas in place. In *Hogan v. Commissioner*[7] it was held that when the owner and operator of a producing Oil and Gas Lease assigns and transfers it for cash and the payment of an overriding royalty, he retains an economic interest in the oil and gas in place which will be depleted by production.

The owner of all or a part of the working interest is also deemed to hold an economic interest. Although the application of this rule would seem very clear, such is not always the case. The facts of *Pearl Oil Co.*[8] illustrate the type of troublesome situations which may arise.

In *Pearl Oil Co.* the petitioner sold a lease and as partial consideration received a note. The purchaser subsequently failed to pay the note and the petitioner obtained a compromise judgment to have the balance due on the note paid out of oil and gas production. The court held that the judgment did not give the petitioner an economic interest in the oil and gas but merely

---

[6] See *Greensboro Gas Co.*, 30 BTA 1361 (1934); *William v. Comm.*, 36-1 USTC § 9178 (5th Cr. 1936).

[7] 44-1 USTC § 9217 (5th Dir. 1944).

[8] 40 BTA 147 (1939).

described the method for payment of the judgment.

In connection with carried interest arrangements, the question of whether or not the carried party is deemed to have an economic interest during the payout period has been the subject of much uncertainty. However, it is now generally accepted that the carried party does not hold an economic interest during payout and is not entitled to depletion nor is he required to report income from production during this period attributable to his carried interest.

Another area of some confusion involves net profits interest. It has been held that the acquisition of a net profits interest as consideration for an indispensable contribution of the use of real property adjacent to the oil deposits results in the creation of an economic interest.[9] However, when received for services or capital disassociated from the lease or minerals in place, or for services not connected with actual development of the lease, a net profits interest does not constitute an economic interest in oil and gas in place.[10]

In connection with shareholders and corporations, despite indirect ownership, shareholders of corporations are generally not entitled to depletion because of production from corporate owned properties. A limited exception to this rule is available for Subchapter S corporations. However, as a result of the rules governing taxation of Subchapter S corporations (Section 1371 et. seq. of the Code) the benefit of percentage depletion is effectively locked into the corporation because of the limitations imposed as to the flow through of deductions.

The question of whether or not production payments give rise to an economic interest was affected by changes in the Code made by the Tax Reform Act of 1969. Prior to the enactment of the Tax Reform Act of 1969 production payments, if payable only out of proceeds from the sale of oil if

---

[9] See Commissioner v. Southwest Exploration Co., 56-1 USTC § 9304 (U.S. Sup. Ct. 1956).

[10] See Helvering v. O'Donnel, 303 US 370 (U.S. Sup. Ct. 1938); Kirby Petroleum v. Comm., 326 US 599 (U.S. Sup. Ct. 1946).

and when produced, were treated as giving rise to an economic interest, thereby entitling the holder thereof to claim depletion.[11] However, as a result of changes made to Section 636 of the Code by the Tax Reform Act of 1969, a production payment is now generally treated as a non-recourse loan not giving rise to an economic interest in the oil and gas in place and, therefore, not subject to depletion.

After it has been determined that depletion will be available with respect to certain income, the next logical question becomes, "When will such deduction be allowed?" Stated generally, the deduction for depletion is allowable in the year of receipt of the income, and time of the receipt of the income will depend on the method of accounting, cash or accrual, used by the recipient of the income for Federal income tax purposes.

Either of two types of depletion may be available under the Code — cost or percentage depletion. Cost depletion is computed by dividing the cost of the mineral interest by the estimated recoverable reserves, and then multiplying by the number of units sold during the taxable year. Total deductions for cost depletion may not exceed the basis of the interest (initial cost plus capitalized leasehold items). The taxpayer has the burden of establishing his basis in the interest. Although this would seem to be an easy task, such is not always the case. For example, where the surface also is acquired none of cost may be allocated to minerals unless the minerals were previously severed or the taxpayer is able to establish a reasonable basis of allocation of cost between the surface and minerals.

Also, for purposes of computing cost depletion, estimated recoverable reserves include not only proven reserves but also "probable" or "prospective" reserves.

The second type of depletion which may be available is percentage depletion. Simply stated, percentage depletion is a deduction provided by statute equal to a certain specified percentage of gross income from oil and gas production. The method of computation of percentage depletion is set

---

[11] See *Lee v. Comm.*, 42-1 USTC § 9375 (5th Cir. 1942); *Anderson v. Helvering*, 40-1 USTC § 9479 (U.S. Sup. Ct. 1940).

forth in Sections 613 and 613A of the Code. Generally, the deduction for percentage depletion is claimed when it exceeds the deduction allowable for cost depletion. In addition to other limitations, a taxpayer must be entitled to deductions for cost depletion to be entitled to claim percentage depletion. That is to say, the taxpayer must have an economic interest in the oil and gas in place. The deduction for percentage depletion is not, however, dependent on the taxpayer's basis in his economic interest. Basis, however, is reduced, but not below zero, by the full amount of deductions for percentage depletion actually claimed.

§ 6.02(b) Depletion after the Tax Reduction Act of 1975. The Tax Reduction Act of 1975 brought about major changes in the area of percentage depletion. Basically, it repealed the deduction for percentage depletion subject, however, to several important limitations (see Section 613A of the Code).

Exemptions from repeal are:

(1) Regulated Natural Gas — Domestic natural gas produced and sold by the producer before July 1, 1976, subject to Federal Power Commission jurisdiction, the price for which has not been adjusted to reflect to any extent the increase in liability of the seller for tax caused by the repeal of percentage depletion. By definition this exemption is no longer important.

(2) Natural Gas Sold Under a Fixed Price Contract — Domestic natural gas sold by the producer under contract, in effect on February 1, 1975, and at all times thereafter before such sale, under which the price for such gas cannot be adjusted to reflect to any extent the increase in liability of the seller for tax as a result of repeal of percentage depletion. Price increases after February 1, 1975 are presumed to take into account increases in tax liability unless shown to the contrary.

(3) Geothermal Deposits — Located in the United States or in a possession of the United States.

(4) Independent Producers and Royalty Owners Exemption — by

far the most important exemption.

Under the Tax Reduction Act of 1975 the exemptions for regulated natural gas, natural gas sold under a fixed price contract, and geothermal deposits percentage depletion continued to be available at a rate of 22 percent. Additionally, no exemption was provided for foreign oil and gas production.

As previously mentioned, the independent producers and royalty owners' exemption to the repeal of percentage depletion is by far the most important. In general, this exemption provides for a continuation of percentage depletion at a rate of 22 percent for a taxpayer's depletable oil quantity, as defined, provided neither the taxpayer nor any related party is engaged in refining or retail marketing of oil and gas products. The depletable oil quantity was initially 2,000 barrels of domestic oil per day, or to the extent the taxpayer elects, an equivalent quantity of domestic natural gas.

The depletable oil quantity is subject to allocation among the taxpayer, his spouse and minor children, and among commonly controlled corporations and trusts. However, no allocation is required among partners and a partnership. Under the provisions of the act, the depletable oil quantity is reduced from 2,000 barrels per day in 1975, by 200 barrels per year after 1975, to 1,000 barrels per day for 1980 and thereafter.

Under Section 613A of the Code, aided by the Tax Reduction Act of 1975, percentage depletion remains limited to 50 percent of net income from a property and, in the aggregate, is further limited under the independent producer and royalty owner exemption to 65 percent of the taxpayer's net taxable income (in each case, before the deduction for depletion). The 65 percent limitation is computed before any carryback of net operating losses or capital losses; an unlimited carryover applies to depletion disqualified under this new 65 percent limitation.

As previously mentioned, allocation of the depletable oil quantity is required among a taxpayer, his spouse and minor children. The requirement for allocation of the depletable oil quantity among controlled corporations applies where one corporation owns more than 50 percent of the stock of another corporation, or where five or fewer individuals, directly or indirectly, own more than 50 percent of the stock of two or more corporations.

Again, the independent producer and royalty owner exemption does not apply if either the taxpayer or any related party is engaged in refining or retail marketing of oil or gas products. A party is not considered operating a refinery unless total runs on any day during the taxable year exceed 50,000 barrels. A party is not considered engaged in retail marketing of oil or gas products unless sales exceed $5,000,000 per year in the aggregate. And, finally, an individual taxpayer will not be considered related to a corporation or partnership carrying on a refining or retail oil or gas business unless he owns five percent or more of the stock of the corporation, or of the profits or capital of the partnership. Additionally, a taxpayer is not considered related to a corporation solely by reason of being a partner in a partnership with such corporation.

The independent producer and royalty owner exemption does not apply to property which was "proven" at the time of acquisition of the property or of acquisition of an interest in a partnership holding such property. For this purpose, "proven" properties are defined as oil and gas properties, the principal value of which has been demonstrated by prospecting, exploration or discovery work. The principal value of the property has been demonstrated by prospecting, exploration or discovery work only if at the time of transfer any oil or gas has been produced by the taxpayer or from the property transferred; or prospecting, exploration or discovery work indicate that it is probable that the property will have gross income from oil and gas from such deposit sufficient to justify development of the property; and the fair market value of the property at the time production commences, excluding actual expenses for equipment and intangible drilling and development costs, is less than twice the value of the property at the time of the transfer.

It would seem that failure to allow the exemption after a transfer of a "proven" property may not be unreasonable since, typically, cost depletion will be available on the fair market value of the property.

The independent producer and royalty owner exemption does, however, continue to apply after the "transfer of a proven property", in general, so long as a single "tentative oil quantity" (initially 2,000 barrels per day) is required to be allocated between the transferor and transferee. Thus, percentage depletion would continue: (1) on a transfer from husband to wife so long as they are married; (2) on a transfer from father to minor son so long as the son remains a minor; (3) on a transfer from a corporation to its wholly owned subsidiary so long as both remain members of the same controlled group or under common control; or (4) on a transfer from a father to a trust for the exclusive benefit of his minor son so long as the son remains a minor but possibly limited to the extent that the revenue, without any reserve for depletion, is actually distributed to the son.

No exemption to the transfer of proven property rule is available for a transfer from an individual (or partnership composed of individuals) to a controlled corporation (as for example, a transfer pursuant to Section 351 of the Code), since there is no provision for allocation of the tentative oil quantity in such situation. Transfers to revocable trust are not considered transfers for purposes of the proven property rules.

The method for computing depletion may be illustrated by the following example:

(1) Computation of percentage depletion, 22 percent of the gross income, not to exceed 50 percent of the net income from the property.

Production sales		$ 114
Less cash expenditures:		
Royalty payments	$ 14	
IDC	20	
Operating costs	15	
Equipment costs	5	54
Net Cash		$ 60
Gross income (after royalty)		$ 100

Less:

IDC	$ 20	
Operating costs	15	
Depreciation of equipment	25	$ 60
Net before depletion		$ 40
Percentage depletion:		
22 percent	$ 22	
50 percent of net income from property ..	20	
65 percent of total taxable income before depletion (assumes no other income or deductions)	24	
Lesser of above	$ 20	

(2) Cost depletion:

Production x remaining leasehold cost
Reserves
10 bbl

$$\frac{10\ bbl}{200\ bbl} \times \$300 = \$\ 15$$

200 bbl

(3) Taxable income from property:

Allowable depletion (percentage)	$ 20
Taxable income from property	$ 20

§ 6.03 Sublease versus Sale

An area of somewhat subtle distinction involves the difference between a sale or a sublease of oil and gas properties, although the different tax treatment afforded is hardly subtle. Basically, a transaction will be treated as a lease or a sublease where the owner of operating rights in the oil and gas property assigns all or a portion of such rights to another person and retains a continuing nonoperating interest. For example, if X (the owner of a working interest), assigns all of the working interest to Y and retains an overriding royalty, X has made a sublease.

An assignment of any type of interest other than operating rights will not be treated as a lease or sublease. Thus, for example, if X assigns an overriding royalty retaining the working interest, the transaction will not be treated as a sublease since the working interest or operating rights were not assigned.

If a grantor assigns all or a fraction of the working interest retaining two or more interests, one of which is a continuing nonoperating interest, the transaction will be treated as a lease or sublease. Basically, as long as no operating rights are retained and a continuing nonoperating interest is retained (which may be in addition to other nonoperating interests), the transaction will be lease or sublease. Thus, for example, where X assigns all of the working interest held by him and retains an overriding royalty and a production payment, the transaction will be treated as a sublease.

A transaction will be treated as a sale or exchange under various circumstances if the consideration received for the transfer is cash or the equivalent of cash. When the owner of any kind of property interest assigns all of his interest or a fractional interest identical, except as to size, with the interest retained, the transaction will be treated as a sale or exchange. For example, if X, holding all of the working interest, assigns all of the working interest for cash or other property or assigns an undivided part of the working interest for cash or other property, the transaction will be treated as a sale or exchange. Similarly, when the owner of a working interest assigns any type of continuing nonoperating interest in the property and retains the working interest, we have a sale. Thus, for example, when X, holding all of the working interest, assigns to Y an overriding royalty interest for cash or other property, the transaction will be treated as a sale of the overriding royalty interest by X to Y.

We also have a sale when the owner of any type of continuing property interest assigns that interest and retains a noncontinuing interest in production. Thus, for example, when X, holding a royalty or net profits interest, conveys such interest for cash or other property and retains a production payment, the transfer by X will be deemed a sale.

As previously mentioned, the tax treatment afforded a transaction classified as a sale differs greatly from the tax treatment of a sublease. Income from a sale or taxable exchange will constitute capital gain, unless the taxpayer is a dealer in oil and gas properties.

The effect of capital gains treatment may be illustrated by the following example:

Proceeds	$ 100
Basis (capitalized costs)	20
Gain	80
Long Term Capital Gain deduction (see Code, § 1202)	(40)
Taxable income	$ 40

Cash or other consideration derived from a lease or sublease is considered lease bonus and will be taxable as ordinary income, which may be subject to depletion.

The effect of lease bonus treatment may be illustrated by the following example:

Proceeds (bonus)	$100
Cost depletion (if any)	— 0 —
Taxable income	$ 100

Section 613A(c) of the Code, added by the Tax Reduction Act of 1975, fails to expressly state whether percentage depletion is available with respect to advance royalties and lease bonuses. However, since percentage depletion allowed by the independent producer and royalty owner exemption require actual production during the year, it seems clear that lease bonuses and advance royalties are not subject to percentage depletion in the year of receipt in the absence of production. Further, it is unclear whether a percentage depletion deduction on a lease bonus would be available if there is actual production in the year of receipt or in a subsequent year. Hopefully, this issue will be cleared up by regulations.

The IRS has adopted a formula for computing cost depletion on lease bonuses which provide as follows:

$$B \frac{(A)}{(A+R)} = \text{Cost depletion}$$

B = Adjusted basis of depletable property immediately before lease or sublease
A = Bonus or advance royalty
R = Royalty to be received in the future

Several problems are presented, however, when attempting to apply this formula. First, no deduction for cost depletion is available without a depletable basis which the taxpayer has the burden of establishing. This may be difficult, for example, where the lease is granted by the fee owner and the minerals were not acquired separately. In addition, difficulty may be presented in establishing estimates of future royalties to be received, particularly in an unproven area. There may be, however, some authority for treating all bonus payments as a recovery of capital. See *Murphy Oil Co. v. Burnet*,[12] where the court stated as follows:

> "... if the bonus and expected royalties together are not found to exceed the capital investment of the Lessor, the entire bonus received in advance of royalties must be treated... as a return of capital, since, in that case, the expected royalties added to the bonus, are, by hypotheses, sufficient to return no more than the Lessor's capital."

§ 6.04 Tax Free Receipt of Economic Interest Under the "Pool of Capital" Doctrine and Recognition of Income on Farmouts[13]

For many years it has been well recognized that one contributing property or services to the "pool of capital" necessary for the development of an oil and gas property in return for an economic interest in the same property does not recognize taxable income.[14] This illustrates, somewhat like the deduction for intangible drilling and development costs and also the deduction for percentage depletion, an area where rules involving oil and gas taxation provide results more favorable than those generally available in other areas. Within very narrow limits, geologists, lease brokers, accountants and lawyers are generally accepted to be within the doctrine.

In order to fall within the "pool of capital" doctrine, the economic

---

[12] 3 USTC § 1002 (U.S. Sup. Ct. 1932).

[13] For a further discussion of this subject, see § **4.03(g)**.

[14] See GCM 22730, 1941-1 CB 214.

interest received must be in the property with respect to which services were performed. The arrangement for receipt of the economic interest from the beginning must require rendition of services for an economic interest. In general, the services to be provided must be of a character which, if not provided by the one receiving the economic interest, would have to have been performed by the Lessee. The services also must relate to the acquisition, exploration and development of the property. In addition, under the "pool of capital" doctrine, expenses incurred by the one rendering services in consideration for the economic interest must be capitalized. And, finally, no further services may be required once the primary services have been performed and the economic interest obtained. Again, the doctrine is generally thought to apply where services and capital are provided or where only services are provided.

Some questions may be raised regarding the continuation of the "pool of capital" doctrine where only services are involved in view of the decisions in *United States v. Frazell*,[15] and *James A. Lewis Engineering, Inc. v. Comm.*[16], and the enactment of Section 83 of the Code, as a part of the Tax Reform Act of 1969.

In *United States v. Frazell*, the Fifth Circuit considered whether a geologist who was to receive in exchange for services an interest in an oil venture should be required to recognize income equal to the fair market value of such interest where such interest was subsequently transferred to a controlled corporation. After considering the application of Section 351 of the Code as well as U.S. Treasury Regulations Section 1.721-1(b)(1) and GCM 22730, the court concluded that the taxpayer geologist should be required to recognize income on the receipt of the free interest, represented by stock in the controlled corporation.

In the decision in *James A. Lewis Engineering, Inc. v. Comm.*, the taxpayer, a petroleum engineer who received an interest in future

---

[15] 65-1 USTC § 9125 (5th Cir. 1965).

[16] 65-1 USTC § 9122 (5th Cir. 1965).

production in exchange for services rendered in installing and supervising a waterflood program, was required to recognize income equal to the fair market value of the interest in future production received at the time of receipt. In considering the application of GCM 22730, the Fifth Circuit concluded that the "pool of capital" doctrine was applicable only through the drilling and development stage, which is considered to be complete when the casing and Christmas tree have been installed, and not in connection with a subsequently undertaken waterflood project.

Additionally, Section 83 of the Code provides that property received as compensation for services is includable in taxable income at its fair market value.

Irrespective of the above discussed authorities, there is apparently still support for the continued application of the "pool of capital" doctrine.[17]

A concept included in the "pool of capital" doctrine involves the recognition of income on farmouts. Prior to 1977, it was generally felt that the typical farmout situation was included in the "pool of capital" doctrine with the result that neither the farmor nor the farmee was required to recognize income. However, the Internal Revenue Service has taken the position in Rev. Rul. 77-176[18] that such is not the case, at least in instances where the party receiving the farmout earns acreage outside the drillsite spacing unit.

Rev. Rul. 77-176 involved a transfer of the full working interest in a drillsite, subject to an overriding royalty interest convertible to a working interest at payout, plus one-half of the working interest in the surrounding acreage by "taxpayer Y" to "taxpayer X" upon completion or abandonment of a test well to be drilled by taxpayer X.

In the ruling, the Internal Revenue Service took the position that taxpayer X is entitled to a full deduction for the drilling costs incurred since he earns the full working interest in the drillsite during the payout period.

---

[17] See *Cline*, 67 TC No. 72 (1977); Rev. Rul. 77-176, IRB 1977-19, 14.

[18] IRB No. 1977-19.

354

Further, the Internal Revenue Service took the position that the "pool of capital" doctrine applies with respect to the drillsite and that taxpayer X does not have taxable income as a result of receipt of the interest in the drillsite. However, they also stated that the "pool of capital" doctrine applies only to receipt of the interest in the same "property" to which taxpayer X made the contribution by drilling and that the interest received by taxpayer X in the surrounding acreage is "different" from the interest in the drillsite and is a separate property.

Under the position taken by the Internal Revenue Service in the ruling, taxpayer X must value the interest earned in the surrounding acreage and include such value in taxable income. Taxpayer Y must compute gain or loss on disposition of the interest in the surrounding acreage and treat the value of such interest as additional capitalized cost in the overriding royalty in the drillsite.

The ruling is understood to assume that an association taxable as a corporation was not created and that an election out of Subchapter K, the partnership income provisions, was made.

Another problem which may be presented in connection with farmouts involves the continued application of the independent producers and royalty owners exemption to the repeal of percentage depletion. Proposed U.S. Treasury Regulations on the transfer of proven properties may preclude percentage depletion on property earned on a farmout in some cases.

The proposed regulations provide that a transfer is deemed to occur on the day on which a contract to transfer the property becomes binding on both parties or, if no contract is made, on the date of actual transfer. The regulations include an example where "F is to drill a well on E's unproven property and is to own the entire working interest" until payout when 50 percent reverts to E. The example concludes that F is entitled to percentage depletion because the "transfer" occurred on the date of the agreement when the property was unproved, not on completion of the well. E is held to also be entitled to percentage depletion on the 50 percent reversionary interest. However, another example provides that where an option to purchase property is acquired and later exercised, the "transfer" occurs on the date of exercise. And, a "transfer" is defined to include a "lease", "sublease" or "assignment".[19]

These regulations may be interpreted to mean that where a farmout does not include a present assignment or sublease or if the assignee is not obligated to drill the well, the transfer occurs upon completion of the well when the property will be "proven" and the transferee will not be entitled to percentage depletion. The same problem would seem to be present where the assignment is to be made only on completion of a "productive" well even if the assignee has agreed to drill the well, as he cannot reasonably agree to drill a productive well.

The "proven property" problem can be solved and the "excess acreage" problem under Rev. Rul. 77-176 reduced by use of a present assignment and sublease form of farmout. Under a present assignment form of farmout agreement the valuation of the excess acreage would be made before the well is drilled when the value would be less, rather than on completion when the value would be much greater if the test well is productive. And, of course, the assignment would be before the property is proven for purposes of the proven property rule.

The present assignment and sublease need not be in recordable form and may terminate after a stated primary term unless drilling or production continues. Provision could be made for recordable instruments to be provided only if production results.

For a sample form, see Appendix B to Chapter Four.

The "excess acreage" problem can be solved by use of a "tax partnership"; that is, by use of an organization subject to the partnership income tax provisions of the Code. This may be either a formal partnership or a joint operating agreement with the provision stricken for the election out of Subchapter K. Use of the tax partnership will eliminate the

---

[19] See Proposed Regulations § 1.613A-3(h).

recognition of income under the rule that income is not to be recognized upon the transfer of property (leases by one party, drilling funds by the other) for an interest in a partnership.[20] Each party would be deemed to own merely a right to share in partnership revenues. However, use of a tax partnership does involve the complications of filing partnership income tax returns, use of a common partnership taxable year for determining income and loss, and general adherence to the partnership income tax rules.

For a sample tax partnership, see Appendix B to Chapter Four.

§ 6.05 Deduction of Intangible Drilling and Development Costs

Under the U.S. Treasury Regulations, any taxpayer who owns operating rights in oil and gas properties, and who incurs intangible drilling and development costs, may elect to expense or capitalize such costs.[21] If an election is made to capitalize intangible drilling and development costs, such costs may nevertheless be expensed if the well is abandoned as dry or not commercially productive. The election to expense intangible drilling and development costs currently must be made in the first taxable year in which such costs are incurred, and once made is binding on the taxpayer for all subsequent years.

In situations where a tax partnership is being used the election to deduct intangible drilling and development costs must be made at the partnership level. Such is the case whether a formal partnership is involved, or merely operations under a Joint Operating Agreement with the provision for election out of Subchapter K of the Code stricken.

The election to deduct currently is available with respect to "intangible drilling and development costs". This term is defined by U.S. Treasury Regulations as any cost incurred which in itself has no salvage value and which is "incident to and necessary for drilling of wells and the preparation

---

[20] See § 721 of the Code.

[21] U.S. Treasury Regulations § 1.612-4.

of wells for production of oil and gas."[22] Expenditures expressly included by the regulations in the definition of intangible drilling and development costs are "wages, fuel, repairs, hauling, supplies, etc." used in the drilling, shooting and cleaning of wells; in the clearing of ground, draining, road making, surveying, and geological work as are necessary in preparation for drilling of wells; and in the construction of such derricks, tanks, pipelines and other physical structures as are necessary for the drilling of wells and the preparation of wells for production of oil and gas. Not included, however, are the costs of installing facilities for the treatment or storage of oil or gas.

The definition includes cost of installation of tangible equipment placed in the well although the cost of the equipment itself must be capitalized and depreciated. However, only costs of installing equipment through the Christmas tree are included. The costs of deepening a well would also qualify as intangible drilling and development costs.

The election to deduct intangible drilling and development costs is limited to the working interest owned during the payout period. U.S. Treasury Regulations provide that an operator may deduct as an expense the intangible drilling and development costs paid or incurred by him, only to the extent the costs are attributable to his share of the total of all operating mineral interests in the well. Intangible drilling and development costs attributable to the portion of the mineral interest held by others must be capitalized. An exception to this general rule recognized by the Internal Revenue Service interprets the limitations set forth in the regulations to mean that, if a taxpayer contributes a disproportionate amount to the development of a property in exchange for an interest in the property, he may deduct that portion of the intangible drilling and development costs which equals his share of the working interest income during the payout period even though the full working interest may not be held following payout.

---

[22] U.S. Treasury Regulations § 1.612-4(a).

The definition of intangible drilling and development costs includes cost to the operator of any drilling or development work done for him by contractors under any form of contract, including turnkey contracts. Therefore, generally, intangible drilling and development costs includes amounts paid pursuant to a turnkey (fixed price) drilling contract. However, controversies have arisen where a turnkey contract has been entered into with the same party from which the lease was acquired. At least two cases support full deduction of the turnkey drilling price in such cases.[23] However, one has denied full deduction, principally on procedural grounds.[24] The position of the Internal Revenue Service as set forth in Rev. Rul. 73-211,[25] is that an amount equal to an arms-length turnkey price is deductible.

Another problem that arises in connection with intangible drilling and development costs involves what is sometimes referred to as purchased intangibles. A taxpayer cannot purchase a deduction for intangible drilling and development costs. The application of this concept is illustrated by the decision in *Platt v. Comm.*,[26] where the taxpayer who acquired an interest in a lease with a well drilling thereon was not permitted to deduct intangible drilling and development costs incurred to the depth of the well as of the date of acquisition.

With respect to the timing of deductions for intangible drilling and development costs, if a taxpayer properly elects to deduct intangibles, the deduction is allowable in the taxable year in which such costs are incurred by an accrual basis taxpayer, or in the year in which such costs are paid by a cash basis taxpayer. For this purpose, the operator is generally treated as the agent of the nonoperating working interest owners. Therefore, timing of

---

[23] See *C.F. Hedges, Jr.*, 41 TC 695 (1964), and *L.L. Stanton*, 26 TCM 191 (1967).

[24] See *Charles M. Bernuth*, 73-1 USTC § 9132 (CA-2 1973).

[25] 73-1 CB 303.

[26] 53-2 USTC § 9560 (7th Cir. 1953).

the deduction is dependent upon payment by the operator to the contractor.

Prepayments of intangible drilling and development costs to the contractor will be deductible in accordance with rules set forth in *Pauley v. U.S.*[27]; Rev. Rul. 71-579[28].

The Internal Revenue Service initially took the position that prepayments to drilling contractors would not be deductible until completion of the well.[29] However, the Internal Revenue Service was then reversed on facts identical to those set forth in Rev. Rul. 170 by the decision in *Pauley v. U.S.*[30] Finally, in 1971, the Internal Revenue Service revoked Rev. Rul. 170, and adopted a ruling conforming to the decision in the *Pauley* case — Rev. Rul. 71-252[31].

The Internal Revenue Service, in general, followed the position that a cash basis taxpayer required under a drilling contract with a driller to make a payment to the driller by a certain time, may deduct the amount of such payments at the time paid even though the well is not drilled until the following year. The drilling contract can be either turnkey, footage or daywork. A turnkey contract is not necessary.

The National Office of the Internal Revenue Service and local agents have attempted to limit the application of Rev. Rul. 71-252 by imposing additional requirements and by making narrow interpretations of the facts of the ruling. Although it is difficult to know for certain what facts must be present before a prepayment can be deducted, the following would be

---

[27] 63-1 USTC § 9280 (DC Cal. 1963).

[28] 1971-1 CB 146.

[29] 1971-2 CB 225.

[30] See Rev. Rul. 170, 1953-2 CB 741.

[31] 63-1 USTC § 9280 (DC Cal. 1963).

advisable:

(1) The contract should be with the actual driller who is unrelated to the operator and the other co-owners unless the contract is a turnkey contract, in which case the contractor need not be the actual driller but should be unrelated to the owner. (A pre-payment that only goes to the operator under an operating agreement probably will not be deductible because the operator is only an agent for the owners.)

(2) If the contract does not cover the drilling of the entire well, the remaining interest must, nevertheless, be obligated to the drilling of the well.

(3) The contract should relate to the drilling of a well on a specific property and should provide for a definite commencement date, which should not be more than one year, at the most, from the date of the contract.

(4) The contract by its original term must require the prepayment to be made on or before a certain date, and if the prepayment is not being made prorata by all owners, must set forth the amount to be paid by or on behalf of each owner.

(5) Many agents insist that it must be shown that the drilling contractor required, requested or at least "bargained for" the prepayment, and that it was not volunteered by the operator without some business benefit being received in return such as, for example, rig availability.

(6) Some preparation toward the drilling of the well should be shown in the year of payment, such as the securing of a drilling permit, the staking of the location, or dirt work at the drilling site.

§ 6.06 Associations Taxable as Corporations

Joint operations may result in the creation of an "association taxable as a corporation". The result of association treatment is that the so-called "corporation" is subject to income tax on its own income. Additionally, a shareholder of a "regular corporation" is not entitled to deductions for losses incurred by the corporation until the stock becomes worthless or he sells or exchanges his stock at a loss. Cash distributions from a corporation to its shareholders out of "earnings and profits" are treated as taxable income to the shareholders. The features of double taxation of corporate profits and lack of flow through of losses dictate that typically association treatment is to be avoided.

The tax status of a joint operating agreement is dependent upon the existence of a "joint profit objective". The existence of the right to take production in kind avoids the finding of a joint profit objective.[32] Existence of a joint profit objective may result in association treatment unless the agreement is modified to have more non-corporate characteristics than corporate characteristics under the regulations.[33]

Pre-existing contracts and calls on production may result in the Internal Revenue Service taking the position that a joint profit objective is present since the co-owners do not actually have the right to take production in kind. Typically, joint operating agreements with the right to take production in kind result in the creation of tax partnerships. However, in a recent Internal Revenue Service National Office Technical Advice Memo, where a lease was acquired subject to a pre-existing production sale contract, thereafter subdivided, and the owners entered into a model form joint operating agreement reserving the right to take production in kind, the Internal Revenue Service took the position that a joint profit objective was actually present as the co-owners did not actually have the right to take their share of production in kind. Thus, an association taxable as a corporation may be created unless the model form joint operating agreement is sufficiently modified to avoid a sufficient number of the other corporate characteristics.

More common, perhaps, is the acquisition and subdivision of a lease

---

[32] See IT 3930, 1948-2 CB 161 and IT 3948, 1949-1 CB 126.

---

[33] See U.S. Treasury Regulations § 301.7701-2(a)(1).

which is subject to a call on production. Apparently the Internal Revenue Service National Office has not yet taken a position as to whether a joint profit objective would be present in such a situation. The conservative approach in such a situation, however, would be either to get the call released or sufficiently modify the model form joint operating agreement so as to insure avoidance of association treatment.

The effects of partnership classification are that:

(1)   Income and deductions from the partnership level flow through to each "partner".

(2)   The partnership entity makes its own tax elections.

(3)   A partnership may be used to avoid capitalization of intangibles under a carried interest situation or in an arrangement providing for division of costs on a tangible-intangible basis.

Section 761 of the Code provides for an election to be excluded from the partnership income tax provisions of the Code (Subchapter K). To be entitled to make such election out of Subchapter K, there must be a partnership and not an association taxable as a corporation.

The effects of an election out of Subchapter K are as follows:

(1)   There is no requirement to file partnership income tax returns.

(2)   Tax elections are made at the co-owner level.

(3)   For Federal income tax purposes, the sharing of expenses and income is prorata.

Chapter Seven

EDITOR'S COMMENTS

As a Landman, you are not expected to originate drilling prospects for your company. However, you may be the only representative of the company your Lessors ever meet. They will have numerous questions concerning what your company is doing, with many turning on how oil and gas is found. And, to do your job, you must be able to explain these techniques to the landowner.

If you work for a larger company, you will frequently be a part of the explorationist team that helps sell a proposed prospect to "committee". To properly support your geological or geophysical co-member, you must be able to understand his part of your team presentation.

In "Exploration Techniques", Bill Sengel will walk you through the oil finding process. You should be fascinated by his tracing of the origin of oil and gas, where and how it accumulates, and how the explorationist finds these accumulations of hydrocarbons.

From Sengel's discussion, you will discover:

• How an explorationist prepares surface and subsurface maps to search for "traps" where oil and gas may have accumulated (§ **7.03**).

• How logs are used to prepare structure and isopach maps and cross sections (§ **7.04**).

- How the explorationist actually applies these techniques in reaching his decision to drill (§ **7.05**).

- The role of the geophysicist (§ **7.06**).

— L.G.M.

Chapter Seven

EXPLORATION TECHNIQUES

By E.W. "Bill" Sengel *

§ 7.01 Introduction

This chapter will pertain to probably one of the most important segments of the oil industry — exploration for hydrocarbons. Without exploration for, discovery of and production of hydrocarbons, both liquid and gaseous, there would be little reason for the industry to exist.

In this chapter we will discuss, to some extent, the origin and migration of hydrocarbons, the trapping mechanisms, and some of the techniques used by explorationists to find the hydrocarbon accumulations.

In an attempt to understand the complexities involved in subsurface geological conditions, explorationists spend considerable time delving into historical geology. Historical geology is the compilation of the efforts of many men and carries the history of the earth back in time through two billion years. It endeavors to trace the events that affect the physical history of the earth: the origin and evolution of life, the distribution of the lands and

---

* Editor's note: Sengel, retired in 1970 after 30 years' service with Schlumberger Well Services, now serves as a consulting well log analyst. A guest lecturer at the University of Oklahoma, Oklahoma State University and the University of Arkansas, Sengel is a member of the Society of Professional Well Log Analysts and an honorary life member of the Oklahoma City Geological Society.

seas during various geologic periods and the development of the rocks of different ages. Geologic time is so vast in scope that major changes in the earth's surface have been, and still are, hardly perceptible in the lifetime of an individual.

Attempts to decipher the geological events of the remote past are based on the fact that these same physical forces are at work on the earth at the present time; for example, limestone banks and reefs being built at various places in the oceans, deltas forming at river mouths, sediments being deposited, sand bars being formed by wave action, fault movements, earthquakes, etc.

Knowledge of the functioning of these forces, plus interpretation of the rocks and fossil remains, has enabled the building of a chronological record of the past. As in human history there are gaps where no information is available. We may never know the complete history of the earth, despite the fact that geologists and paleonthologists of all nations have devoted over a century of study to the subject.

Geologically speaking, the mountains of today are relatively young, and many of them exist where seas have repeatedly covered the area. There has been a succession of uplifting, erosion or destruction, submergence by seas and deposition of sediments. These occurrences were never coincidental at all places on the earth, so seas and mountains were always changing places. Plants and animals in huge quantities were destroyed and buried during these changes.

§ 7.02 The Oil Formation Process

§ 7.02(a) **Origin of Oil.** When one fills his tank with gasoline or uses any one of a number of petroleum products, he marks the end of a fascinating drama whose scenes may span many, many millions of years.

Many people have studied the when, where, how and under what conditions crude oil was formed. There are probably as many theories as there are students. No one theory seems to account for all the facts, which suggests oil may have been formed in more than one way.

The one theory that most authorities seem to agree on, though, is that oil was formed from organic matter — the remains of plants and/or animals. The organic matter decomposes and at the same time is acted on by bacteria. This may have dispensed with all material other than fats, oil substances, etc. These substances, through modifications not yet fully understood, were changed into liquid and gaseous hydrocarbons. There has been much laboratory and field work to substantiate this theory.

Bacteria are simple single-celled microorganisms which reproduce or multiply with extreme rapidity. They occur abundantly and are widely distributed every place in nature where there are moisture, minerals and/or organic matter. They can live and be active in any environment where petroleum ocurs regardless of temperature, salinity or depth.

As oil is formed it may adhere to the matrix material as a film, or accumulate in a small hole as a droplet. It may be released so it is free to move by various means. Bacteria may dissolve or decompose the matrix material and release the oil to form droplets, or they may build up on the surface of the matrix and push the oil off. Oil in small holes may be pushed out of the hole by a gas build-up in the hole which expels the oil. In any event the oil forms droplets which are free to move.

It is very likely that the destruction of the myriad of plants and animals occurred in an environment where much water was involved. Consequently they probably settled into or were covered by mud.

§ **7.02(b) Migration of Oil.** It is probable that decomposition and bacterial action began during the mud stage and progressed as more and more material was deposited.

Compaction of these sediments caused a marked reduction of pore space and may be a predominate cause of movement of water and hydrocarbons upward and outward. They would continue to move until such time as they entered porous rocks known as reservoirs. If the reservoir rock had some sort of trapping condition the hydrocarbons would accumulate and form an oil pool.

§ 7.02(c) Reservoir Rocks. The various types of rocks that comprise the outer portion of the earth are classified as igneous, metamorphic and sedimentary. Only the sedimentary rocks are of major importance in petroleum geology.

The most common types of sedimentary rocks are shales, sandstones, conglomerates and carbonates (limestone and dolomite). Of these, reservoir rocks are usually sandstones, conglomerates and carbonates.

Sandstones are composed mostly of quartz grains and may vary from clean types to those in which clay and/or silt particles tend to clog the openings between grains.

The older sandstones are generally more deeply buried. They will be more compacted and cemented with calcareous, dolomitic, siliceous, iron oxide or other materials. Consequently, they generally have less porosity and permeability than younger sandstones.

Conglomerates are consolidated gravels composed of pebbles of various sizes held together by the same type of cementing materials as sandstones.

Some reservoirs occur in carbonate rocks, limestones and dolomites. Many forms of marine life extract lime from sea water and use it to build their shells or body structures. After death, the organic matter decays and the calcareous skeletal material accumulates to form limestone. Limestone also forms as a chemical precipitate. Dolomite can form by precipitation, but most commonly is formed by chemical change which occurs when magnesium replaces some of the limestone.

Of the various kinds of sedimentary rocks, shales predominate, probably 80 percent of the total. They are finely divided particles of older rocks (compacted muds) which were deposited in the still waters of seas and lakes. As many shale beds contain disseminated organic residues, they are believed to be some of the most important source beds for petroleum and natural gas.

Shales generally are highly compacted and are not considered as reservoirs unless there are some unusual conditions. If they are fractured or contain fissures and are in close proximity to sandstone or limestone reservoirs, they may function as part of the connecting drainage channels.

§ **7.02(d) Types of Traps.** Sedimentary rocks are generally deposited in a reasonably orderly fashion. The succession of layers depends on the type and quantity of material and the water action in which the material is deposited. Some of these layers will be of reservoir type rocks, and if they are in the vicinity of source rocks they may have hydrocarbons migrate into them. These hydrocarbons will continue to migrate until something stops their movement and causes them to accumulate. The stoppages are called traps.

Subsequent movements in the earth's crust cause these layers to be bent, tilted, broken, folded, raised, lowered, etc., which interrupts their continuity and causes the traps that stop the movement of hydrocarbons and allow accumulation in the porous spaces. This creates an oil or gas pool which is the target of explorationists (see Figures 14-26).

Traps are broadly divided into two categories — structural and stratigraphic. Structural traps are related to folding, bending or tilting of the formations, and stratigraphic traps are related to interruptions of deposition. There are many variations and combinations of these traps.

Examples of structural traps are anticlines, synclines, domes and monoclines (see Figures 15-18).

Some examples of stratigraphic traps are unconformity traps, pinchouts, channel fills and offshore bars (see Figures 19-26).

These diagrams are very much idealized in order to illustrate the types of trapping mechanisms. In nature they can be and generally are more complex. There can be any and all kinds of combinations, depending on the earth's crust movements and activities.

The primary function of the exploration geologists is to find one or more of these traps and cause a hole to be drilled.

§ 7.03 Exploration Techniques in General

Explorationists use many methods in searching for traps. They make maps of the surface using present topography, outcrops, quarries,

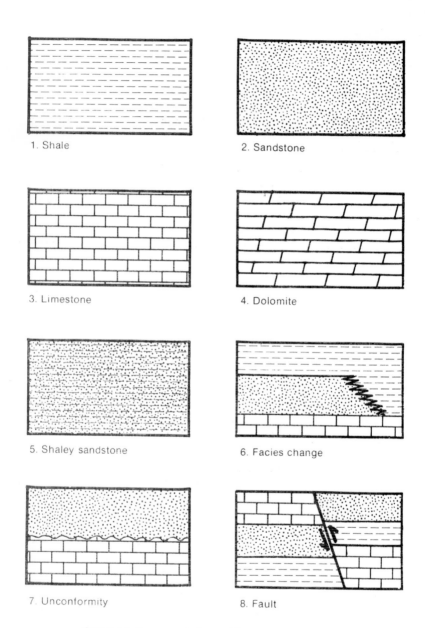

1. Shale

2. Sandstone

3. Limestone

4. Dolomite

5. Shaley sandstone

6. Facies change

7. Unconformity

8. Fault

Figure 14. Some symbols used in geological mapping.

372

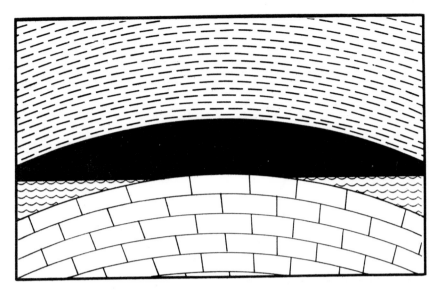

Figure 15. Anticline.

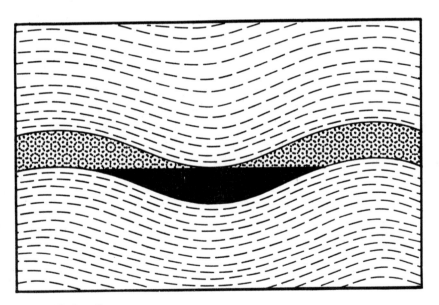

Figure 16. Syncline.

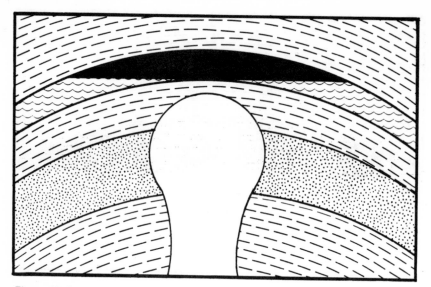

Figure 17. Salt Dome.

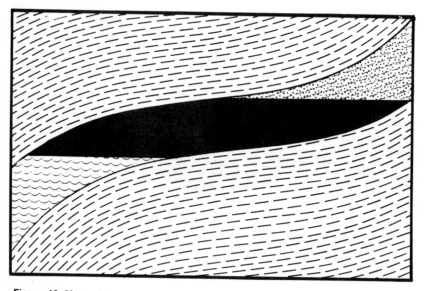

Figure 18. Monocline.

374

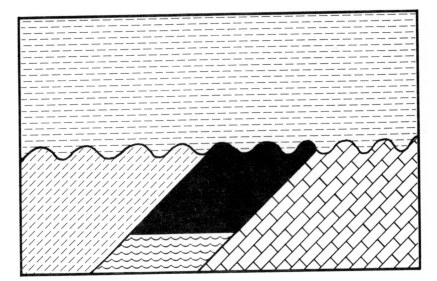

Figure 19. Angular Unconformity.

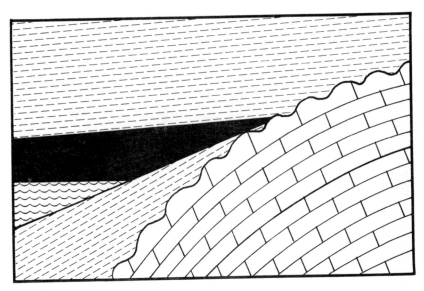

Figure 20. Pinchout.

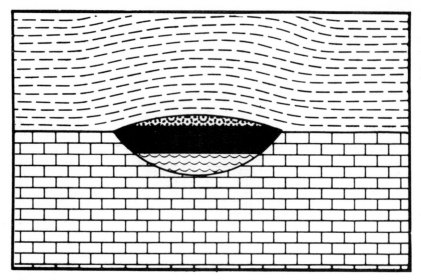

Figure 21. Channel Fill.

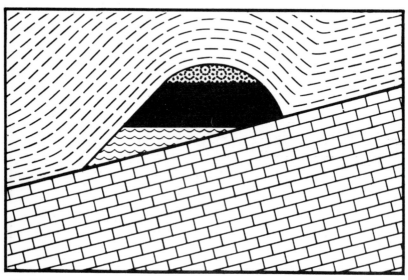

Figure 22. Offshore Bar.

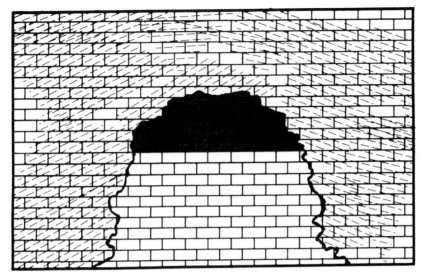

Figure 23. Reef Limestone.

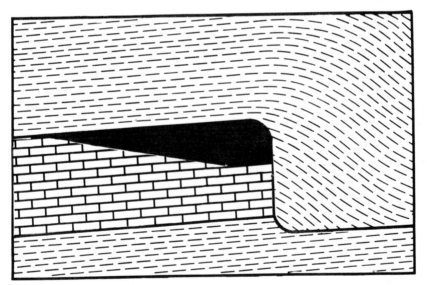

Figure 24. Truncated Limestone.

377

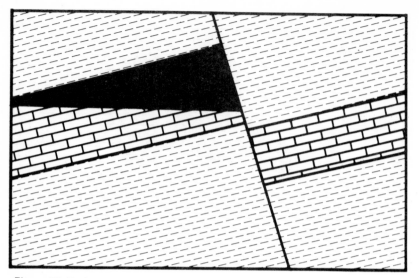

Figure 25. Faulted Limestone.

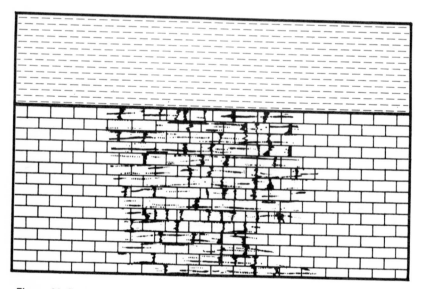

Figure 26. Fractures and crevices in limestone.

roadcuts, aerial photographs, vegetation, soil surveys, etc. These maps often help to determine the types of formations, thicknesses, aerial extent and attitude in relation to other formations. This information can be combined with other data and extrapolated to depth.

Another type of map that can be constructed uses information from core holes, well logs in existing drill holes (both productive and dry), geophysical surveys, gravity measurements, magnetic measurements, etc. to obtain information relative to formations at depth. These are called subsurface maps and may lead to the drilling of a hole subsequently finding a producing zone (or being abandoned).

Two of the primary sources of information used to make subsurface maps are well logs in existing holes and geophysical data obtained by geophysical crews using seismograph equipment.

For many years it has been the practice of operators to run one or more logs in drill holes. Consequently there are many thousands of logs in every oil and gas province available to the geologist.

Well logs are records of the drill hole. They are measurements of formation characteristics such as resistivity, spontaneous potential, sound travel time, density, radioactivity, hydrogen content and others, plotted versus depth of the hole. They are a continuous recording so that any part of the borehole may be studied (see Chapter Eight on well logging).

The particular well log that is generally used by geologists for correlation work is the one that is run from surface to total depth of the well. In older wells this will be the Electrical Log, which is comprised of a spontaneous potential curve and several resistivity curves (see Figure 27).

The spontaneous potential is a measurement of voltage variations in the mud column. There are many factors that cause and affect the spontaneous potential curve, but they will not be discussed here. The primary cause is that the water in the mud column and the salty water in the porous formations create a battery effect in the mud column that can be measured with proper instrumentation. This curve is used primarily to help locate reservoir type rocks and for correlation.

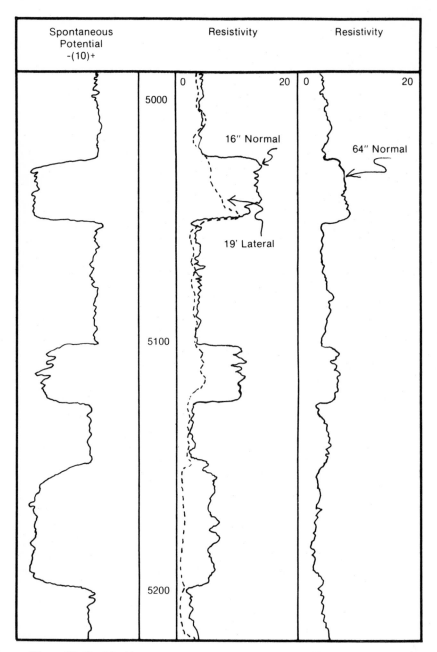

Figure 27. Electrical Log.

380

Resistivity is the ability of a formation to impede the flow of electricity through itself. Important factors that affect the resistivity of a formation are the presence or absence of porosity, the amount and type of water in the pore spaces, and the presence or absence of hydrocarbons. Resistivity is used to help identify and evaluate possible producing zones.

In more recent wells there are Induction-Electrical Logs and Dual Induction Logs. These are comprised of a spontaneous potential curve and various resistivity or conductivity (inverse or resistivity) curves. Quite often a gamma ray curve is recorded in the same log track with the spontaneous potential curve. This is a measurement of the natural radioactivity of the formations and can help to identify reservoir rocks.

Technically, these newer type logs are considerably different from the older logs. The primary difference is in accuracy of evaluation of the formations, but for correlative purposes they provide the same information (see Figures 28-29).

§ 7.04 Use of Logs

With the many logs available to the geologist, he has several options at his disposal:

(1) He may make a *structure map* using contours to connect points of equal elevation, thus indicating the ups and downs of the selected zone (see Figure 30).

(2) He may make an *isopach map* using contours to connect points of equal thickness of the selected zone. This indicates where the thicker part of the zone should be found (see Figure 31).

(3) He may make a *cross section*, correlating the logs across the area of interest. This gives a diagram showing vertical and horizontal variations of the rocks through a particular geologic interval (see Figure 32).

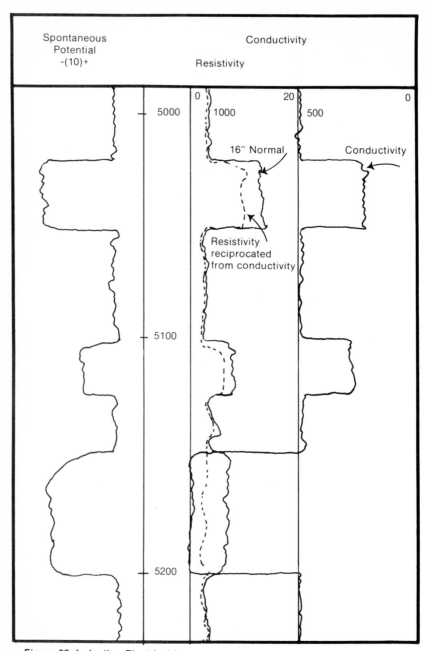

Figure 28. Induction Electrical Log.

382

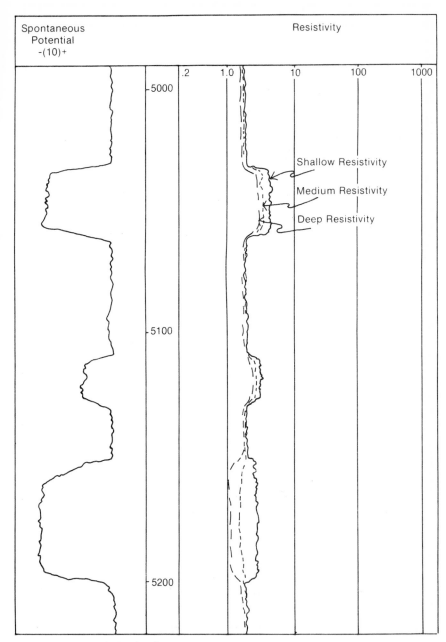

Figure 29. Dual Induction Log.
This log uses logarithmic scales for resistivity.

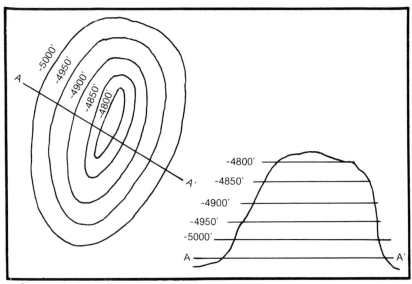

Symmetrical anticline.
Both sides are fairly equal slope.

Asymmetrical anticline.
One side steeper than the other.

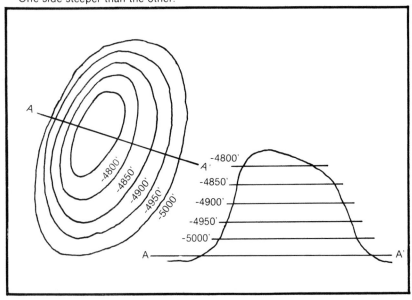

Figure 30. Structure Map.
Contour lines indicate variations in elevation.

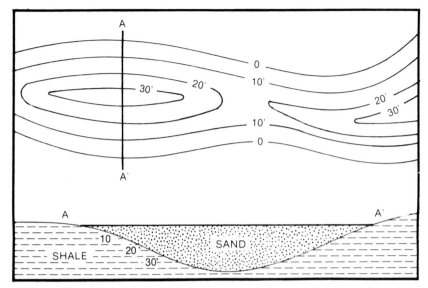

Channel fill sand.

Offshore bar sand.

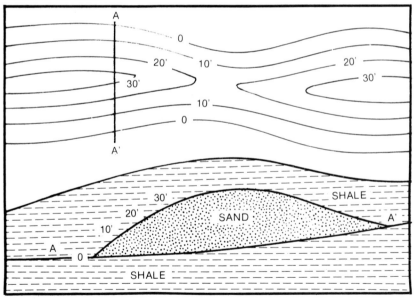

Figure 31. Isopach Map.
Contour lines indicate variations in sand thickness.

385

Preparation of the data to be used is very important. The well location and elevation must be accurate. The interval to be mapped must be selected. The logs must be correlated properly. The tops and bottoms of

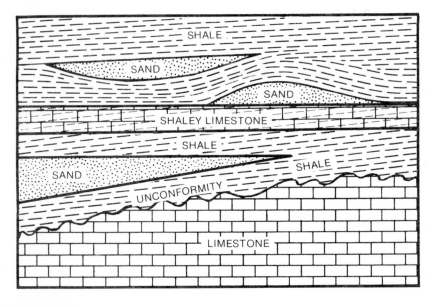

Figure 32. Schematic cross section.
Indicates vertical and horizontal variations
across the selected interval of interest.

the selected formation must be accurately determined. A common depth datum is selected, usually sea level. The data must be tabulated in such a manner that it is readily usable.

§ 7.05 Examples

§ 7.05(a) **Structure Contour Map.** The geologist wants to develop a prospect in a quarter section where he has six logs available. The logs are in dry holes, but two of the wells had shows on drill stem tests. The logs are located as shown in Figure 33. The logs with tops of the sand of interest (Zone B) are shown in Figure 34. The depth of the top with the well elevation subtracted, to give sub-sea level depth is shown for each log. In Figure 35, the sub-sea level depth is shown for each well and contours have been drawn to accommodate these depths, resulting in an apparently symmetrical anticline. From this map the geologist would recommend drilling at locations X and Y. The sand should be found considerably higher structurally than Wells 3 and 4, which had shows. If the drilling of either location confirms the map and production is found, the geologist will use the data from the new well to refine the map.

§ 7.05(b) **Isopach Contour Map.** A company has acquired leases on the southeast quarter of the section in Figure 36. There are logs in four dry holes drilled several years ago. Recent drilling in the northwest quarter has found some good production. Logs are available on the wells as indicated. All 13 logs are shown in Figure 37. The producing zone is the B sand.

The geologist suspects a northwest-southeast trending channel fill sand, and will endeavor to develop some drillable locations on the acreage.

The resulting map is shown in Figure 38. This indicates there are several good prospects at W, X, Y and Z. Subsequent drilling will be necessary to prove or disprove the theory.

§ 7.05(c) **Cross Section Map.** Wells in adjoining sections are shown in

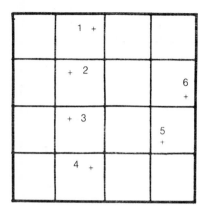

Figure 33. Locations of dry holes with logs available.

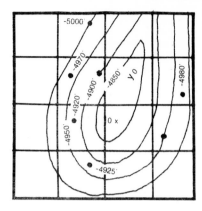

Figure 35. Contours drawn from information from dry holes.

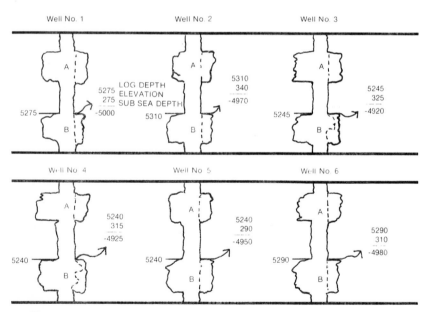

Figure 34. Logs from the dry holes.
Tops of Zone B have been picked and sub sea depths determined.

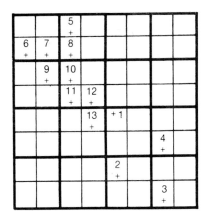

Figure 36.
Wells with logs available.

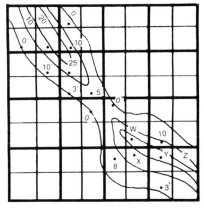

Figure 38.
Contours from log information.

Logs from Wells

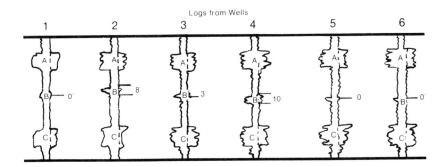

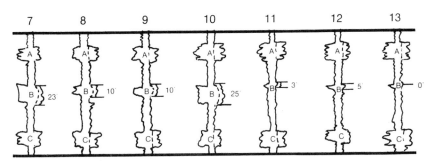

Figure 37. Logs available from all wells.
Thickness of Zone B has been determined.

389

Figure 39. Well 3 is an old shallow dry hole. The others are dry holes with wells 2 and 4 having shows in Zone E, but non-productive.

The logs in the wells are shown in Figure 40 with lines connecting tops of correlative zones. Indications are that Zone E should be present in Well 3 and higher structurally than in the two wells that had shows. Well 3 becomes a good deepening prospect or a new well at the same location.

Figure 41 shows wells drilled at different times by different operators. Wells 1 and 5 are dry holes, and wells 2, 3 and 4 are producers from a shaley sand zone.

A company has acquired the wells and the entire section. The geologist wants to develop drilling prospects. A cross section is prepared from the logs in the wells. Figure 42 shows the cross section and indicates the sand should be thicker between wells 3 and 4, suggesting possibly a channel sand trending northwest-southeast. Subsequent drilling will provide information to refine the map and possibly develop a good field.

§ 7.05(d) Comments. It must be kept in mind that the figures and diagrams are very idealized. The intent and purpose is to illustrate a situation and/or technique. In actual practice, the geologist's job is much more complicated. "Mother Nature" does not readily give up her secrets. Practically all subsurface work is done indirectly, using measurements of the earth's characteristics and properties and/or information implied from other sources. The geologist does an extremely competent job in the face of such odds.

Many times he will be faced with the task of finding an exploratory drilling location in an area where there are few, if any, well logs available. Maybe no wells have been previously drilled, so there is no information available. In a situation such as this he must rely entirely on other sources of data.

§ 7.06 Geophysics
Geophysics is the study of the earth by measuring its physical properties. These include measurements of seismic reflections, the

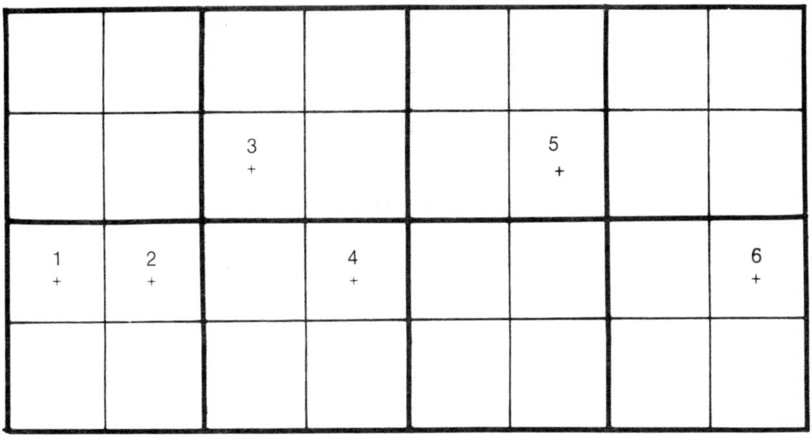

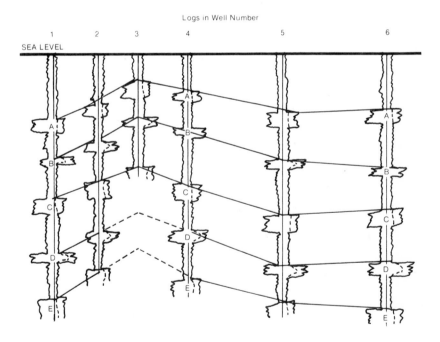

Figure 39. Locations of the dry holes with logs available.

Logs in Well Number

SEA LEVEL

Figure 40. Cross section made from the logs in the dry holes.

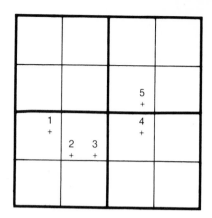

Figure 41.
Locations of the wells with logs.

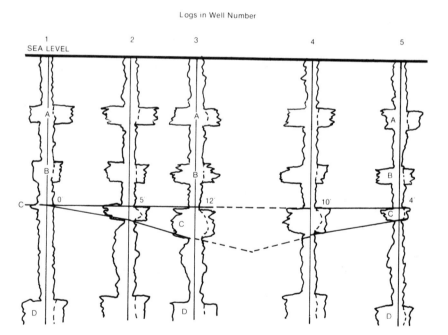

Figure 42. Cross section made from logs in the wells.

392

magnetic field and the gravitational field.

The purpose of a seismic survey is to map geologic structure. Measurements are made of the time it takes a seismic wave to travel from a source of energy (explosive, compressed air or gas, vibroseis, etc.) down to a selected reflection formation and back to the recorder at the surface. (See Figure 43.) Variations in time are related to the depth and attitude of the formation. This is interpreted to indicate structural variations.

Magnetic surveys are measurements of the earth's magnetic field, the object of which is to locate magnetic material concentrations or of determining the depth of basement rocks.

Gravity surveys are measurements of the gravitational field at different places to determine the distribution of densities. This is related to rock types.

Seismic surveys are performed by a geophysical field crew, working in straight lines criss-crossing the area of interest. They will work with truck mounted drilling equipment, recording trucks, geophones (instruments to detect the seismic waves), explosives, miles of wire and cables, etc.

Shallow shot holes are drilled along the selected line, recording trucks are positioned at optimum locations, geophones are placed on the ground at selected intervals and connected to the recording truck, and explosives are detonated at proper times at the bottom of the shot holes. This provides the energy to supply the seismic wave, and recordings are made as indicated in Figure 44. Many, many of these recordings are made. They may then be attached side by side to provide complete coverage across the area of interest.

Geophysicists and geologists trained in their use then study these recordings and make interpretations as to structure variations.

If wells are drilled as a result of this work the information from the wells is used to confirm or revise the interpretation.

In many areas, for example marine areas, geophysics provides the major source of information.

Geophysics does not, with few exceptions, locate oil or gas directly. It primarily helps detect structure of rock layers, folds, domes, faults, etc.,

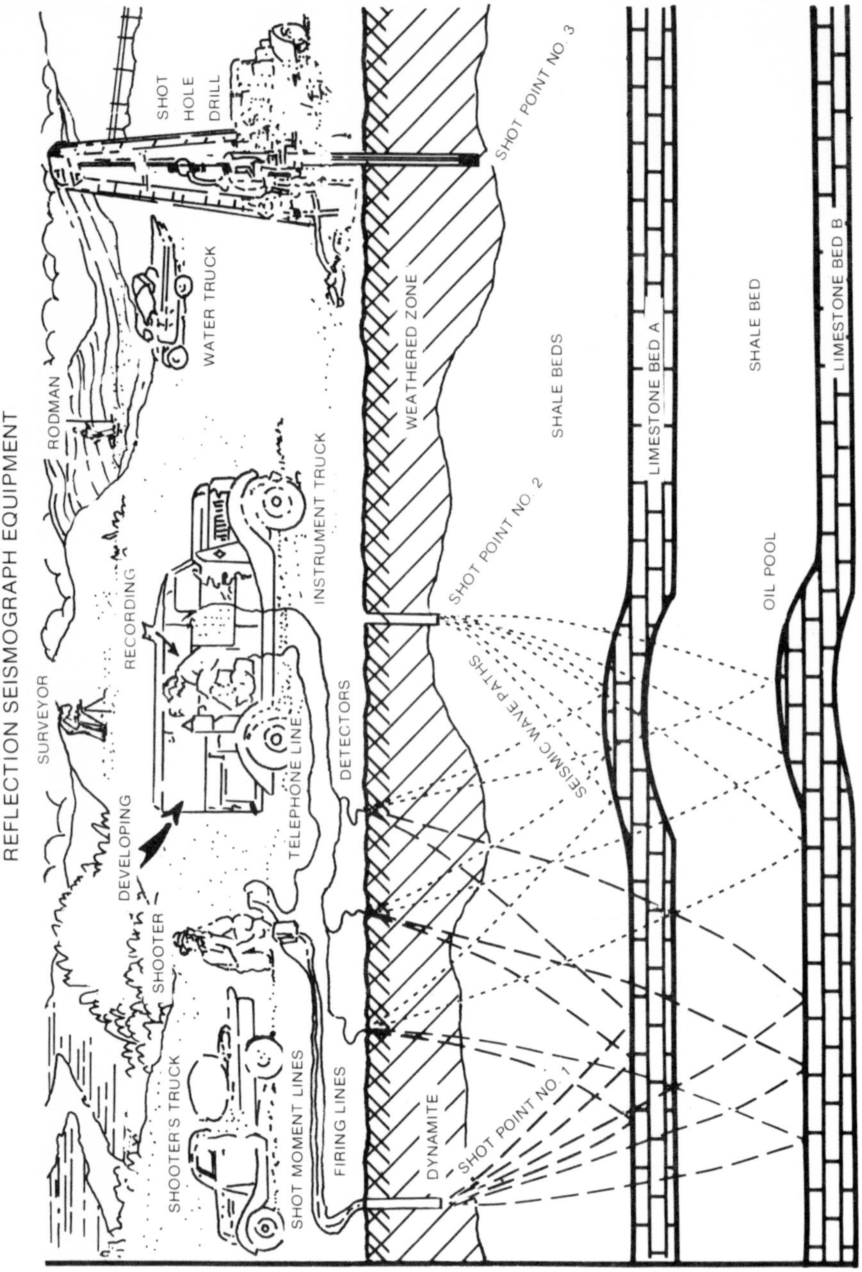

Figure 43. Equipment used to make a reflection seismograph survey.

394

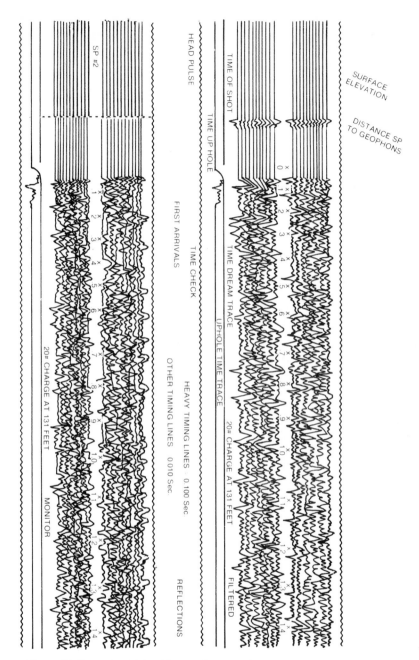

Figure 44. Records obtained from seismograph surveys.

395

which is extremely useful by explorationists in their search for traps.

§ 7.07 Summary

The purpose of this chapter has been to present to you an insight into the why and how explorationists search for hydrocarbons.

The fact that hydrocarbons were formed, and migrated into reservoirs that were somehow interrupted or changed so that the oil or gas could accumulate, is all important. This is the only reason that oil or gas fields exist.

The methods used by exploration geologists are many and varied. Two of the major sources of information have been presented. Geologists will use these in many ways, as varied as their experience and knowledge. Theirs is an extremely competent profession, continually pitting their skills against a very difficult and uncooperative adversary.

The fact that the industry has continued to grow and prosper for over 100 years is evidence of their successes.

Generally, after the exploration geologist has found a field, it is turned over to the production department to develop and produce to maximum efficiency.

The explorationist then turns his attention to finding new fields by starting the whole procedure all over again.

Chapter Eight

EDITOR'S COMMENTS

In Chapter Seven, you saw the importance of logs to the explorationist in his search for oil and gas. In this chapter, Bill Scales discusses "Electrical Well Logging" and its importance to the oil industry, explaining the type of electrical well logs available and what they can tell you.

Not only oil companies, but bankers and governmental agencies as well, rely on electrical logs as one of their chief sources of information concerning proposed or present oil and gas operations.

Here are points covered in this chapter that should be of particular interest to you as a Landman:

• How electrical well logs are actually used in the industry (§ **8.02**).

• How electrical logs are used to determine rock "porosity", and which logs are most effective under which conditions (§ **8.06**).

• How electrical logs are used to determine water saturation (§ **8.07**).

• The use of logging programs (§ **8.09**).

Scales, at the end of this chapter, also provides a highly useful schedule of available commercial logging services and a summary of types and primary uses of the various electrical well logs (Figures 55-57).

— *L.G.M.*

Chapter Eight

ELECTRICAL WELL LOGS

By Bill M. Scales *

§ 8.01 Introduction

Electrical logs have been used in the petroleum industry for over a half century. Well logs have provided information to major companies, independent operators, geologists, engineers, government agencies, financial institutions and other interested parties on millions of wells drilled in search of hydrocarbons.

Any record of subsurface formation properties as they are related to depth can be considered a well log. For example, a record of drilling time in terms of the minutes per foot required to drill at a particular depth is a well log, but since it is not recorded from a wireline it is not an electrical log. It

---

* *Editor's note:* Scales, operations manager of the Houston Hill Estate, Fort Worth, served for years with Welex, a Halliburton Company, where his duties included log analysis and direction of Welex's Customer Log Seminar Program. A guest lecturer at a number of colleges and universities on the subject of electric logging, Scales' professional honors have included serving as president of the Petroleum Club of Fort Worth, the Petroleum Engineers Club of Fort Worth and the Fort Worth Chapter of the Society of Professional Well Log Analysts, as well as offices in the Fort Worth Geological Society and as vice-chairman of the Society of Petroleum Engineers.

does, however, provide useful information. The fastest drilling formations are referred to as "drilling breaks" and usually occur in porous formations.

Electrical logs are *electrical properties* of a well recorded in terms of depth. Other formation measurements such as travel time of sound (acoustic-velocity log), radioactive measurements (density log, neutron log), and other wireline measurements are usually lumped together under the broad term of electrical logs.

Another term sometimes substituted for electrical logs and having the same meaning is "open hole" logs. Since electrical logs measure the resistivity of sub-surface rocks, they cannot be run inside casing. Casing conducts an electrical current; therefore, it is necessary to run electrical logs before casing is set in the bore hole. Usual logging points are at the time the well reaches total depth (TD), before pipe is run or the well plugged, and before intermediate casing strings are set.

Four or five major logging companies perform most of the logging services in the United States. Dozens of independent companies also perform logging services, but their areas of operation are localized and the wells they log are generally shallow.

Many of the logs available to oil operators are results of major oil company research. The logging companies acquire the patent rights and basic research information on new logging systems, through royalties and trades, from the majors. Further developments by the service companies are usually necessary before a new log is commercial.

§ 8.02 Use of Logs — In General

People within the petroleum industry use electrical well logs for different purposes. For example, a person in exploration will be most likely to use logs for correlations, depth control and lithology (rock type) determinations. Petroleum engineers would be more apt to use them for porosity determinations, water saturation calculations and other reservoir studies. Landmen may use them as a cure for insomnia. A partial list and description of uses is as follows.

§ 8.03 Correlations[1]

Using electrical well logs for well to well correlations is very important. It is probably the most common use of logs. Explorationists locate zones of interest, certain formation tops or "markers" in one well and compare them with the same "markers" in a nearby well. Using this information they can prepare maps of depths of formations, thickness of beds and oil-water contacts. Cross sections of areas can be constructed using logs such as the simplified example in Figure 45. A working cross section using logs from wells drilled over an area covering about a mile is in Figure 46. An examination of the logs revealed that close comparisons could be made of several different zones and the cross section was constructed. Correlations were made using similarities of log responses and distinctive markers which occur from one well to the next.

Correlating zones from one well to another well aids in determining whether a well is running high, low or flat to the other. The relative positions of the zones is usually easy to find. Determine the depth of the "marker" in each well. Subtract the respective elevation, which is shown on the log headings, from the "marker" depth. The smallest number would represent the highest well.

§ 8.04 Lithology Determinations

Logging programs can be designed for reasonably accurate rock composition interpretations. These lithology determinations involve cross-plotting of density, neutron and acoustic-velocity logs and are usually done with the aid of computers. With some knowledge of the minerals involved, complex lithology interpretations can be made. Coal and sulphur can also be located with a program of this type, and such a program is offered as a commercial service.

§ 8.05 Depth Determination

The depths of sub-surface formations as they are reflected on an

---

[1] *Editor's note:* For additional material on the correlation process, see Chapter Seven.

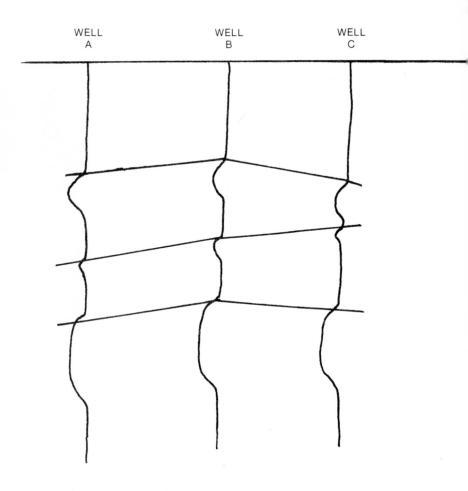

Figure 45. Simplified Correlations.

The simplified correlations above show how markers are used from one well to another. The same curve characteristics appear on each well.

402

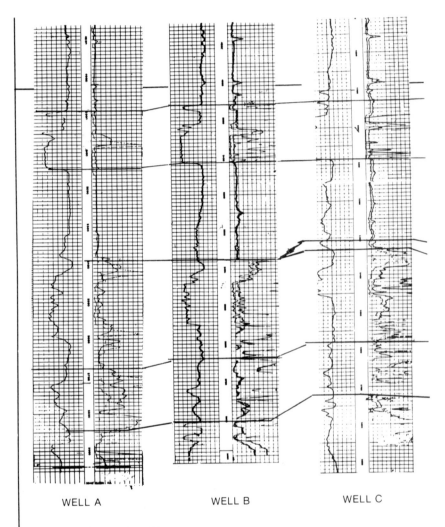

WELL A   WELL B   WELL C

Figure 46. Correlation of Logs.

    The above logs illustrate how correlations can be made using electrical logs. Some of the zones were correlated using the SP curve (left track) and others were made on resistivity curves (right track).

electrical well log are generally accepted as very accurate. Since a log is a graph of depth and its relation to a property of rocks, such as resistivity, depth becomes very important. The depth scale is in a narrow column near the center of the log in which the depths are listed. It is usually referred to as the depth column. In most areas of the United States the logs are run on a five-inch (log) to 100-feet (borehole) scale, which is referred to as the detail section, and a two-inch (log) to 100-feet (borehole). In areas of complex geology, a third scale is sometimes run being one-inch (log) to 100-feet (borehole). The five-inch scale is used for calculations. The one-inch and two-inch scales are used for correlations.

§ 8.06 Porosity Determinations

§ 8.06(a) In General. A rock must be porous if it is to contain hydrocarbons, so it is important to be able to measure the amount of storage space a formation has for fluids.

The porosity of a rock is the ratio of void volume within the rock to the bulk volume of rock. This void space or porosity is most often expressed as a percentage. In logging terminology, the Greek letter phi ("ϕ") is used as the symbol for porosity. A rock with one-tenth of its volume being void space would be described as ϕ = 10%.

There are a number of logs which measure formation properties which can be related to porosity. At least one type of porosity log is run on almost every well. Since an accurate porosity is hard to obtain in areas of complex lithologies, it is not unusual to run two or more porosity logs.

The porosity logs available in most areas include the following: (1) microlog or contact log, (2) microlaterolog or FoRxo, (3) neutron log, (4) sidewall neutron, (5) gamma ray[2], (6) density logs, (7) density-neutron combination, (8) acoustic-velocity or sonic logs, and (9) caliper log[2].

A review of each of these porosity logs follows.

---

[2] The gamma ray and caliper do not measure porosity, but are usually run simultaneously with porosity determining devices.

§ 8.06(b) Microlog or Contact Log. The microlog or contact log is one of the oldest types of porosity logs, being introduced in the early fifties. This log is a pad mounted resistivity device. The pad, which is forced against the wall of the hole during logging, has two short spaced electrode arrangements, each having a different depth of investigation. One curve is influenced by the resistivity of the mud cake on the wall of the hole and the other curve looks a few inches deeper at what is called the invaded zone — the area flushed with mud filtrate (see Figure 47). The result is a log with two curves. The recorded measurement is in ohm-meters.

In a permeable zone, drilling mud will try to enter the formation. Only the filtrate is able to enter the rock and the solids in the mud form a mud cake on the wall of the borehole reducing the size of the hole to smaller than bit size. The microlog, in these zones of interest, will indicate that there is permeability present by a positive separation between the two curves. The longer spaced curve will be reading higher resistivities than the short spaced curve. In zones of low permeability or no permeability, the curves will almost track each other or there will be negative separation.

The presence of porosity and permeability is easier to see on the microlog than it is to estimate or calculate. Although porosity estimates can be made under certain conditions, the microlog is usually not depended on for accurate porosities. The presence of permeability can be detected with the microlog but cannot in any circumstances be used for quantitative permeabilities.

The microlog is still run today, but not so often as the only porosity log. It is recorded in combination with other logs. An example is shown in Figure 48.

Micrologs should be run in fresh muds, i.e., not in a salt mud. When the mud is salty, the invaded zone resistivity can be measured with the microlaterolog.

§ 8.06(c) Microlaterolog or FoRxo. The microlaterolog device is designed for logging in salt base drilling muds. Salt muds are commonly used in the Permian formations of west Texas and New Mexico, in most of

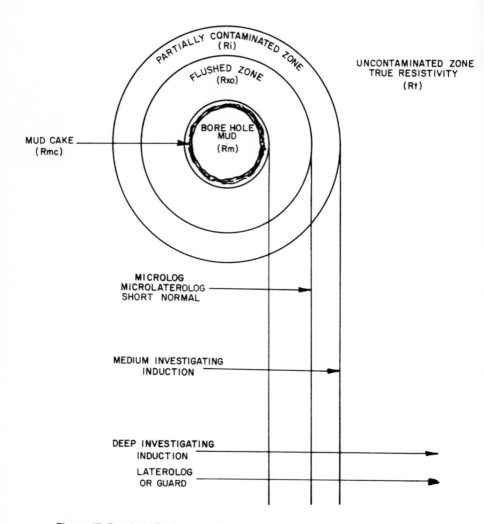

Figure 47. Borehole Environment Schematic.

The borehole schematic illustrates the problem of mud filtrate invasion and the depths of investigation of different logging devices. Mud cake is formed by filtrate entering permeable rocks and the mud solids plate against the wall of the hole forming the cake. As the filtrate enters the rock it flushes out the native fluids. This area is known as Rxo in logging terminology. The deep investigating logs are designed to measure formation which is uncontaminated by drilling fluids.

406

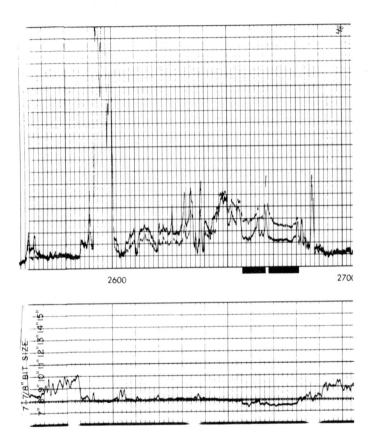

Figure 48. Microlog Example with Caliper.

The positive separation. which indicates porosity and permeability. is shown in the blocked section. 2656-80. Notice the mud cake build-up in this zone on the caliper. The hole is washed out around 2700 and 2575. The zone from 2590-2600 is hard limestone. Negative separation is shown from 2600' to 2642. denoting a lack of permeability.

407

Kansas and in other parts of the country where surface casing does not cover salt zones. The recorded values are in ohm-meters.

Since electrical current flows easily in very low resistivity muds, it is necessary to focus the current into the formation with the microlaterolog. It measures only a few inches back into the rock where mud filtrate has invaded (see Figure 47). The resistivity measurement of the invaded zone, where native fluids are flushed out and replaced with mud filtrate, is known in logging terminology as Rxo. With knowledge of the resistivity of the invaded zone (Rxo) and the resistivity of the mud filtrate (Rmf) a porosity estimate can be made. Interpretation charts relating Rxo and Rmf to porosity are available from every logging company. Moveable hydrocarbons can also be computed using a logging program which includes the microlaterolog. The moveable oil plot (MOP) requires a minimum of three logs. The program would require a porosity log such as the density log, the sidewall neutron log or the sonic log, the microlaterolog, and a deep investigating resistivity log which responds to formation uncontaminated by mud filtrate.

Porosity is usually indicated by movement of the microlaterolog curve to the left as shown in Figure 49.

§ 8.06(d) Neutron Log. The neutron log is also one of the older porosity tools. For many years it was the only porosity tool other than the above mentioned microlog devices.

Neutron tools or sondes contain a neutron source which bombards the formations with neutrons. The radiated formation contains atoms which collide with the neutrons, and as the neutrons are slowed they are captured by hydrogen atoms which in turn release gammas which are measured by a gamma detector. Hydrogen atoms in a rock are generally associated with water or oil in the pore spaces. The gammas released by hydrogen are empirically related to porosity.

The neutron curve is usually presented in the right tracks of the log grid and a gamma ray curve is presented in the left track giving the gamma ray neutron name that is so familiar. Neutron curves indicate porosity with

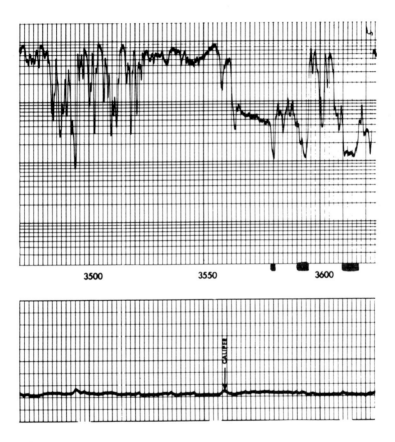

Figure 49. Microlaterolog or FoRxo.

The microlaterolog can be used as a porosity tool in salt base muds. The blocked intervals illustrate how porosity would appear on the log with deflections to the left. The zone 3526-56 has extremely low porosity. The caliper in the left track indicates no mud cake as is generally the case in brine muds.

deflections or "kicks" to the left. The highest porosity ideally would be indicated by the farthest deflection to the left.

A zone with low porosity would cause the neutron to deflect to the right side of the log. There would be less oil or water in a zone of low porosity than there would be in a zone of high porosity. Therefore, less hydrogen would be present and the neutron curve would respond with deflections to the right. Shales appear on the neutron log as porous intervals with kicks to the left. Shale porosity is not considered to be productive since the pore spaces are not interconnected, and as a result there is no permeability. Shales generally have very high porosities and the neutron responds with large deflections to the left.

Gamma ray-neutron logs are capable of logging in open holes or in cased holes. Since they are capable of logging through casing, they can be used in old wells on which no electrical logs are run. They are also used to tie in open hole logs with casing collars and thus insure accuracy in perforating.

Neutrons are more accurate in porosity determinations when they are decentralized and the neutron source is near the borehole. Since casing is centralized, it tends also to centralize the neutron tool resulting in less dependable quantitative work.

§ **8.06(e) Sidewall Neutron.** The development of the sidewall neutron devices has contributed to a much improved porosity log. A neutron source and one or more gamma detectors are mounted on a pad that is extended from the body of the tool on an articulated arm. The pad is pressed against the borehole with a great deal of pressure so that the source and detector are completely decentralized. This decentralizing technique, coupled with solid state electronics and better calibration methods, results in an excellent porosity tool.

Curves on the sidewall neutron log are presented linearly and an increase in porosity would be indicated with a deflection to the left, as on the older neutron logs. Several curves are presented on the sidewall neutron logs. In the track left of the depth column, the gamma ray and the caliper are recorded. The right tracks usually show two curves, one being the raw neutron count data and the other is a direct reading of porosity expressed in percentage. The porosity read-out is made by an on-board computer module that converts the raw neutron count data into porosities. It is not necessary to perform any calculations. A porosity scale is shown on the log, and it is a simple matter to read off porosities. These porosities are corrected for rock types.

§ 8.06(f) Gamma Ray. It was mentioned earlier that a gamma ray curve was recorded with the neutron logs and placed in the left track of the log. The gamma ray measures the natural radioactivity of the subsurface rocks. All rocks have some radioactive material which will emit gamma rays. Natural gamma ray emission occurs even when no logging tool is in the borehole to measure it. The gamma radiation measurement helps with the identification of lithology. They are also widely used in correlations of zones in one well to the same zones in another well.

The gamma ray is presented so that rocks of low radioactivity appear as deflections to the left. Rocks capable of producing hydrocarbons are recognized as having low radioactivity or a low gamma count. Limestones, dolomites and sandstones would fall in the low gamma radiation category. An exception to this would be when fluid in the spaces is radioactive such as in Permian sandstones in west Texas and New Mexico. Other rocks low in radioactivity would include anhydrite, gypsum and halite (salt). The rocks with low radioactivity will appear with the gamma ray curve moving to the left.

Rocks with high radioactivity levels will appear with the gamma curve moving to the right. Shale, potassium minerals and igneous rocks are in the high radioactivity category. Shales are very common and would appear on a log as high gamma count and the gamma ray curve will deflect to the right.

Rocks of a complex nature that contain minerals of low and high radioactivity will produce a log curve that falls in the middle range. For example, a shaley sand would appear on the gamma ray log in the mid-range between the low radioactivity and the high radioactivity range.

The gamma ray can be run inside casing or in the open hole. In casing it is commonly run with a magnetic collar locator for tying in casing collars and formation depths. It is used in this manner to insure perforating accuracy. In open hole logging, the gamma ray is usually run simultaneously with other logs. Its main purpose when used in this manner is to distinguish between shales and possible productive rocks such as sandstones, limestones and dolomites.

§ 8.06(g) Density Logs. The density log, which is also run simultaneously with the gamma ray, has come on since the late sixties as the most popular porosity log. The density log measures the electron density of the formation. With knowledge of the rock type in the formation, the measured density can be related to porosity.

The density log equipment is very similar to that of the sidewall neutron. It is a pad-mounted tool which contains two detectors and a source. Instead of a neutron source, it utilizes a gamma emitting source. The pad containing the source and detectors is forced against the borehole. As the formation is bombarded with gamma rays, the gammas are scattered and this gamma scattering property of a formation is directly related to formation density. Porosities are electronically computed and are recorded on the log. The direct reading porosity curve is similar to the porosity read-out on the sidewall neutron log. A linear porosity scale is placed on the log with the low porosities to the right and the higher porosities to the left.

Density logs are commonly run with a gamma ray curve and a caliper presented in the left track. Density logs can be run in empty holes (gas drilled) or in boreholes containing mud. Although it is sometimes necessary to attempt to obtain a log through casing, they are much more reliable in uncased holes. Micrologs are sometimes run in combination with the density when logging an uncased borehole containing drilling fluid. The microlog cannot be used in casing or in an empty hole.

Excellent porosity determinations are made from the density log under most conditions when lithology characteristics are known. Density logs are especially preferred in high porosity sands and in shaley sands.

§ **8.06(h) Density-Neutron Combination.** Gas in the pore spaces of a formation can cause density porosity calculations to be misleading. When gas is present, the porosity readings from the density logs are optimistic.

When gas is present in pore spaces the neutron porosities also can be misleading. The neutron is primarily a hydrogen-seeking device. Since there is relatively less hydrogen in a gas zone than in water or oil zones, the neutron tends to indicate lower porosities in gas filled pore spaces.

The density and neutron are presented in porosity values. Gas can be located by looking for zones where the density porosity is higher than the neutron porosity. In water or oil the neutron porosity and the density porosity read approximately the same values. In shales the neutron usually indicates very high porosities, much higher than the density. Generally, the only times the neutron will read lower porosities than the density is when gas is present in the pore spaces (see Figure 49).

§ **8.06(i) Acoustic-Velocity or Sonic Logs.** The acoustic-velocity or sonic devices consist of one or more transmitters and two or more receivers. The transmitters emit a sound pulse, and the travel time of sound is measured at the receivers. The recorded value is the travel time in microseconds per foot. The log is a record of travel time ($\triangle$ t) through a vertical distance of rock. One foot measurements are most often taken; however, two-foot and three-foot spacings are also available. The travel time of sound through rock can be related to porosity. As with the sidewall neutron and the density, knowledge of lithology is necessary.

The acoustic velocity log responds to low porosities with the curve deflecting to the right. It responds to higher porosities with curve excursions to the left. As on all previously discussed porosity logs, the high porosities appear as a kick to the left. The lower porosities appear as kicking to the right.

The acoustic-velocity log is most dependable in the more compacted formations. It is not considered at its best in shallow unconsolidated high porosity sands.

Attempts are sometimes made to use the tool as a porosity device in

413

cased holes. The open hole results are considered more reliable. In no case can it be used in empty holes

Gamma rays and calipers are usually run simultaneously, and they are placed in the left track. Other combinations are sonic-neutrons, which are run to locate gas. The response is much like the density-neutron combination in gas, where the neutron porosities are lower than the density derived porosities. The neutron porosity values will be lower than those of the acoustic-velocity.

In some areas a combination induction log and acoustic-velocity log are run. Its purpose is to obtain resistivity measurements and porosities with one trip in the borehole.

§ **8.06(j) Caliper Log.** A caliper log is run simultaneously with all of the above mentioned porosity logs.

The caliper provides a continuous record of hole diameter. Whenever the caliper indicates hole size that is smaller than bit size it is assumed that mud cake is present. Mud cake is formed when mud filtrate invades a rock, leaving mud solids plated on the borehole. For a zone to be capable of taking mud filtrate, it must have some permeability and porosity. It is widely believed that if a formation is permeable enough to take a fluid, it will be permeable enough to produce a fluid.

The caliper will also indicate hole enlargements or wash-outs. Sharp wash-outs opposite porosity kicks on the porosity logs usually mean that the porosity reading is unreliable and is probably a spurious event.

Hole volumes can be calculated from a caliper. This information is used to calculate the amount of cement needed to cement the casing in the hole. Possible packer seats for drill stem tests can be picked from a caliper. A section with uniform hole sizes and little or no enlargement is usually selected for the best chance of obtaining a good packer seat.

§ **8.06(k) Selecting the Porosity Log.** Porosity ranges, as seen in various oil provinces, can be from nearly zero to as high as approximately 40 percent. Many Louisiana-Texas Gulf Coast sands have porosities in the 30-

38 percent range. Hydrocarbons can also be produced from sandstones in the 10-20 percent range, such as those found in hard rock areas. These porosities are considered to be on the low side and would probably have to be fractured.

Limestones and dolomites can vary from almost zero to as high as 25-30 percent. Some of the ultra-deep wells in west Texas have an average matrix porosity of two percent and are prolific producers. In wells where the matrix porosities are low and the production good, secondary porosities are usually found. Vugs and fractures are secondary porosities which are a result of geologic changes.

Primary porosity is of genetic origin.

Shales have porosity which may measure in the 40 percent range. Pore spaces in shales are not interconnected; therefore, there is no permeability. Shales are not considered to have effective porosities.

The measurement of porosity is very important because porosity is so necessary for production. Without porosity measurements, the evaluation of storage space in a reservoir would be worthless.

The selection of a porosity log or porosity logs to be used in a logging program can be a problem. There are a number of factors to consider. On close-in wells or field development wells, many operators prefer the same type log that was used on the offset well. This is especially true if the offset well turned out good. If an accoustic-velocity log was run on the successful offset, then the acoustic-velocity log would likely be run on the new well. People in the petroleum industry tend to follow the leader when the leader was successful. The advantage of following the same program is the feeling that a closer comparison can be made from one well to another. In complex lithologies, accurate porosities are difficult to determine. Under these conditions it would be wise to consider two or more porosity devices. Computer derived cross plots generally can furnish the porosity information needed. A cross plot of the sidewall neutron and the density logs is commonly used in complex lithologies.

In any well where there is even a remote chance of the presence of gas, the density-neutron combination should be considered. Lower neutron log porosities and higher density log porosities in gas zones have resulted in numerous unexpected gas wells.

The larger logging service companies have facilities in or near most hydrocarbon producing areas. They have knowledgeable, trained personnel who are familiar with the problems of their area. Making recommendations for logging programs is part of their service. They will also price these recommended programs.

§ 8.07 Water Saturations

§ 8.07(a) In General. The use of electrical logs to determine the water saturation of the sub-surface rocks is a common and an important function of log calculations. In fact, almost all calculations using electric logs would be to determine the porosity and the water saturation of the rocks. Knowledge of porosity and water saturations aids in the decision on whether to plug a well or to attempt a completion. Well site calculations for the two parameters are made on practically every well drilled in search of hydrocarbons.

The water saturation of a rock is the ratio of the volume of water within the pores of the rock to the total pore volume. The logging symbol for water saturation is Sw, and it is usually expressed in percentage. A rock with the total pore volume filled with water would be described Sw = 100%, in logging terminology. If water occupies half of the pore spaces it would be described as Sw = 50%. The other half of the pore spaces would be assumed to contain hydrocarbons.

Nearly all formations contain some water so log calculations may indicate water saturations of 100 percent down to very low values. Rarely are sub-surface formations found that contain no water.

Sometimes water saturations are such that they cause some confusion.

In Gulf Coast areas some sandstones may calculate as high as 60 percent water and produce hydrocarbons water free. In other areas, water saturations of 45 percent may be too high and 100 percent water may be produced. Critical saturations depend on the rock type and the grain size. Local knowledge is needed when water saturations are above 40 percent. On wildcat wells, where there is no information available on critical saturations, a log analyst may recommend a drill stem test on zones of interest.

The reason for calculating a log for water saturations is because it is the formation waters that conduct an electric current. Electricity can pass through rocks because of the formation salt water they contain. But for a few exceptions, dry rocks are excellent insulators. Oil and gas are good insulators, and pure water is also a good insulator. Sub-surface formations have measurable resistivities because of the water in their pore spaces. The resistivity of a formation will depend mainly on the resistivity of the water in the rock, the amount of water in the rock, and the arrangement of the pore spaces.

§ 8.07(b) Water Resistivity. It was mentioned earlier that pure water is a good insulator. If salt is added to the water it becomes more conductive. An electrical current can be conducted through water by ions formed when salt goes into solution in the water. The saltier the water, or when more ions are in solution, the more conductive the water will be.

Temperature also has an effect on the ability of water to conduct a current. In higher temperatures, the ions move easier, and the water becomes more conductive. Since the temperature in a borehole increases with depth this becomes an important factor and a measured bottom hole temperature will appear on the log heading.

The logging symbol for formation water resistivity is Rw, and it is expressed in ohm-meters. A knowledge of Rw is necessary when calculating water saturations.

§ 8.07(c) Resistivity. Another factor necessary for the calculation of water saturation is the resistivity of the uncontaminated rock or formation. This measurement is made in the borehole by logging tools made for this purpose.

Since the terms resistivity and conductivity have been used in several connections, and since the terms will be used very often in this chapter and in future work with electrical logs, this may be a good time to further discuss the terms.

The resistivity of a material is the degree to which it resists the flow of an electric current. The reciprocal of resistivity is conductivity. The conductivity of a material is the degree to which it conducts the flow of an electrical current. Since the terms are reciprocal, a material with low electrical resistance would conduct a current easily, thereby having a high conductivity. A material with high electrical resistance would not conduct a current easily, thereby having low conductivity.

The units of resistivity on an electrical log are

$$\frac{ohms \times meter^2}{meter}$$

and are usually referred to as ohm-meters. Conductivity measurements are presented in milli-ohms/m. A scale of these units can be found at the top and bottom of a log. Curve deflections to the right indicate an increase in resistivity. Resistivity measurements of sub-surface formations have been the key measurements since the time of the first logs.

The old electrical logs or surveys measured the resistivity of sub-surface rocks using varied spacings and configurations of electrodes. Even though the "ES", as the older electrical logs were often called, is rarely considered in a modern logging program, it still has some importance. Most of the wells drilled before the middle fifties were logged with a log of this type. Therefore, the early day electrical logs are the only records available in thousands of wells.

The logging symbol for the resistivity of a sub-surface formation which is uncontaminated by drilling fluids is Rt. Rt is sometimes referred to as true resistivity. It will be seen later that a measurement of Rt is necessary for log calculations.

Earlier in the discussion of micro devices, the resistivity of the rock flushed with mud filtrate was discussed. The logging symbol for the resistivity of the mud filtrate filled rock is Rxo. The symbol for the resistivity of partially mud filtrate flushed rock is Ri, or resistivity of the invaded zone. Figure 47 presents a schematic of the borehole environment indicating the resistivity measurement relationship.

§ 8.07(d) Types of Resistivity Electrical Logs.

§ 8.07(d)(1) Induction Log.
The most commonly used log for measuring the resistivity of uncontaminated formation is the induction log. Many of the induction surveys run are comprised of four curves. These curves are the induction resistivity and conductivity, the short normal, and the spontaneous potential (SP).

The induction log is a multiple coil device that investigates deep in the formation. Focusing is accomplished so that borehole materials, the invaded zone and nearby formations do not influence its measurements in most cases.

The induction system works well where the borehole is empty, or contains a mud at least three times more resistive than the formation waters. It is more reliable in formations with low resistivities than in high resistivity formations. Generally, it measures good values in zones with five feet or more bed thickness. Resistivity increases are shown with curve deflections to the right (see Figure 50).

§ 8.07(d)(1)[a] Short Normal.
The short normal electrode system has a shallow depth of investigation. It is designed to measure the resistivity of the formation a few inches beyond the borehole. In permeable zones this area of investigation is generally invaded with mud filtrate.

Earlier electrical logs also had a short normal and the curve helped in transition to induction devices, especially in correlations from one well to another.

In addition to being one of the geological correlation curves, the short normal can be used for determination of bed thickness, and in some instances, for porosity estimates. As with all resistivity curves, it kicks to the right with an increase in resistivity.

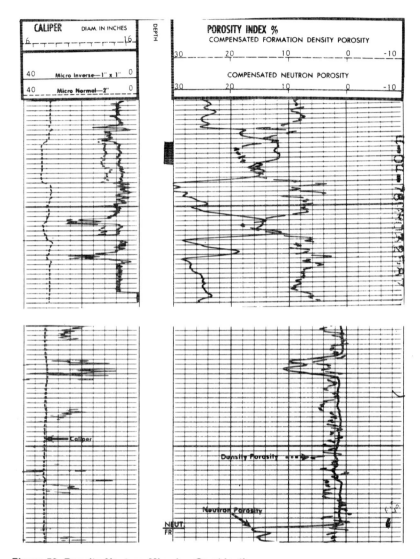

Figure 50. Density-Neutron-Microlog Combination.

The caliper and microlog curves are placed in the left track. Mudcake is indicated on the caliper and positive separation on the microlog in the blocked section. This indicates the presence of permeability.

In the right track, the neutron curve indicates lower porosity than the density curve in the blocked section. A porosity scale is shown at the top of the example. The density curve shows 19 percent porosity and the neutron curve 12 percent. The zone produced gas and is typical of a gas zone. Just under the blocked zone, the neutron and density curves are reading the same porosity. This could be oil or water.

The bottom section of log shows a "tite" section with very little porosity or permeability.

420

§ 8.07(d)(1)[b] Spontaneous Potential (SP). The SP curve is a measurement of naturally occurring potentials in the borehole. Several things have to be present in the borehole to cause these relative potential changes. There must be a difference in the salinity of the borehole fluid (fresh mud) and the water (salty) in the formation, the presence of shale, and a permeable bed.

Naturally occurring potentials in the borehole are measured and shown in the track left of the depth column as the SP curve. These potentials are present even when the logging tools are not in the borehole. Measurement units are in millivolts.

In shales and other non-permeable beds the SP curve stays to the right side of the track, about two or three divisions left of the depth column. The curve kicks to the left when ionic permeabilities are encountered. Some permeability must be present to cause the kick to the left. However, it does not have to be enough permeability for commercial production (see Figures 51 and 52).

The curve is often used for correlations. It is also used as a flag or signal that some permeability is present when the curve kicks to the left. This movement of the curve to the left would indicate a possible zone of interest. Another use of the SP curve is that of estimating formation water salinities.

Since it is necessary to have a salinity contrast between the formation waters and the borehole fluid, it is easily understood why no SP's are recorded in brine muds. The salinities of the borehole fluids and the formation waters are nearly the same. Brine muds are frequently used in west Texas, New Mexico, Kansas, and in several other areas. In some areas, such as the Rocky Mountains, fresh waters are in the formations and fresh muds are in the borehole. Here again there is no salinity contrast and the SP is of little value.

§ 8.07(d)(2) Dual Induction Log. The newest approach to induction logging is the dual induction system. Its general appearance is quite different from the induction curve discussed above. Dual induction logs are presented on a logarithmic grid, in contrast to the older induction logs which were run on a linear grid. Two induction curves are employed that

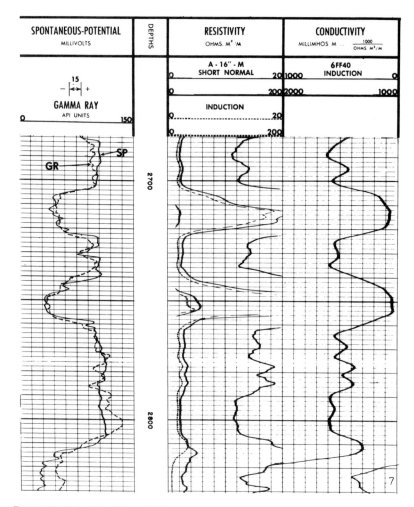

Figure 51. Induction-Electric Log.

This example, from the Woodbine formation of east Texas, illustrates how an induction log will appear in hydrocarbon zones and in a water zone. It also shows the responses of the gamma ray and the SP curve in sands and shales.

With kicks to the left, the SP curve indicates zones of possible interest at (1) 2704-22, (2) 2742-59 and (3) 2810-TD. These zones also appear as low radioactivity on the gamma ray. The zones where the SP is deflected to the right and where the gamma ray indicates high radioactivity are shales.

The zones of possible interest have porosities in the range of 30 percent. Zone 1 is gas productive. Zone 2 is oil productive and Zone 3 is water. With the porosities constant. Zones 1 and 2 indicate the presence of hydrocarbons because of the higher resistivities. Zone 3, with the low-resistivity indicates salt water.

The conductivity curve is recorded in the far right track. Divide its values into 1,000 and the result should be the same as the induction resistivity which is seen in the track right of the depth column.

422

have different depths of investigation. The medium investigation curve is designed to measure formation resistivity out from the borehole but in an area influenced by the invasion of mud filtrate (see Figure 47). The deep investigating curve is designed to read deeper into the formation in uncontaminated formation. A third curve is used and it is a laterolog type curve. The readings of this curve are more influenced by the borehole and the immediate surrounding flush zone. In areas where invasion of mud filtrate is a problem, complicated calculations can be made using the values of the three curves to determine a reasonable true resistivity.

Increases in resistivity are shown by the deflection of the curves to the right. The logarithmic scale will record very high resistivities without going off-scale. However, induction devices are at their best in low resistivities and are not considered very accurate in values of over 100 ohm-meters.

§ **8.07(d)(3) Laterologs or Guard Logs.** It was mentioned earlier that induction logs could be run in empty boreholes or when the mud in the borehole is approximately three times more resistive than the formation waters. It was also mentioned that in many areas wells are drilled with salt muds. In salt muds, resistivities of the mud and formation waters are very nearly equal. In these wells, where the mud is salty, the induction logs do not give reliable resistivities. True resistivities are determined by the running of focused electrode devices called laterologs or guard logs. Salt muds are more conductive than the surrounding formations and the measuring currents will take the path of least resistance. The laterolog devices have guard electrodes that force the measuring current into the formation. The resulting measurement is true resistivity.

§ **8.07(d)(4) Combination Tools.** The trend in recent years has been to "stack" logging devices and run more than one survey simultaneously. Valuable rig time is saved in this manner. Some of the combinations are for the measurement of porosity. The density neutron log is a combination of porosity tools. Others combine a resistivity device and a porosity device and are able to obtain true resistivity and porosity in one run (see Figure 53).

In some areas — mostly along the Gulf Coast — the induction-acoustic-velocity combination is run. This combination provides resistivity

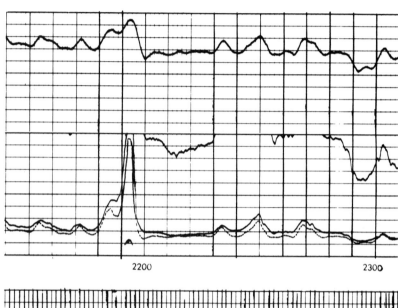

2200   2300

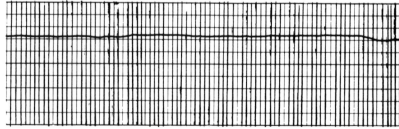

Figure 52. SP Curve Appearance in Tite Zones.

    The SP curve is in the left track. In non-permeable or tite zones the curve will be ranging to the right of the track approximately three divisions from the depth column. The example illustrates the shale base line. The rocks from 2200-2300 are shales and limey shales. The resistive zone just above 2200 is a tite limestone. The SP curve will deflect to the left in permeable rocks.

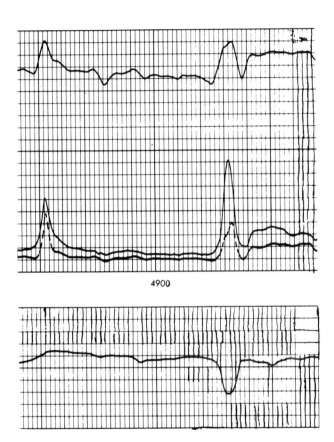

4900

Figure 53.

The SP curve is in the left track and it deflects to the left 4926-36. This kick to the left indicates a zone of interest. Above this zone the SP is straight and the low resistivities are typical of shale. At the top of the example note the then tite streak.

425

information from the induction and porosity information from the travel time of sound as measured by the acoustic-velocity.

The surveys are recorded on a split grid. Resistivity information is presented on a logarithmic scale and the travel time of the acoustic velocity is linear.

In addition to the SP, induction and acoustic-velocity, a computed curve is recorded and referred to as Rwa for apparent water resistivity. The Rwa technique is used to examine a great number of possible productive zones in a short time. It is a back calculation of water resistivity using a variation of basic logging equations.

Salt mud combinations are also available. A gamma ray-neutron and laterolog type device are stacked and run simultaneously.

§ 8.08 Archie Equations

The basic logging equations were developed by G.E. Archie and are referred to as the Archie Equations. Archie presented the relationships in a paper published in 1942 in the transactions of AIME entitled "The Electrical Resistivity Log as an Aid in Determining Some Reservoir Characteristics". Most interpretations performed with open hole logs use the basic Archie Equation which is expressed as follows:

$$Sw = \sqrt{\frac{F \times Rw}{Rt}}$$

where

Sw = Water saturation

F = Formation factor which is related to porosity

$F = \dfrac{1}{\varnothing_m}$ m is cementation factor which ranges from 1.3 in high porosity to 2.3 in low porosity rocks.

For mathematical convenience, m = 2 is often used.

Rw = Resistivity of the formation water

Rw is measured from produced waters in nearby areas. Tabulations of Rw are available in most areas.

Rt = True resistivity as measured by induction or laterolog devices

Values from porosity logs and resistivity logs are placed in the Archie Equation for calculation.

All of the logging service companies provide charts for solving the Archie Equations. Many electrical well logs can be calculated in a straightforward manner; however, many logs present difficulties that require log specialists with years of experience.

§ 8.09 Logging Programs

Logging programs are usually designed to provide formation resistivity and formation porosities which can be used in the Archie Equation. Usually a minimum program would consist of a resistivity survey and one porosity log. More complete programs would require two or more porosity logs and perhaps even a change in the resistivity device. The selection of a suitable program usually depends on previous experience in an area, knowledge of rock types to be logged and even personal preference. Since the selections will vary from area to area, it is again suggested that a logging program be recommended by logging service companies working in the area of interest. They can also provide pricing information on the various programs available.

A schedule of available commercial logging services and a brief description of each are presented in Figures 54-56.

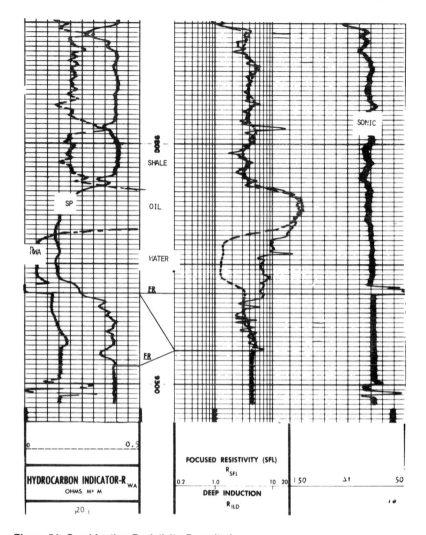

FOCUSED RESISTIVITY (SFL)
R_{SFL}

HYDROCARBON INDICATOR-R$_{WA}$
OHMS. M² M

DEEP INDUCTION
R_{ILD}

Figure 54. Combination Resistivity-Porosity Log.

The combination induction-sonic log was recorded in one run in the borehole. The logging tool is pulled upward and the SP starts moving to the left at 9260 indicating the presence of some porosity and permeability. At 9218 the SP curve moves back to the right as it leaves the zone of interest and goes back into shale. The sonic curve is on the far right and it verifies the presence of porosity 9218-60 as it reads a △ t of 80 microseconds, which calculates approximately 20 percent porosity.

The induction curve is presented in the logrithmic track as the dashed or interrupted curve. It reads very low resistivities in the bottom of the zone because it is filled with conductive salt water. The oil/water contact is at 9238. At this point the resistivity values increase because there are hydrocarbons present and therefore less salt water.

The Rwa curve in the left track is a quick look method using Archie Equations. Apparent water resistivities are calculated and recorded on the log. Zones with Rwa values calculating higher than the norm possibly contain hydrocarbons.

This is a very complicated and sophisticated sonde. The sonic electronics are in the top of the instrument. Note that the sonic log does not log the bottom 30' of borehole.

428

COMMERCIAL OPEN HOLE LOGGING SERVICES

Arranged by Type and Company

REF. NO.	WELEX	SCHLUMBERGER	DRESSER-ATLAS	GO INTERNATIONAL
1.	Electric Log	Electrical Log	Electrolog	Electrical Log
2.	Induction Electric Log	Induction Electrical Log	Induction Electrolog	Induction-Electrical
3.	Dual Induction Guard Log	Dual Induction Laterolog	Dual Induction Focused Log	
4.	Guard Log	Laterolog-3, Laterolog-7	Laterolog	Laterolog
5.	Contact Log	Microlog	Minilog	Micro-Electrical Log
6.	FoRxo Log	Microlaterolog	Micro-Laterolog	Micro-Laterolog
7.		Proximity Log	Proximity Log	
8.	Acoustic Velocity Log	Sonic Log	Acoustilog	Sonic Log
9.	Compensated Acoustic Velocity Log	BHC Sonic Log	BHC Acoustilog	Sonic Log (Borehole Compensated)
10.	Fracture Finder Log	Amplitude Log	Fraclog	Amplitude
11.	Micro-Seismogram Log	Variable Density Log	Variable Amplitude Density Log	Seismic Spectrum
12.	Density Log	Formation Density Log	Densilog	Density Log
13.	Compensated Density Log	Compensated Formation Density Log	Compensated Densilog	Density Log (CDL)
14.	Simultaneous Gamma Ray-Neutron Log	Gamma Ray-Neutron Log	Gamma Ray-Neutron Log	Gamma Ray-Neutron Log
15.	Side Wall Neutron Log	SNP Neutron Log	Epithermal Sidewall Neutron Log	Sidewall Neutron Log

Figure 55

429

RESISTIVITY LOGS

PRIMARY USAGE	
Correlation True Resistivity	1. THE ELECTRIC LOG usually consists of an SP curve and three resistivity curves. Electrode spacings of resistivity curves range from approximately 16"-18" (normal) to 16'-22' (lateral). It is used for correlation and was formerly used with the aid of charts to approximate formation resistivity. This log has been replaced, in general, by the Induction Electric Log because of its better response characteristics in thin beds.
Correlation True Resistivity	2. THE INDUCTION ELECTRIC LOG is usually an SP curve, 16"-18" (normal) resistivity curve, induction conductivity curve (about 40" spacing), and the induction resistivity curve. It is used as described above under (1). In beds of about 6' or greater thickness, the induction resistivity value is usually very close to the value of formation resistivity. It is used when Rmf / Rw is greater than 4.
Correlation True Resistivity Invaded Zone	3. THE DUAL INDUCTION LOG contains SP, deep and shallow investigation induction resistivity curve values and a very shallow investigation guard type resistivity measurement. The three resistivity measurements provide a basis for evaluating Rt, Ri and Di.
Correlation True Resistivity	4. THE GUARD LOG is a recording of resistivity measured with one spacing of guard electrode length. It may be run with SP or gamma ray. It is used when Rmf / Rw is less than 3. Guard Electrode length ranges from 60" to 120" depending on the company and tool type.
Identify Permeable Zones Resistivity of Flush Zone	5. THE CONTACT LOG is a record of two resistivity measurements made by two very shallow investigation pad mounted electrode systems. "Positive separation" of the two curves usually is an indication that mudcake is present between the pad and the rock. It is used when Rmf / Rw is greater than 5, and is usually recorded with caliper. The two resistivity curves provide a basis for determining Rxo and mudcake thickness.
Resistivity of Flush Zone Porosity Moveable Oil	6. THE FoRxo LOG is a single resistivity measurement from a pad mounted guard type electrode. In invaded rocks with thin mudcake, the FoRxo measures Rxo directly. It is often used in conjunction with Guard or Induction Logs to identify permeable zones containing "moveable oil". It is used in fresh or salt mud and is recorded with a caliper.
	7. THE PROXIMITY LOG is similar to (6) above, but somewhat deeper investigation

Figure 56

ACOUSTIC LOGS

Porosity Travel Time

8. THE ACOUSTIC VELOCITY LOG records the time required for an acoustic compression wave to travel through one foot of rock. This value of time (interval transit time) is related to porosity. The log is usually recorded with a Gamma Ray Log and Caliper.

Porosity Travel Time

9. THE COMPENSATED ACOUSTIC VELOCITY LOG is similar to (8) above, but some of the extraneous responses related to irregular borehole geometry are removed.

Locate Zones of Fracture

10. THE FRACTURE FINDER LOG is a measure of the amplitude of the acoustic signal that has traveled through the rock around the borehole. The presence of fractures has been shown to reduce the amplitude of the received acoustic signal.

Locate Zones of Fracture

11. THE MICRO-SEISMOGRAM LOG utilizes a method of recording the total acoustic signal instead of one feature of it as in (8), (9) and (10) above. The Micro-Seismogram Log is usually used in conjunction with the Fracture Finder Log.

RADIOACTIVE LOGS

Porosity Bulk Density

12. THE DENSITY LOG records the value of formation bulk density. The tool emits a beam of gamma rays and the number of those gamma rays that reach the detector is related to formation density. The log is recorded with Gamma Ray and Caliper.

Porosity Bulk Density

13. THE COMPENSATED DENSITY LOG is similar to (12) above but compensated to minimize the effects of mudcake thickness variations in the tool response.

Porosity

14. THE SIMULTANEOUS GAMMA RAY NEUTRON LOG is a record of the response of the detector near the neutron source. From this it is possible to approximate formation porosity. Also recorded is the natural gamma radiation of the formation.

Porosity

15. THE SIDEWALL NEUTRON LOG is recorded in terms of formation porosity. It is a refinement of (14) above and, apparently, more reliable and is insensitive to hole diameter variations.

Figure 57

431

INDEX

A

B

C

437

439

CONTRACTS

D

E

Traps — see "Geology and Geophysics", above
Unconformities — see "Geology and Geophysics — Traps", above

F

FARMOUT AGREEMENTS
Abandonment
        Duty to Plug and Abandon — see "Assignor's Obligations", below
        Take Over Right — see "Notice", below
Acceptance
        Assignment — see "Assignment", below
        Farmout Agreement .... § 4.03(i)
Access .... § 4.03(f)
Area of Mutual Interest — see "Leases", below
Assignability .... § 4.03(i)
Assigned Interest — see "Rights Assigned and Reserved", below
Assignee's Obligations
        Access — see "Access", above
        Delay Rentals — see "Delay Rentals", below
        Drilling — see "Drilling Obligations", below
        Duty to Plug .... § 4.03(f)
        Indemnification — see "Indemnification", below
        Information — see "Testing and Reporting", below
        In General .... §§ 4.03(d), 4.03(f)
        Notice — see "Notice", below
        Regulatory Compliance .... § 4.03(f)
Assignment
        Acceptance .... § 4.03(h)
        Area Covered — see "Leases", below
        Assignment of Farmout — see "Assignability", above
        Depths — see "Depth Limitations", below
        Description — see "Leases", below
        In General .... §§ 4.03(d), 4.03(h)
        Prior to Abandonment — see "Notice", below
        Rights Assigned — see "Rights Assigned and Reserved", below
        Substances — see "Substances Covered", below
        Tax Consequences — see "Tax Consequences", below
        Time for Acceptance .... § 4.03(h)
Assignor's Obligations
        Assignment — see "Assignment", below

G

Limitations on Grant — see CONVEYANCES; MINERAL CONVEYANCES AND RESERVATIONS
Oil and Gas Leases — see OIL AND GAS LEASES
Reservations — see CONVEYANCES; MINERAL CONVEYANCES AND RESERVATIONS
Term Grants — see CONVEYANCES; MINERAL CONVEYANCES AND RESERVATIONS

H

HABENDUM CLAUSE
Conveyances — see CONVEYANCES; LAND OWNERSHIPS
Mineral Conveyances — see MINERAL CONVEYANCES AND RESERVATIONS; MINERAL OWNERSHIP
Oil and Gas Leases — see OIL AND GAS LEASES
HOMESTEAD ACT — see FEDERAL LANDS; LAND OWNERSHIP
HOMESTEAD AND COMMUNITY PROPERTY — see CANADIAN PETROLEUM LAND OPERATIONS; CONVEYANCES; CURING TITLE DEFECTS; JUDICIAL ACTIONS; LAND OWNERSHIP; MINERAL CONVEYANCES AND RESERVATIONS; MINERAL OWNERSHIP

I

IDC DEDUCTION — see TAXATION
ILLEGAL PURPOSE — see CONTRACTS
IMPLIED COVENANTS — see OIL AND GAS LEASES
INDIAN LANDS
Adverse Possession of — see ADVERSE POSSESSION
Canada — see CANADIAN PETROLEUM LAND OPERATIONS
In General .... §§ **1.03(a), 2.02(a)**
INTANGIBLE DRILLING AND DEVELOPMENT COSTS — see TAXATION
INTERSTATE TRANSPORTATION AND SALE OF NATURAL GAS — see GAS CONTRACTS

J

JOINT OPERATING AGREEMENTS — see OPERATING

L

462

N

467

470

471

476

478

Curative Instrument, Quit Claim Deed as — see CURING TITLE
    DEFECTS
Effect of — see CONVEYANCES
Patent, Conveyance Prior to Perfecting Right to — see FEDERAL
    LANDS; PATENTS

R

RAILROADS
    Canadian Grants — see CANADIAN PETROLEUM LAND
        OPERATIONS
    Federal Grants to — see FEDERAL LANDS
    Rights of Way
        Conveyances Adjoining — see CONVEYANCES; MINERAL
            CONVEYANCES AND RESERVATIONS
        Oil and Gas Leases Adjoining — see OIL AND GAS LEASES
RECORD CHECK — see TITLE EXAMINATION
RECORDING
    Acknowledgment as Prerequisite for .... § **2.02(c)(7)**
    Bona Fide Purchaser, Effect on — see BONA FIDE PURCHASER;
        NOTICE
    Constructive Notice, Recording as — see NOTICE
    Curative Statutes, Recording as Prerequisite for Operation of — see
        CURING TITLE DEFECTS
    In General .... §§ **2.02(c)(1), 2.02(c)(7)**
    Notice, Recording as — see NOTICE
    Unrecorded Claims
        In General .... § **2.01**
        Validity of .... § **2.01**
REFORMATION — see JUDICIAL ACTIONS
RELICTION — see RIPARIAN LANDS AND COASTAL WATERS
RENTAL DIVISION ORDER — see DELAY RENTALS
RENTAL STIPULATION — see DELAY RENTALS
RENTALS — see DELAY RENTALS
RESCISSION AND REFORMATION — see JUDICIAL ACTIONS
RESERVATIONS — see CONVEYANCES; MINERAL CONVEYANCES
    AND RESERVATIONS
RESTRICTED LANDS AND RESTRICTED INDIANS — see INDIAN
    LANDS
REVENUE RULING 77-176 — see TAXATION

S

STATUTE OF FRAUDS — see CONTRACTS; CONVEYANCES
STATUTE OF USES — see LAND OWNERSHIP
SUBSTITUTE WELL — see FARMOUT AGREEMENTS
SUPPORT AGREEMENTS
SURFACE OWNERSHIP
Canada — see CANADIAN PETROLEUM LAND OPERATIONS
Defined — see MINERAL CONVEYANCES AND RESERVATIONS;
    MINERAL OWNERSHIP
SURVEY SYSTEMS
Canada — see CANADIAN PETROLEUM LAND OPERATIONS
In General — see DESCRIPTION

T

TAXATION — see also FARMOUT AGREEMENTS; OPERATING
    AGREEMENTS
Canada — see CANADIAN PETROLEUM LAND OPERATIONS
States — see CONVEYANCES
Tax Liens — see CURING TITLE DEFECTS
U.S. Federal Income Tax
    Association Taxable As a Corporation — see "Partnerships",
        below
    Capital Gains Treatment — see "Sales, Exchanges and
        Subleases", below
    Compensation for Services — see "Services", below
    Corporations
        Subchapter S — see "Subchapter S Corporations", below
    Depletion
        Carried Interest — see "Lessee's Right to Claim", below